International Association of Geodesy Symposia

*Chris Rizos, Series Editor*
*Pascal Willis, Series Associate Editor*

# International Association of Geodesy Symposia

*Chris Rizos, Series Editor*
*Pascal Willis, Series Associate Editor*

---

Symposium 104: Sea SurfaceTopography and the Geoid
Symposium 105: Earth Rotation and Coordinate Reference Frames
Symposium 106: Determination of the Geoid: Present and Future
Symposium 107: Kinematic Systems in Geodesy, Surveying, and Remote Sensing
Symposium 108: Application of Geodesy to Engineering
Symposium 109: Permanent Satellite Tracking Networks for Geodesy and Geodynamics
Symposium 110: From Mars to Greenland: Charting Gravity with Space and Airborne Instruments
Symposium 111: Recent Geodetic and Gravimetric Research in Latin America
Symposium 112: Geodesy and Physics of the Earth: Geodetic Contributions to Geodynamics
Symposium 113: Gravity and Geoid
Symposium 114: Geodetic Theory Today
Symposium 115: GPS Trends in Precise Terrestrial, Airborne, and Spaceborne Applications
Symposium 116: Global Gravity Field and Its Temporal Variations
Symposium 117: Gravity, Geoid and Marine Geodesy
Symposium 118: Advances in Positioning and Reference Frames
Symposium 119: Geodesy on the Move
Symposium 120: Towards an Integrated Global Geodetic Observation System (IGGOS)
Symposium 121: Geodesy Beyond 2000: The Challenges of the First Decade
Symposium 122: IV Hotine-Marussi Symposium on Mathematical Geodesy
Symposium 123: Gravity, Geoid and Geodynamics 2000
Symposium 124: Vertical Reference Systems
Symposium 125: Vistas for Geodesy in the New Millennium
Symposium 126: Satellite Altimetry for Geodesy, Geophysics and Oceanography
Symposium 127: V Hotine Marussi Symposium on Mathematical Geodesy
Symposium 128: A Window on the Future of Geodesy
Symposium 129: Gravity, Geoid and Space Missions
Symposium 130: Dynamic Planet - Monitoring and Understanding…
Symposium 131: Geodetic Deformation Monitoring: From Geophysical to Engineering Roles
Symposium 132: VI Hotine-Marussi Symposium on Theoretical and Computational Geodesy
Symposium 133: Observing our Changing Earth
Symposium 134: Geodetic Reference Frames
Symposium 135: Gravity, Geoid and Earth Observation
Symposium 136: Geodesy for Planet Earth
Symposium 137: VII Hotine-Marussi Symposium on Mathematical Geodesy
Symposium 138: Reference Frames for Applications in Geosciences
Symposium 139: Earth on the Edge: Science for a sustainable Planet
Symposium 140: The 1st International Workshop on the Quality of Geodetic Observation and Monitoring Systems (QuGOMS'11)
Symposium 141: Gravity, Geoid and Height systems (GGHS2012)
Symposium 142: VIII Hotine-Marussi Symposium on Mathematical Geodesy
Symposium 143: Scientific Assembly of the International Association of Geodesy, 150 Years

More information about this series at http://www.springer.com/series/1345

# IGFS 2014

Proceedings of the 3rd International Gravity Field Service (IGFS), Shanghai, China, June 30 – July 6, 2014

Edited by

Shuanggen Jin
Riccardo Barzaghi

Springer

*Volume Editor*
Shuanggen Jin
Shanghai Astronomical Observatory
Chinese Academy of Sciences
Shanghai
China

Riccardo Barzaghi
Politecnico di Milano, DICA
Piazza Leonardo da Vinci 32
Milano
Italy

*Series Editor*
Chris Rizos
School of Civil & Environmental Engineering
University of New South Wales
Sydney
Australia

*Associate Editor*
Pascal Willis
Institut national de l'Information
géographique et forestière
Direction de la Recherche
et de l'Enseignement
Marne-la-Vallée
France

ISSN 0939-9585 ISSN 2197-9359 (electronic)
International Association of Geodesy Symposia
ISBN 978-3-319-39819-8 ISBN 978-3-319-39820-4 (eBook)
DOI 10.1007/978-3-319-39820-4

Library of Congress Control Number: 2016952239



Printed on acid-free paper

This Springer imprint is published by Springer Nature
The registered company is Springer International Publishing AG Switzerland

# Preface

The accurate and precise estimation of the gravity field of the Earth is nowadays required in many geodetic and geophysical investigations. The recent satellite missions devoted to the observation of the gravity field of the Earth have strongly improved the resolution and precision of the estimated global geopotential models. Global mass redistributions in the Earth environment can be observed and modeled through gravity from space and can improve the knowledge of the Earth system and climate changes. A unique height system can be estimated for the whole Earth, which is fundamental in, e.g., evaluating sea level variations. The new fields also allow innovative investigations of the solid Earth giving new details of the crust and mantle and variation over time. These improvements in the estimation of the global geopotential models also require updated/new methods in modeling the higher frequency of the gravity field and denser local data coverage to achieve 1-cm geoid accuracy, which is likely to be required in a few years for practical applications. The 3rd International Gravity Field Service (IGFS) General Assembly that focused on the above issues, including methods for observing, estimating, and interpreting the Earth gravity field as well as its applications, was successfully held in Shanghai, China from June 30 to July 6, 2014.

These proceedings contain 24 peer-reviewed papers presented at the 3rd IGFS General Assembly, which was organized by the International Gravity Field Service (IGFS), Commission 2 of the International Association of Geodesy (IAG), and Shanghai Astronomical Observatory (SHAO), Chinese Academy of Sciences. The IGFS is an official IAG Service which coordinates and harmonizes the activities of other "Level 1" gravity-related services, namely, the Bureau Gravimetrique International (BGI), the International Geoid Service (IGeS), the International Center for Earth Tides (ICET), the International Center for Global Earth Models (ICGEM), and the International Digital Elevation Model Service (IDEMS). Over 130 participants from 25 countries attended this assembly. There were 80 oral papers and 30 posters presented in the 6 days of the assembly. The list of participants and paper titles can be found at the IGFS2014 website http://202.127.29.4/meetings/igfs2014. The scientific sessions were centered on:

**Session 1 Gravimetry and Gravity Networks**
Conveners: Sylvain Bonvalot and Dan Roman

**Session 2 Global Geopotential Models and Vertical Datum Unification**
Conveners: Michael Sideris and Jiancheng Li

**Session 3 Local Geoid/Gravity Modeling**
Conveners: Urs Marti and Riccardo Barzaghi

**Session 4 Satellite Gravimetry**
Conveners: Roland Pail and Shuanggen Jin

**Session 5 Mass Movements in the Earth System**
Conveners: Rene Forsberg and Shuanggen Jin

**Session 6 Solid Earth Investigations**
Conveners: Carla Braitenberg and Rene Forsberg

We express our gratitude to all those who have contributed to the successful 3rd IGFS General Assembly, particularly the conveners who devoted a lot of time in organizing attractive sessions and scheduling the program of the assembly and the associate editors of these proceedings who played a leading role in the peer review process until the final acceptance for publication. Most important, sincere thanks were given to the IAG proceedings editor Dr. Pascal Willis who advanced and kept on track the publication of these proceedings. Meanwhile, we would like to thank all reviewers, who are listed in this volume as an appreciation of their dedication.

In addition, we wish to thank the local organizing committee (LOC) members of the Satellite Navigation and Remote Sensing Group, Shanghai Astronomical Observatory, Chinese Academy of Sciences, who hosted the Assembly. Beyond responsibility for the website, registration, technical support, and all kinds of other arrangements, the LOC organized a visit to the Sheshan Astronomical Observatory, including a guided tour to the astronomical museum and to the 65-m radio telescope. Shuanggen Jin and his team (Rui Jin, Guiping Feng, Attaullah Khan, Tengyu Zhang, Xuerui Wu, Andres Calabia, Xuechuan Li, Yi Yang, Fang Zou, Nasser Najibi, Yang Zhou, and Xin Zhao) have done more than their share in bringing the IGFS General Assembly to success.

Finally, we would like to gratefully thank the Springer Publisher for their processes and cordial cooperation to publish this proceeding.

Shanghai, China
Milano, Italy
July 2015

Shuanggen Jin
Riccardo Barzaghi

# Contents

# Part I

# Gravimetry and Gravity Networks

# Quality Assessment of the New Gravity Control in Poland: First Estimate

P. Dykowski and J. Krynski

**Abstract**

The new gravity control in Poland is based on absolute gravity measurements. It consists of 28 fundamental stations and 168 base stations. Fundamental stations are located in laboratories; they are to be surveyed in 2014 with the FG5-230 gravimeter. Base stations are monumented field stations; they were surveyed in 2012 and 2013 with the A10-020 gravimeter. The part of the new gravity control consisting of base stations is the subject of the quality assessment.

Besides absolute gravity measurements the vertical gravity gradient was determined at all 168 base stations. Nearly 350 single absolute gravity measurement setups and vertical gravity gradient determinations performed provide valuable and comprehensive material to evaluate the quality of the established gravity control. Alongside the establishment of the base stations of the gravity control, multiple additional activities were performed to assure and provide a reliable gravity reference level. These activities concerned regular gravity measurements on monthly basis with the A10-020 on the test network at Borowa Gora Geodetic–Geophysical Observatory, calibrations of metrological parameters of the A10-020 gravimeter and scale factor calibrations of LCR gravimeters, participation with the A10-020 in the international (ECAG2011, ICAG2013) and regional comparison campaigns of absolute gravimeters.

Careful analysis of the data gathered throughout the project resulted in the estimation of the Total Uncertainty budget for the A10-020 gravimeter on each of 168 base stations. It provides a reliable quality assessment of the new gravity control in Poland.

**Keywords**

A10 absolute gravimeter • Gravity control • Uncertainty

## 1 Introduction

The technical project for the modernization of gravity control in Poland was developed at the end of 2011 (Krynski et al. 2013). The new gravity control in Poland (PBOG14) is based in principle on absolute gravity measurements. The PBOG14 consists of 28 fundamental stations located in laboratories and 168 base stations (1 in 1,850 km$^2$) which are monumented field stations. Fundamental stations are to be surveyed in 2014 with the FG5-230 of the Warsaw University of Technology. Base stations were surveyed in 2012 and 2013 with the A10-020 absolute gravimeter of the Institute of Geodesy and Cartography (IGiK), Warsaw. They consist of 57 ASG-EUPOS (active GNSS network) eccentric stations, 78 POGK98 (previous gravity control) stations, 4 EUVN (European Vertical Reference Network) stations, 22 POLREF (densification of European Reference Network) stations, and 7 new stations (Fig. 1).

P. Dykowski (✉) • J. Krynski
Institute of Geodesy and Cartography, 27 Modzelewskiego St, 02-679 Warsaw, Poland
e-mail: przemyslaw.dykowski@igik.edu.pl


S. Jin, R. Barzaghi (eds.), *IGFS 2014*, International Association of Geodesy Symposia 144,
DOI 10.1007/1345_2015_46

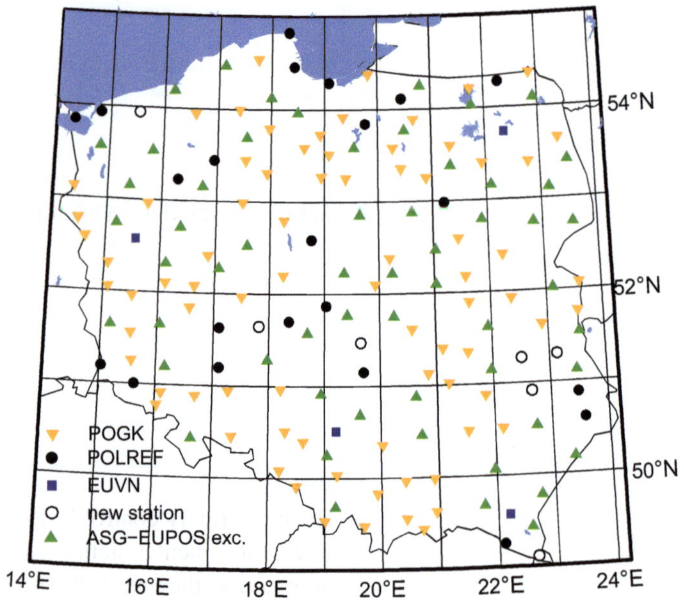


**Fig. 1** PBOG14 stations

Surveys at the base stations (absolute gravity and vertical gravity gradient) were carried out from July 2012 till December 2013. Absolute gravity was determined with the A10-020 in six campaigns. Also in six campaigns vertical gravity gradient with LaCoste&Romberg (LCR) gravimeters was determined. Each campaign lasted no longer than 2 weeks. The acquired data has been processed by April 2014 and the performance of the A10-020 gravimeter has initially been analyzed.

The results of numerous test measurements with the A10-020 absolute gravimeter in laboratory as well as in various field conditions conducted since 2008 on the test network at Borowa Gora Geodetic–Geophysical Observatory (Krynski and Sekowski 2010; Sekowski and Krynski 2010; Dykowski et al. 2012) were fundamental for the specification of the project objectives as well as the measurement methodology. Important role played also the results from the survey of the Finnish First Order Gravity Control in 2009–2010 (Mäkinen et al. 2010) and tests with a group of LCR gravimeters equipped with modern feedback systems (Dykowski 2012). All the activities described above before and during the project including calibrations and absolute gravity comparison campaigns were to provide the most reliable gravimetric level yet in Poland.

Although the A10 absolute gravimeters are widely used for more than a decade, no study on the Total Uncertainty (T.U.) budget estimation for any A10 gravimeter was reported and published. Total Uncertainty budget estimated for A10 gravimeters when processing data from ICAG (International Comparison of Absolute Gravimeters) and ECAG (European Comparison of Absolute Gravimeters) absolute gravity comparison campaigns was incomplete; moreover it was in most cases based on the Micro-g LaCoste Inc. provided values. For the first time an experimental T.U. value was presented for the A10-020 gravimeter for ICAG2013. It was based on the results of a previous research (Dykowski et al. 2013). The material collected during the project on the modernization of gravity control in Poland is a unique source of information on the performance of the A10-020 gravimeter. It makes possible to provide a complete and reliable estimation of T.U. budget.

## 2 Supplementary Activities During the Establishment of PBOG14

The basic activities during the establishment of the new gravity control concerned absolute gravity determinations with the A10-020 gravimeter, and vertical gravity gradient measurements with the LCR gravimeters needed for reducing the measured gravity values to a benchmark level. Multiple additional activities were carried out throughout the project to ensure accuracy and reliability of the determined gravity values. They concerned calibrations of the laser and frequency standards of the A10-020 gravimeter, calibrations of the LCR gravimeters used, regular repeatable measurements with the A10-020 at Borowa Gora Observatory, and the most crucial for the gravity reference level estimation: the participation in international, regional and local comparison campaigns of absolute gravimeters.

### 2.1 Calibrations of the A10-020 Laser and Frequency Standards

The calibrations of the components of the A10-020 gravimeter were planned to cover all absolute gravity surveys to be conducted within the project, i.e. those on PBOG14 stations,

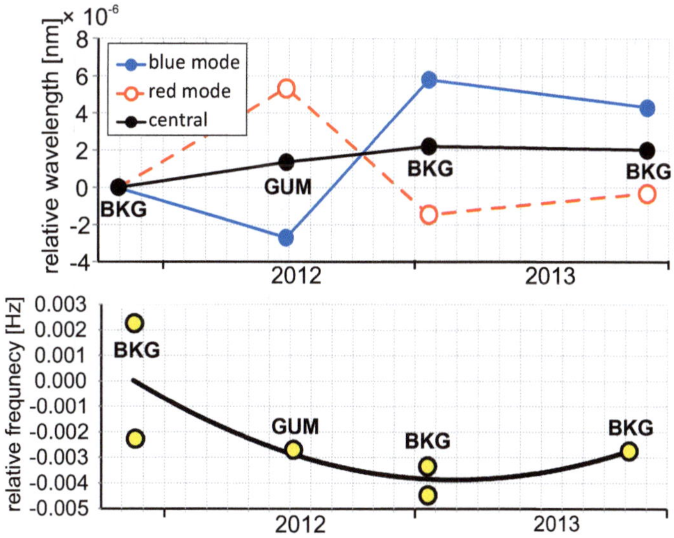


**Fig. 2** Laser (*top*) and frequency standard (*bottom*) calibration data throughout the establishment of PBOG14

those during AG comparison campaigns, and those during regular A10 gravimeter surveys at Borowa Gora Observatory.

The ML-1 laser of the A10-020 gravimeter was calibrated four times during the course of the measurement part of the project: three times at the Bundesamt für Kartographie und Geodäsie (BKG) which is in close cooperation with IGiK in terms of absolute gravity determinations, and once at the Central Office of Measures (GUM) – Polish metrological institute. The results of laser calibrations are shown in Fig. 2 (top) with distinction to the red, blue mode and the central frequency. The central frequency of the laser drifted slightly in the course of the project, resulting in $\sim$3.3 μGal change in the observed gravity value. This value was considered significant enough to be included into the calculations of gravity values determined with the A10-020. Laser wavelengths were linearly interpolated between consecutive calibration data for each day of absolute gravity surveys.

The frequency standard of the A10-020 gravimeter was also calibrated four times: three times at BGK and once at GUM (Fig. 2 – bottom). The total frequency variation in the course of the project results in less than 0.005 Hz ($\sim$1.0 μGal in terms of gravity variation). For the calculation of gravity determinations with the A10-020 these calibration data were interpolated with second degree polynomial for each day of absolute gravity surveys.

## 2.2 Calibrations of LCR Gravimeters

Three LCR gravimeters with modern feedback systems were used for the vertical gravity gradient determinations on all PBOG14 stations. They were calibrated twice on the Central Gravimetric Calibration Baseline in Poland, i.e. in June 2012 and July 2013. Each calibration was performed on at least three spans of the baseline of minimum 30 mGal gravity difference. As the vertical gravity gradient determination requires a measurement range of no more than 400 μGal the scale factor was determined with far greater accuracy than needed. The effect of LCR calibration on the vertical gravity gradient determined was found negligible.

## 2.3 Regular Absolute Gravity Measurements with the A10-020 at Borowa Gora Observatory

In the course of the project, gravity was regularly determined with the A10-020 on monthly basis on three stations at Borowa Gora Geodetic-Geophysical Observatory: two laboratory stations A-BG and BG-G2, and the field station, 156. The measurements were performed with the same methodology as on PBOG14 stations. The results obtained were used for quality assessment of the A10-020 absolute gravimeter performance during the project. Figure 3 presents the results of each single setup of absolute determinations of gravity on the laboratory station A-BG and the field station 156. The standard deviation of the determined gravity on the field station equals 5.1 μGal. The value was used as the representative "system model" uncertainty of the A10-020 in field conditions for the T.U. estimates on PBOG14 stations. This value was further used as "System Model" uncertainty instead of 10 μGal suggested by the manufacturer (Micro-g Inc. 2008). The scattering shows also the variability of the determined gravity at a station when no correction due to local hydrology is applied. Statistics of the results on the test network are presented in Table 1.

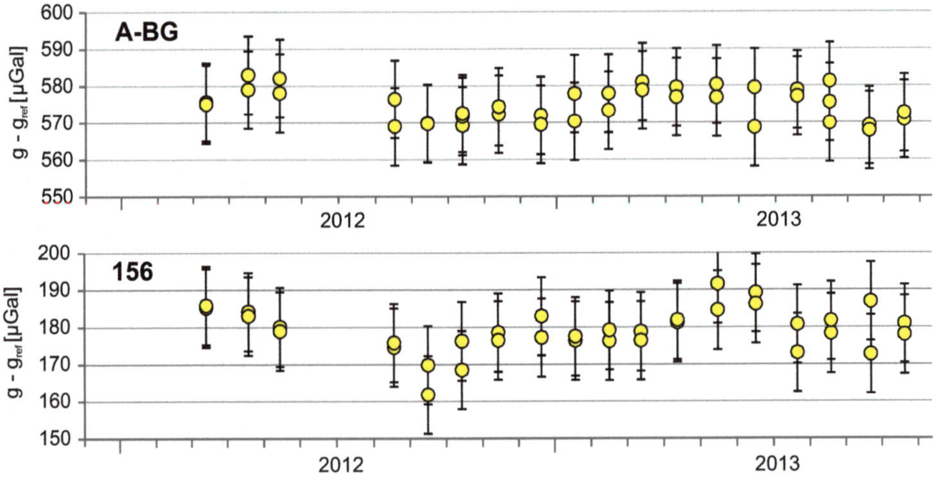


**Fig. 3** Results of gravity survey with the A10-020 at Borowa Gora Observatory test network; $g_{ref} = 981,250,000$ μGal

**Table 1** Statistics of absolute gravity determinations with the A10-020 at Borowa Gora Observatory test network ($g_{ref} = 981,250,000$ μGal) (μGal)

| Station | | | |
|---|---|---|---|
| | $g - g_{ref}$ | Std dev. | Max–Min |
| A-BG | 574.8 | 3.7 | 12.1 |
| BG-G2 | 439.8 | 5.9 | 19.8 |
| 156 | 179.0 | 5.1 | 22.1 |

## 2.4 Absolute Gravimeter Comparison Campaigns

An essential part of the establishment of the gravity control concerned the participation of the A10-020 gravimeter in local, regional and international absolute gravimeter comparison campaigns. All comparisons were planned in a way that would allow repeated control of the gravity reference level provided with the A10-020 for the project of the modernization of the gravity control. The A10-020 participated in one regional AG comparison at Wettzell Observatory in January 2013 organized by BKG. Besides A10-020 four FG5 gravimeters participated in the comparison. The resulting offset for the A10-020 was determined as −7.0 μGal. The A10-020 also participated in ECAG2011 (Francis et al. 2013) and ICAG2013, both carried out in Walferdange, Luxemburg, in November 2011, and 2013, respectively. The observed offsets were −5.8 μGal (ECAG2011) and −4.7 μGal (ICAG2013). For the establishment of the gravity reference level for the PBOG14 a single offset value was calculated, based on international comparison campaigns as a weighted average with the weights proportional to the number of AG participating in the comparison campaign. The value used is −5.0 μGal with the standard deviation of 1.1 μGal.

The A10-020 also participated in two local comparison campaigns with the FG5-230 of Warsaw University of Technology. The campaigns were carried out at the gravimetric laboratory at Borowa Gora Observatory in December 2012 and May 2013 giving offsets for the A10-020 of +9.6 and +7.0 μGal, respectively. In the ECAG2011 the FG5-230 had an unusual offset of −11.9 μGal and the A10-020 had a −5.8 μGal offset (Francis et al. 2013), therefore the local campaign results are consistent within a few μGal. Yet as the FG5-230 did not participate in ICAG2013 and because of the unusual offset during ECAG2011 it was decided not to include these comparisons in the final gravity control data processing.

## 3 Vertical Gravity Gradient Determinations with LCR Gravimeters

Test measurements with LCR gravimeters conducted in the framework of the independent project in May 2012 (Dykowski et al. 2012) were used to check the methodology for the determination of vertical gravity gradient specified in the technical project. For nearly 330 measurements (2 on each station) performed on all PBOG14 stations the estimated measurement error for a single survey (Fig. 4a) was below 10.0 μGal (assumed in the technical project) with the average of 4.0 μGal and the standard deviation of 2.1 μGal. It was calculated as a standard deviation of LCR readings on repeated measurement levels. The difference between gravity reductions to a benchmark level from two independent vertical gravity gradient determinations (Fig. 4b) were smaller than 10.0 μGal (assumed in the technical project) with the average of 4.3 μGal. Reductions of the gravity to a benchmark level were calculated using a second degree polynomial approximation at the measurement height. Vertical gravity gradient values determined on PBOG14 stations using linear approximation vary from −3.68 to −2.75 μGal/cm with the average of −3.09 μGal/cm.

The uncertainty of the determined vertical gravity gradient at the station was estimated as the half of the difference

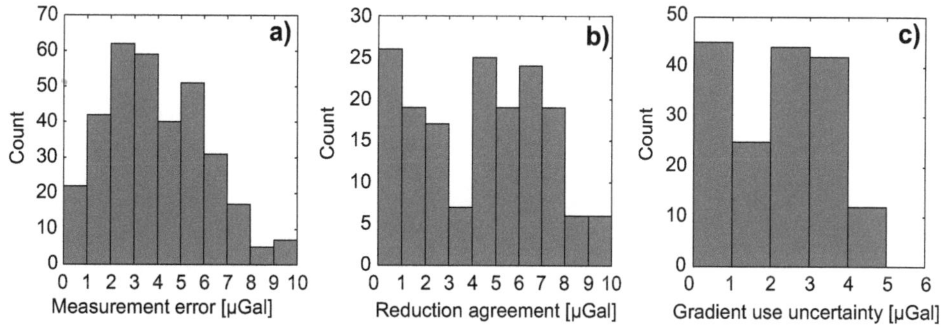


**Fig. 4** Histograms of vertical gravity gradient determinations: (**a**) measurement errors, (**b**) reduction agreement, (**c**) vertical gravity gradient uncertainty

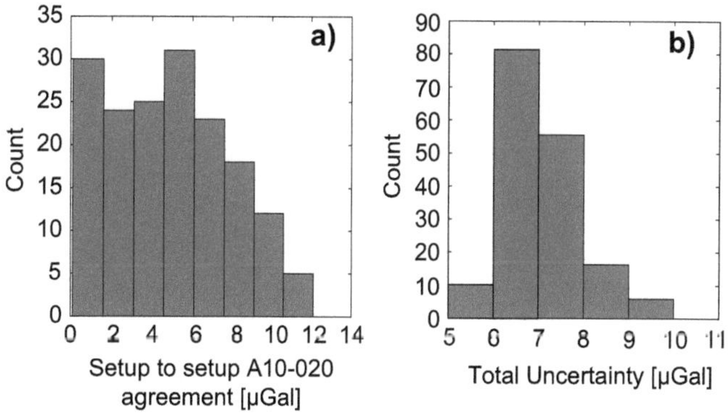


**Fig. 5** Histograms of absolute gravity determinations: (**a**) setup to setup agreement on PBOG14 stations, (**b**) Total Uncertainty values on PBOG14 stations

between two reductions of gravity from the level of determination with the A10-020 to a benchmark level with the use of a second-degree polynomial. Values of the estimated uncertainty are presented in Fig. 4. They are all smaller than 5.0 μGal (Fig. 4c) and their average is 2.2 μGal (Fig. 4c).

## 4 Absolute Gravity Measurements with the A10-020

At least two independent setups of the A10-020 absolute gravimeter were performed at each PBOG14 station accordingly to the measurement methodology applied (Krynski et al. 2013). A total number of nearly 350 single setups were performed on 168 PBOG14 stations. Setup to setup agreement at PBOG14 stations is presented in Fig. 5a. Its values vary from 0.2 to 12.0 μGal with the average of 5.0 and 3.0 μGal standard deviation. On five stations the setup to setup agreement exceeded 10 μGal (assumed as a limit in the technical project) yet these station were accepted as correct as their T.U. values were good.

## 5 Total Uncertainty Budget Estimation

A crucial part of the project concerned a reliable estimation of the A10-020 Total Uncertainty budget for a complete evaluation of the quality of PBOG14 gravity control. Total Uncertainty has been expressed by three main terms: $\sigma_{stat}$ – statistical uncertainty (function of set scatter), $\sigma_{sys}$ – measurement system uncertainty, and $\sigma_{model}$ – correction/reduction models uncertainty.

$$\text{T.U.} = \sqrt{\sigma_{stat}^2 + \sigma_{sys}^2 + \sigma_{model}^2}$$

The values of those terms and their components used for the estimation of the T.U. budget for gravity determination with the A10-020 on PBOG14 stations are presented below. All the values used were estimated based either on the absolute gravity measurements and gravity gradient determinations or on supplementary activities carried out during the project. For example, calibrations became a valuable source of information used for the estimation of the components of the T.U. budget for the A10-020 gravimeter allowing studies on its long and short term stability. At the same time, introducing the results of calibrations reduced the uncertainty estimates for the laser wavelength and frequency of the clock. For the laser the uncertainty was estimated as single calibration uncertainty as the laser frequency values for each measurement were calculated from linear interpolation. For the rubidium clock the uncertainty was calculated from the residual fit of a second degree polynomial to the calibration data. The values contributing to the T.U. estimate are as follows:

- $\sigma_{stat}$ – statistical uncertainty (function of set scatter) (sr);
- $\sigma_{sys}$ – measurement system uncertainty;
  - setup (0.1–6.0 μGal – half of setup-to-setup agreement) (sr);
  - system model (5.1 μGal based on Borowa Gora outdoor station results) (ss);

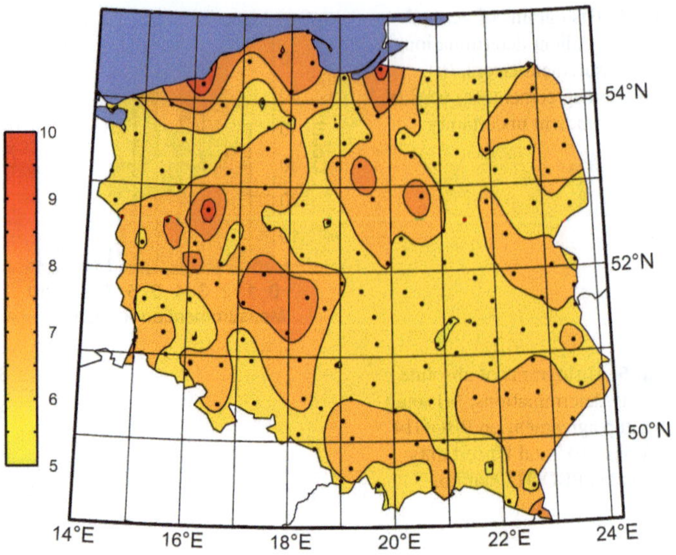


**Fig. 6** Distribution of the total uncertainty of the A10-020 from T.U. values on 168 PBOG14 stations (μGal)

  - laser wavelength (0.7 μGal) and rubidium frequency (0.2 μGal) (ss);
  - Self Attraction Correction (SAC) (0.3 μGal) (ss) – Jiang et al. (2012);
  - Diffraction Correction (DC) (0.6 μGal) (ss) – Jiang et al. (2012);
  - std dev. of A10-020 offset determination (1.1 μGal) (ss);
- $\sigma_{model}$ – correction/reduction models uncertainty;
  - gradient use (0.0–5.0 μGal) (sr);
  - tide and ocean loading (1.5 μGal) – Dykowski and Sekowski (2014) (ss);
  - polar motion (0.05 μGal) (ss);
  - barometric (1.0 μGal) (ss).

The presented uncertainties can be assigned into two groups: systematic ones (marked as "ss") and site related (marked as "sr"). Systematic uncertainties in total give a value of 5.5 μGal and the site related uncertainties vary from 1.6 to 8.4 μGal. For each PBOG14 station T.U. had been estimated based on the root of the sum of the squared values listed above. Values of T.U. estimated for the PBOG14 stations range from 5.7 to 10.0 μGal with the average of 7.1 μGal and a standard deviation of 0.8 μGal. The histogram of T.U values is presented in Fig. 5b. Distribution of estimated T.U. values on PBOG14 stations is presented in Fig. 6. Considering that T.U. values should not exceed 10 μGal (assumed in the technical project) the T.U. values estimated for PBOG14 stations prove to be satisfying as they were larger than 10 μGal only on four stations. At almost 85% PBOG14 stations T.U. values do not exceed 8 μGal. Reliable evaluation of the T.U. for the A10-020 gravimeter is an indication of good quality of the newly established gravity control in Poland. The correctness of the T.U. estimation was visible during the ICAG2013 campaign; even though it was almost twice smaller than the uncertainty specified by the manufacturer, it proved to be slightly larger than the offset estimated during the campaign. A more detailed study on the T.U. value for the A10-020 can be found in Dykowski et al. (2013).

## 6 Conclusions

Supplementary activities performed during the establishment of the modernized gravity control in Poland (PBOG14), i.e. calibrations of the laser and frequency standard of the A10-020 gravimeter, calibrations of the LCR gravimeters used, repeatable measurements with the A10-020 on the test network, and participation in comparison campaigns of absolute gravimeters were essential for ensuring the proper quality of its gravity reference level. Good stability of the components of the A10-020 gravimeter (laser wavelengths, clock frequency) during the project was confirmed, showing, however, a need to perform calibrations on regular basis. Regularly repeated gravity measurements with the A10-020 gravimeter carried on the test network at Borowa Gora Observatory confirmed the stable performance of the A10-020 throughout the realization of the project proving the A10-020 as a trustworthy tool to establish and maintain the gravimetric reference level. The participation of the A10-020 in absolute gravimeter comparison campaigns provided a correction to the established gravimetric reference level of +5.0 μGal for PBOG14 stations. The final stage of the PBOG14 evaluation was the T.U. estimation for each of the absolute gravity station. T.U. of the A10-020 gravimeter estimated during the project is smaller by 20–30% than the one based on the suggestions of the manufacturer. It proofs potentiality of the A10 gravimeter to provide high accuracy gravity determinations in field conditions. A thorough and complete evaluation of the T.U. budget proved that the quality of the PBOG14 established is as good as planned in the technical project.

**Acknowledgements** The project was financed by the Head Office of Geodesy and Cartography in Poland. The authors would like to acknowledge Walyeldeen Godah, Anna Jedrzejewska, Marcin Sekowski and Lukasz Zak of the Institute of Geodesy and Cartography, Warsaw, who participated in gravity measurements during the realization of the project.

## References

Dykowski P (2012) Vertical gravity gradient determination for the needs of contemporary absolute gravity measurements – first results. Rep Geod 92(1):23–35

Dykowski P, Sekowski M, Krynski J (2012) Testing the suitability of the A10-020 absolute gravimeter for the establshment of new gravity control in Poland, IAG Symposia, vol 140. Springer

Dykowski P, Krynski J, Sekowski M (2013) Approach to the A10 gravimeter total uncertainty budget estimation, IAG Symposia, vol 143. Springer

Dykowski P, Sekowski M (2014) Tidal investigations at Borowa Gora Geodetic-Geophysical Observatory, EGU General Assembly 2014, held 27 April – 2 May, 2014 in Vienna, Austria

Francis O, Baumann H, Volarik T, Rothleitner C, Klein G, Seil M, Dando N, Tracey R, Ullrich C, Castelein S, Hua H, Kang W, Chongyang S, Songbo X, Hongbo T, Zhengyuan L, Pálinkás V, Kostelecký J, Mäkinen J, Näränen J, Merlet S, Farah T, Guerlin C, Pereira Dos Santos F, Le Moigne N, Champollion C, Deville S, Timmen L, Falk R, Wilmes H, Iacovone D, Baccaro F, Germak A, Biolcati E, Krynski J, Sekowski M, Olszak T, Pachuta A, Agren J, Engfeldt A, Reudink R, Inacio P, McLaughlin D, Shannon G, Eckl M, Wilkins T, van Westrum D, Billson R (2013) The European comparison of absolute gravimeters 2011 (ECAG-2011) in Walferdange, Luxembourg: results and recommendations. Metrologia 50:257–268. doi:10.1088/0026-1394/50/3/257

Jiang Z, Pálinkâš V, Arias FE, Liard J, Merlet S, Wilmes H, Vitushkin L, Robertsson L, Tisserand L, Pereira Dos Santos F, Bodart Q, Falk R, Baumann H, Mizushima S, Mäkinen J, Bilker-Koivula M, Lee C, Choi IM, Karaboce B, Ji W, Wu Q, Ruess D, Ullrich C, Kostelecký J, Schmerge D, Eckl M, Timmen L, Le Moigne N, Bayer R, Olszak T, Ågren J, Del Negro C, Greco F, Diament M, Deroussi S, Bonvalot S, Krynski J, Sekowski M, Hu H, Wang LJ, Svitlov S, Germak A, Francis O, Becker M, Inglis D, Robinson I, (2012) The 8th International Comparison of Absolute Gravimeters 2009: the first Key Comparison (CCM.G-K1) in the field of absolute gravimetry, Metrologia, 49(6): 666–684. doi:10.1088/0026-1394/49/6/666

Krynski J, Olszak T, Barlik M, Dykowski P (2013) New gravity control in Poland – needs, the concept and the design. Geod Cartogr 62(1):3–21. doi:10.2478/geocart-2013-0001

Krynski J, Sekowski M (2010) Surveying with the A10-20 absolute gravimeter for geodesy and geodynamics – first results. Rep Geod 1(88):27–35

Micro-g LaCoste Inc. (2008) g8 user's manual, March 2008, p 18

Mäkinen J, Sekowski M, Krynski J (2010) The use of the A10-020 gravimeter for the modernization of the finnish first order gravity network. Geoinform Issue 2(1):5–17

Sekowski M, Krynski J (2010) Gravity survey with the A10-20 absolute gravimeter – first results, IAG symposium on terrestrial gravimetry: static and mobile measurements (TG-SMM2010), 22–25 June 2010, Saint Petersburg, Russia, pp 78–83

# Estimability in Strapdown Airborne Vector Gravimetry

David Becker, Matthias Becker, Stefan Leinen, and Yingwei Zhao

**Abstract**

Estimability was introduced as a measure of how much observations can contribute to the estimation process for a linear system. Several publications analyse the estimabilities for integrated navigation systems, comprising a strapdown inertial measurement unit (IMU) and observations from global navigation satellite systems (GNSS). This concept will hereby be adapted to strapdown vector gravimetry, being a special application of a IMU/GNSS system. Simulations are presented showing to what extent different observation types, as positions, velocities and attitudes, can contribute to the determination of gravity disturbances. Knowledge about these characteristics of the system may be useful for the planning of future airborne gravimetry missions.

In particular, the estimability of the deflection of the vertical (DOV) is analysed. During a non-accelerated flight, it is poorly estimable due to its high sensitivity to attitude errors. However, for flight manoeuvres comprising horizontal or vertical accelerations, the DOV estimability can be shown to increase significantly.

**Keywords**

DOV • Estimability • IMU • Observability • Strapdown vector gravimetry

## 1 Introduction

Strapdown airborne gravimetry has been carried out since the 1990s. Kalman filters are commonly used for the estimation process, including system states for gravity disturbances in addition to the standard navigation states (position, velocity and attitude) and sensor-error related states as biases or scale factors (Deurloo 2011; Kwon and Jekeli 2001; Tomé 2002). In order to estimate gravity disturbances, the actual acceleration of the aircraft has to be subtracted from the specific force measured by the IMU. Position and velocity observations taken from two-frequency phase-differential GNSS (PDGNSS) are commonly used for this (Deurloo 2011; Kwon and Jekeli 2001; Tomé 2002). The aircraft's acceleration is then, implicitly or explicitly, gained by numerical differentiation. Furthermore, attitude observations may be available when using multi-antenna GNSS-systems (Cohen et al. 1994; Vander Kuylen et al. 2006).

Hereby we present a systematic analysis, showing to *what extent* such additional observations support the gravity estimation process.

Estimability was introduced for covariance analyses of linear systems by Baram and Kailath (1987). By their definition, a system is estimable, if and only if the covariance in *any* direction of the state space, **u**, reduces through a set of observations. However, not only this binary property (estimable or non-estimable) is of interest, but also the quantification *how much* covariances reduce for a *specific* **u**, or in other words, how much the observations can contribute to the estimation process. This quantitative definition of

D. Becker (✉) • M. Becker • S. Leinen • Y. Zhao
Physical and Satellite Geodesy, Technische Universitaet Darmstadt, Franziska-Braun-Straße 7, 64287 Darmstadt, Germany
e-mail: dbecker@psg.tu-darmstadt.de





S. Jin, R. Barzaghi (eds.), *IGFS 2014*, International Association of Geodesy Symposia 144,
DOI 10.1007/1345_2015_209

estimability, as it is used in Hong et al. (2008) and Moon et al. (2008), will be used throughout this paper. Details can be found in Sect. 3.

An IMU usually contains three accelerometers in perpendicular directions. This enables the determination of the vertical and horizontal gravity components, i.e. the deflection of the vertical (DOV), while classical spring-gravimeters used in airborne gravimetry are limited to scalar gravimetry (cf. Forsberg and Olesen 2010). In Serpas and Jekeli (2005) it is shown, how measured DOVs can contribute to geoid determination. However, its estimability is close to zero during a non-accelerated horizontal flight due to a high sensitivity to attitude errors. In Jekeli and Kwon (1999) and Kwon and Jekeli (2001), a correlation technique for several flights along the same track is used to overcome this problem. In Sect. 4.2 it is shown, that also flight manoeuvres comprising accelerations can lead to a higher estimability of the DOV. In addition to an intuitive explanation, simulation results will be presented.

## 2 System Model and Observation Model

In this section we present the linearised system and observation models, which are used for the further analysis. For detailed derivations of such linearised strapdown models, see Jekeli (2001) and Titterton and Weston (2004). The frames used herein are the body-fixed $b$, navigation $n$, Earth-fixed $e$ and inertial frame $i$. Body-frame and IMU sensor frame are assumed to be identical.

A centralised Kalman filter is used, i.e. all inputs and outputs are connected by a single system model. For reasons of numerical stability, an error-state Kalman filter (sometimes called *indirect* Kalman filter) is used. In vector notation, the 18-dimensional system state is defined as

$$d\mathbf{x} = \{d\mathbf{p}^n, d\mathbf{v}^n, d\boldsymbol{\psi}^n, d\mathbf{b}^b_a, d\mathbf{b}^b_\omega, d\delta\mathbf{g}^n\}^T, \quad (1)$$

being the errors of position, velocity, attitude, accelerometer biases, gyroscope biases and gravity disturbance (three dimensions each). The system state transition from epoch $k$ to epoch $k+1$ is given as

$$d\mathbf{x}_{k+1} = \Phi_k d\mathbf{x}_k + \mathbf{w}_k = (\mathbf{I} + \mathbf{F}_k(t_{k+1} - t_k))d\mathbf{x}_k + \mathbf{w}_k, \quad (2)$$

with Gaussian noise $\mathbf{w}_k = \mathcal{N}(\mathbf{0}, \mathbf{Q}_k)$, time $t$, epoch numbers $k$ and unit matrix $\mathbf{I}$. For better readability, the epoch numbers will be omitted for most formulae. The matrix $\mathbf{F}$ is defined here in a simplified version, neglecting very small effects as e.g. differential changes of coriolis acceleration and transport rate due to small position errors. In Hong et al. (2008) it is shown, that such small changes of the system model do not significantly affect estimabilities.

$$\mathbf{F} = \left[\begin{array}{c|c|c|c|c|c} \mathbf{0} & \mathbf{I} & \mathbf{0} & \mathbf{0} & \mathbf{0} & \mathbf{0} \\ \mathbf{0} & -2\left[\omega^n_{ie}\times\right] - \left[\omega^n_{en}\times\right] & -\left[\mathbf{C}^n_b\mathbf{f}^b_{ib}\times\right] & -\mathbf{C}^n_b & \mathbf{0} & \mathbf{I} \\ \mathbf{0} & \mathbf{0} & \left[-(\boldsymbol{\omega}^n_{ie} + \boldsymbol{\omega}^n_{en})\times\right] & \mathbf{0} & -\mathbf{C}^n_b & \mathbf{0} \\ \mathbf{0} & \mathbf{0} & \mathbf{0} & \mathbf{0} & \mathbf{0} & \mathbf{0} \\ \mathbf{0} & \mathbf{0} & \mathbf{0} & \mathbf{0} & \mathbf{0} & \mathbf{0} \\ \mathbf{0} & \mathbf{0} & \mathbf{0} & \mathbf{0} & \mathbf{0} & \mathbf{0} \end{array}\right], \quad (3)$$

with specific force $\mathbf{f}_{ib}$, Earth rotation and transport rate $\boldsymbol{\omega}_{ie}$ and $\boldsymbol{\omega}_{en}$, and rotation matrix from $b$-frame to $n$-frame, $\mathbf{C}^n_b$. $[\mathbf{a}\times]$ denotes a skew symmetric matrix resembling the cross product: $[\mathbf{a}\times] \cdot \mathbf{b} = \mathbf{a} \times \mathbf{b}$, for vectors $\mathbf{a}$ and $\mathbf{b}$.

Note the identity matrix in the last column of $\mathbf{F}$: Because of the time-differential nature of $\mathbf{F}$, it implies that an error in the estimated gravity disturbance directly affects the first derivative of the velocity error, reflecting a fundamental property of airborne gravimetry: that gravity disturbances cannot be separated from aircraft accelerations, unless additional observations (as GNSS positions) are available.

The gravity disturbance is in that way modelled as a random walk process (cf. Deurloo 2011; Tomé 2002), which corresponds to a first-order Markov process. In the literature, higher-order Markov processes are commonly applied (cf. Deurloo 2011; Kwon and Jekeli 2001). However, the simple model is sufficient here, since the definition of estimability used in the following (see Sect. 3) only accounts for the functional transfer of observation information into the estimated states, while the system noise model is completely neglected.

The linearised observation models for the measurements $z$ are given as $\mathbf{z}_k = \mathbf{H}_k d\mathbf{x}_k + \mathbf{v}_k$, with $\mathbf{v}_k = \mathcal{N}(\mathbf{0}, \mathbf{R}_k)$ being Gaussian noise. For the three observation types: position, velocity and attitude observations, the matrices $\mathbf{H}$ are given as

$$\mathbf{H}_{\text{position}} = \left[\mathbf{I}\ \mathbf{0}\ [\mathbf{l}^n\times]\ \mathbf{0}\ \mathbf{0}\ \mathbf{0}\right] \quad (4)$$

$$\mathbf{H}_{\text{velocity}} = \left[\mathbf{0}\ \mathbf{I}\ \left[-\mathbf{C}^n_b\boldsymbol{\Omega}^b_{ib}\mathbf{l}^b\times\right]\ \mathbf{0}\ (\mathbf{C}^n_b\left[\mathbf{l}^b\times\right])\ \mathbf{0}\right] \quad (5)$$

$$\mathbf{H}_{\text{attitude}} = \left[\mathbf{0}\ \mathbf{0}\ \mathbf{I}\ \mathbf{0}\ \mathbf{0}\ \mathbf{0}\right], \quad (6)$$

where $\mathbf{l}$ is the GNSS-antenna's position w.r.t. the body frame origin (also called *lever arm*), and $\boldsymbol{\Omega}^b_{ib} = \left[\boldsymbol{\omega}^b_{ib}\times\right]$.

## 3 Estimability

In this section, we closely follow the approach of Moon et al. (2008). For an epoch $k$ and the series of $N$ upcoming observations, estimability is a measure of how much the system state estimate can benefit from these observations. For this time-window, the system noise model is neglected for the variance propagation (i.e. $\mathbf{Q}_k = \mathbf{0}$). The *information matrix* of such a series of observations is given as

$$\mathbf{L}_{k,k+N} = \Sigma_{i=0}^{N} \mathbf{\Phi}_{k+i,k}^{T} \mathbf{H}_{k+i}^{T} \mathbf{R}_{k+i}^{-1} \mathbf{H}_{k+i} \mathbf{\Phi}_{k+i,k}, \quad (7)$$

where $\mathbf{\Phi}_{k+i,k} = \mathbf{\Phi}_{k+i-1} \mathbf{\Phi}_{k+i-2} \ldots \mathbf{\Phi}_k$ is used as transition matrix from epoch $k$ to epoch $k+i$. The estimability $\nu$ may then be calculated with respect to the a-priori covariance matrix $\mathbf{P}_k$ and a direction in the state space, $\mathbf{u}$, as

$$\nu(\mathbf{L}_{k,k+N}, \mathbf{P}_k, \mathbf{u}) = \frac{\mathbf{u}^T(\mathbf{P}_k - \mathbf{P}_{k+N})\mathbf{u}}{\mathbf{u}^T \mathbf{P}_k \mathbf{u}}, \quad (8)$$

$$\text{with } \mathbf{P}_{k+N} = (\mathbf{P}_k^{-1} + \mathbf{L}_{k,k+N})^{-1}. \quad (9)$$

This relation of a-priori covariances $\mathbf{P}_k$ and a-posteriori covariances $\mathbf{P}_{k+N}$ yields an estimability $\nu$ between zero (observations do not improve the system state in the direction $\mathbf{u}$) and one (a-posteriori covariances reach zero). Negative values for $\nu$ are not possible, since observations will never increase the system's estimated uncertainty (remember the neglect of the system noise).

For our analysis, we chose $\mathbf{u}$ from a standard basis $\mathbf{e}_i$ of the state space of the system defined in Sect. 2. Thus, we compute estimabilities for system states, in particular for the vertical and horizontal gravity disturbances.

In Moon et al. (2008), an equivalent formulation of $\nu$ can be found, based on a singular value decomposition (SVD) of the information matrix. This formulation is used for the experiments in Sect. 4, due to its higher numerical stability:

$$\text{SVD}: \ \mathbf{U\Sigma V}^T = \sqrt{\mathbf{P}_k} \mathbf{L}_{k,k+N} \sqrt{\mathbf{P}_k}, \quad (10)$$

where $\mathbf{\Sigma}$ is a diagonal matrix with elements $\sigma_i$. Then,

$$\mathbf{D} = \text{diag}(\frac{\sigma_i}{1+\sigma_i}), \ \mathbf{z} = \frac{\sqrt{\mathbf{P}_k}\mathbf{u}}{\|\sqrt{\mathbf{P}_k}\mathbf{u}\|_2}, \quad (11)$$

$$\nu(\mathbf{L}_{k,k+N}, \mathbf{P}_k, \mathbf{u}) = \mathbf{z}^T \mathbf{UDU}^T \mathbf{z}. \quad (12)$$

## 4 Experiments and Results

A simple simulator is used to obtain the matrices $\Phi_k$ along predefined trajectories, each of them flown at 100 m/s. The estimability was calculated for a set of 100 observation epochs, at a rate of 1 Hz. These 100 s intervals reflect a spatial resolution of 5 km (half wavelength). The observation noise is modelled to be Gaussian and non-correlated over time. Note, however, that real GNSS-observations *are* correlated over time in general, in particular when using higher sampling rates.

For all scenarios, we define an a-priori matrix of covariances $\mathbf{P}_k$, being a diagonal matrix of assumed system state variances for a typical navigation-grade IMU: 1 cm for position, 1 mm/s for velocity, 10 arcsec for attitude, 1 mGal for accelerometer biases, 1 arcsec/s for gyroscope drift, and 5 mGal for gravity disturbances (horizontal and vertical channel). The lever arm is arbitrarily defined as $\mathbf{l}^b = (3\,\text{m}, 3\,\text{m}, 3\,\text{m})^T$.

### 4.1 Non-accelerated, Horizontal Flight

The non-accelerated, horizontal flight is the default type of motion in Airborne Gravimetry missions. The estimability is shown for a range of position observation accuracies, see Fig. 1, scenario (a). Note, that even with millimetre-level position observations, the DOV estimability remains close to zero. PDGNSS position observation accuracies can be expected to be on the order of several centimeters to decimeters, depending on the baseline length (distance between aircraft and reference station).

Next, velocity observations are introduced in addition to the position observations. The vertical channel of $\boldsymbol{\delta}\mathbf{g}$ can benefit from these additional observations (Fig. 2). However, PDGNSS-derived velocities can be expected to have an uncertainty of several centimeters per second, not yielding any substantial improvements, if position observations on the 20 cm level or better are also available. For all of the combinations of position and velocity observations shown in Fig. 2, the DOV estimability remains very small ($\nu < 1\%$, no figure shown). Apparently, the attitude uncertainty is the predominating factor for the DOV estimation, which cannot improve through position and velocity observations for the non-accelerated flight.

With high-accuracy roll and pitch observations being available, the DOV gains estimability (Fig. 3, right). Note, however, that such observations are difficult to obtain in practice. For example when using a multi-antenna attitude-GNSS-system, the baseline lengths (distances between any two antennas) are limited to the dimensions of the aircraft (several meters). In such a setup, those systems only provide heading and tilt angle accuracies of several arc minutes (cf. Cohen et al. 1994; Vander Kuylen et al. 2006).

The *vertical* channel remains practically unaffected by attitude observations (Fig. 3, left), meaning that the a-priori attitude uncertainty of the system of 10 arcsec is not the limiting factor for the vertical gravity determination. Also, heading observations will not improve the estimabilities of $\boldsymbol{\delta}\mathbf{g}$ (no figure shown).

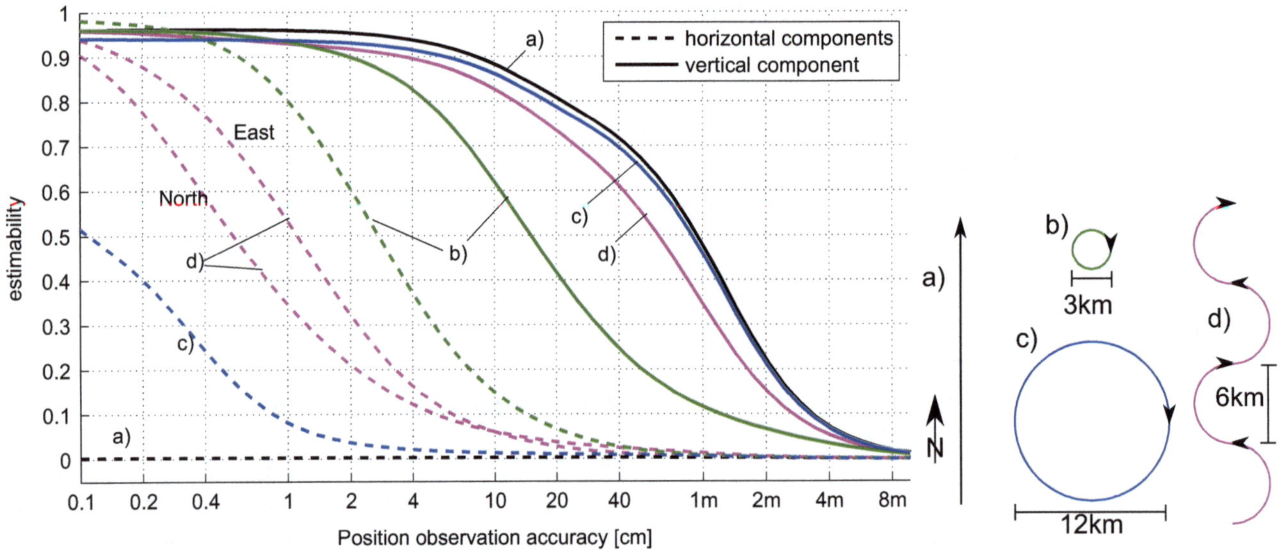


**Fig. 1** Estimabilities of $\boldsymbol{\delta g}$ for different position observation accuracies and different manoeuvres (**a**) to (**d**), at 100 m/s constant speed (ground tracks are shown on the *right*). For symmetry reasons, the scenarios (**a**), (**b**) and (**c**) have identical north- and east-component estimabilities (*dashed lines*)

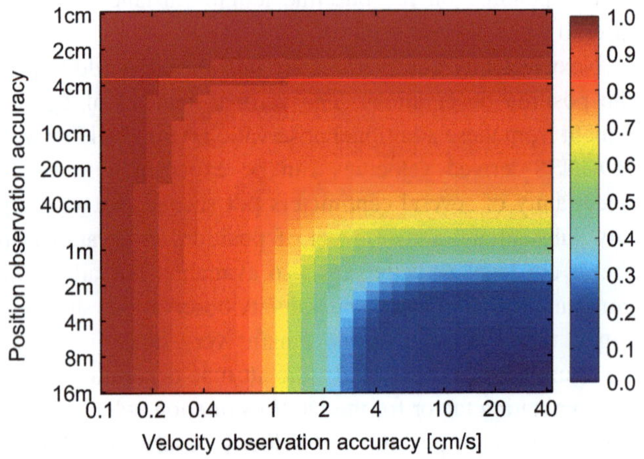


**Fig. 2** Non-accelerated, horizontal flight: vertical gravity disturbance estimabilities for combinations of position and velocity observation accuracies

## 4.2 Flight Manoeuvres

As shown before, the DOV estimability is closely related to misalignments. These can become estimable, even only having position observations, when the aircraft is accelerated, e.g. in vertical direction by non-linear height-changes, or in horizontal direction by course changes. This effect is intuitive, because accelerating into a misaligned direction will yield an offset in position. In the filtering context, this offset will show up as an innovation.

Figure 1 shows the estimabilities for different manoeuvres: straight line (a), horizontal circles of 3 km (b) and 12 km (c) diameter, and a serpentine-shaped trajectory with turns of 6 km diameter (d), all flown at 100 m/s constant speed, with only position observations being available. Apparently, there is a trade-off between the estimabilities of the horizontal and vertical components: Higher aircraft accelerations yield a higher DOV estimability, but also a lower estimability for the vertical component. Thus, the non-accelerated, horizontal flight along a straight line (a) is optimal for the estimation of the vertical component, while a manoeuvre with strong accelerations, as the small circle (b) or the serpentine line (d), is preferable for the determination of the DOV.

Note, that during real flights, actual errors of the IMU sensors tend to be larger for manoeuvres of higher dynamics, as the manoeuvres (b), (c) and (d) shown above. Such effects were completely neglected in this section. Therefore, the findings of this section should be regarded as a theoretic basis for future experiments.

Also note, that the aircraft's attitude, roll, pitch, and yaw, does not affect the gravity's estimability (no figure shown). Consequently, the findings presented hereby are also applicable to non-strapdown systems.

# 5 Conclusions and Future Work

The work presented here shows how flight manoeuvres and different combinations of observations can affect the estimability of gravity disturbances. This type of analysis can be a useful tool for understanding the characteristics of

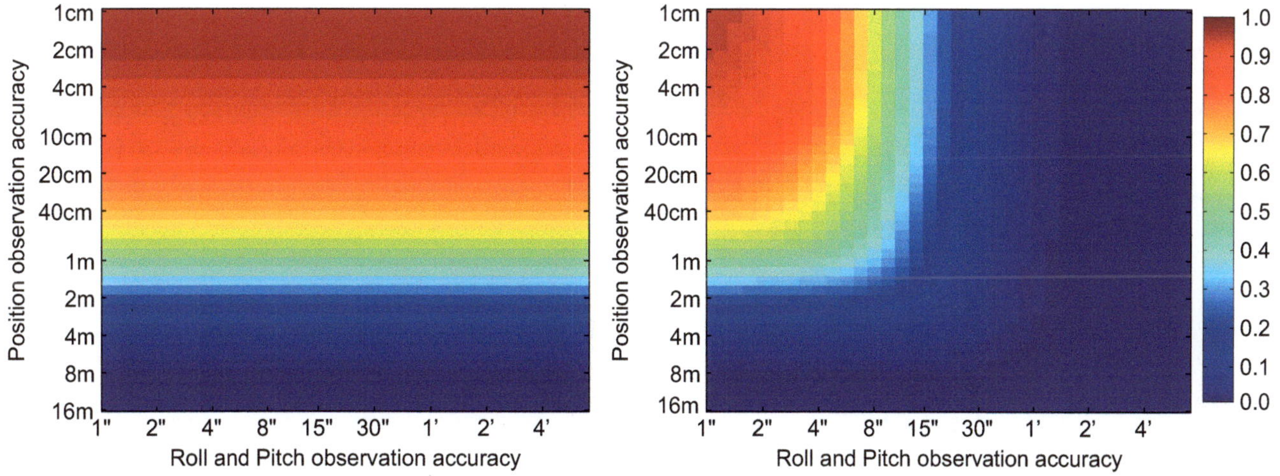


**Fig. 3** Non-accelerated, horizontal flight: gravity disturbance estimabilities for combinations of position and roll/pitch observation accuracies. *Left*: vertical gravity component, *right*: horizontal components (DOV)

a strapdown gravimetry system. This may contribute to the planning process of future missions. Note, that the findings presented here depend on the states' a-priori covariances $\mathbf{P}_k$, which was chosen here for a typical navigation-grade system. For example, a lower-grade system will benefit more from GNSS-derived attitude observations, because the attitude uncertainties are higher.

Real systems may differ from the idealised models presented here, as e.g. the assumption of non-correlated Gaussian noise may not apply, in particular for the GNSS observations. Also, trajectories with higher dynamics, as presented in Fig. 1, may lead to larger errors of the inertial sensors.

An analysis of the sensor bias estimabilities is subject to further research. In particular, the separation of time-varying accelerometer biases and gravity disturbances may increase the long-term stability of the measured gravity.

## References

Baram Y, Kailath T (1987) Estimability and regulability of linear systems. In: 26th IEEE conference on decision and control, 1987, vol 26. IEEE

Cohen CE, Parkinson BW, McNally BD (1994) Flight tests of attitude determination using GPS compared against an inertial navigation unit. Navigation 41(1): 83–97

Deurloo RA (2011) Development of a Kalman filter integrating system and measurement models for a low-cost strapdown airborne gravimetry system. Ph.D. dissertation, Faculty of Sciences, University of Porto, Porto

Forsberg R, Olesen AV (2010) Airborne gravity field determination. In: Sciences of geodesy-I. Springer, Berlin/Heidelberg, pp 83–104

Hong S et al (2008) Observability measures and their application to GPS/INS. IEEE Trans Veh Technol 57(1):97–106

Jekeli C (2001) Inertial navigation systems with geodetic applications. Walter de Gruyter, Berlin

Jekeli C, Kwon JH (1999) Results of airborne vector (3-D) gravimetry. Geophys Res Lett 26(23):3533–3536

Kwon JH, Jekeli C (2001) A new approach for airborne vector gravimetry using GPS/INS. J Geodesy 74(10):690–700

Moon JH et al (2008) Estimability measures and their application to GPS/INS. J Mech Sci Technol 22(5):905–913.

Serpas JG, Jekeli C (2005) Local geoid determination from airborne vector gravimetry. J Geodesy 78(10):577–587

Titterton D, Weston JL (2004) Strapdown inertial navigation technology, vol 17. The Institution of Engineering and Technology, London

Tomé P (2002) Integration of inertial and satellite navigation systems for aircraft attitude determination. Ph.D. dissertation, Faculty of Sciences, University of Porto, Porto

Vander Kuylen L et al (2006) Comparison of attitude performance for multi-antenna receivers. Eur J Navig 4(2):1–9

# A First Traceable Gravimetric Calibration Line in the Swiss Alps

Urs Marti, Henri Baumann, Beat Bürki, and Christian Gerlach

**Abstract**

In order to determine scale factors or more complex calibration functions of relative gravimeters it is usually necessary to calibrate these instruments regularly on points with known gravity values. Especially well suited are points with absolute measurements and with large gravity differences. This implies that gravimetric calibration lines are usually implemented in north–south direction or on stations with big height differences. The latter has the advantage that traveling time can be kept rather short. In 2013 we established a calibration line in the Swiss Alps between Interlaken (altitude 570 m) and Jungfraujoch (altitude 3,500 m). This line consists of 7 absolute stations and several eccentric points. The total gravity difference is more than 600 mGal. All absolute stations are easily accessible by car or are located in immediate vicinity of a station of the Jungfrau railway. Therefore, it is possible to measure the whole line with relative instruments in a closed loop in 1 day. The absolute stations have been chosen in a way that they are accessible during the whole year and that the gravity difference between two neighboring stations does not exceed 150 mGal. So, it is possible to calibrate as well gravimeters with a limited measuring range or with a non-linear scale factor. The expanded uncertainties (k = 2) of the gravity values at the absolute stations are varying between 5 and 6 μGal. All vertical gravity gradients (VGGs) have been determined by relative measurements on three levels above the marker. The newly established calibration line is free to be used by the whole gravity community and we hope that many institutions will profit. It is to our knowledge worldwide the first traceable gravimetric calibration line satisfying the BIPM Mutual Recognition Arrangement (MRA).

The paper describes the details of the establishment of the calibration line, the absolute measurements and a first comparison with relative measurements with three relative gravimeters.

**Keywords**

Absolute gravimetry • Gravimetric calibration line • Switzerland

U. Marti (✉)
Federal Office of Topography swisstopo, Wabern, Switzerland
e-mail: urs.marti@swisstopo.ch

H. Baumann
Federal Institute of Metrology METAS, Wabern, Switzerland

B. Bürki
Geodesy and Geodynamics Lab, ETH Zurich, Switzerland

C. Gerlach
Bavarian Academy of Sciences and Humanities, Munich, Germany





S. Jin, R. Barzaghi (eds.), *IGFS 2014*, International Association of Geodesy Symposia 144,
DOI 10.1007/1345_2015_52

## 1 Introduction

Gravimetric calibration lines play an important role at least since the appearance of the first transportable static relative gravimeters in the 1940s. Their readings (spring lengths, counter units, electrical units) have to be translated into acceleration units. This can be done by relative measurements on stations with known gravity or gravity differences. Modern calibration lines consist at least of two stations with absolute gravity. More points help in the determination of non-linear characteristics of the calibration function. The gravity difference between the extreme points should be as large as possible to cover a large part of the gravimeters measuring range. On the other hand, the absolute stations should be located as close as possible in order to keep the measuring time short. This is in favor of accuracy and cost. Therefore, absolute stations with big height differences are very well suited for the use as calibration lines. One example of such a line is the German calibration line from Garmisch to the Zugspitze (Timmen et al. 2006) with a gravity difference of more than 500 mGal. Another calibration line to be mentioned is the one located in Orangeville north of Toronto with 5 stations and a gravity range of 119 mGal. On this line, all the Scintrex instruments are tested before their delivery to the customer (Scintrex 2012).

We now extended the existing Swiss calibration line between Interlaken and the Jungfraujoch with a total gravity range of more than 600 mGal. The well distributed seven absolute stations allow a fast check and calibration of all types of relative meters.

## 2 Former Works for a Calibration Line in Switzerland

The first gravimetric calibration lines in Switzerland were established in the 1950s near the cities where such instruments were in use at that time (Zurich, Lausanne, Geneva). These lines were usually rather short and covered only a very limited gravity range. For comparisons on a bigger scale, calibration lines in France (usually Paris – Toulouse – Bagnères) and Germany were visited. For some applications the calibrations took place on various lines of the National Gravimetric network (Hunziker 1959).

But already in 1954 the line between Interlaken and the Jungfraujoch was observed for the first time with Worden gravimeters. This line had two intermediate stations in Lauterbrunnen and on Kleine Scheidegg (Hunziker 1959). Due to the limited measuring range of the Worden instrument at this time, an accuracy in the order of 0.1 mGal was reached.

In 1978 the first modern absolute gravity measurements were performed in Switzerland (Marson et al. 1981) where the rise and fall meter of the Istituto di Metrologia G. Colonnetti (IMGC, Torino; today INRiM) was used. Besides of geodynamic purposes, these measurements were foreseen as well for a renewal of the calibration line between Interlaken and Jungfraujoch. These two stations have been measured in 1979 and show a gravity difference of about 605 mGal. In 1980, five additional stations with several eccentric markers were established and observed with three Lacoste&Romberg (LCR) relative meters (Klingelé and Kahle 1987). The gravity difference between two neighboring stations does not exceed 200 mGal in order to allow measurements with instruments of a limited range (e.g. LCR type D meters). The standard deviation between two neighboring stations was in the order of 0.01–0.04 mGal.

In 1999, the first Swiss Absolute Gravity Meter (FG5) was acquired by the Federal Institute of Metrology (METAS) and one of its first applications was the re-observation of the absolute station at Jungfraujoch. Unfortunately, at that time the original station from 1979 was not accessible anymore and a new station in another room (around 10 m away from the original one) had to be chosen. Therefore, the results are not comparable.

In 2008, METAS declared in the frame of the CIPM (Comité International de Poids et Mesures) Mutual Recognition Arrangement (MRA) (BIPM 2003), its first CMC (Calibration and Measurement Capability) in the field of gravity acceleration. The claimed CMC of 4 $\mu$Gal ($k = 1$) could be confirmed in several international key comparisons (e.g. Jiang et al. 2012; Francis et al. 2013).

## 3 The Current Network

In 2009 a renewal of the calibration line between Interlaken and the Jungfraujoch was decided. All the seven original sites of 1979–1980 should get an absolute gravity station. Therefore, a modification of the network was necessary and stations suitable for FG5 measurements (Niebauer et al. 1995) had to be found. At the same time, the original absolute station in Interlaken was moved to a more suitable and stable site out of the groundwater zone. The station at Jungfraujoch had to be moved again because the station from 1999 was no longer useable. The new point was re-built on the exact location of the original point from 1979. But due to massive new constructions, it has to be considered that gravity has changed significantly since then.

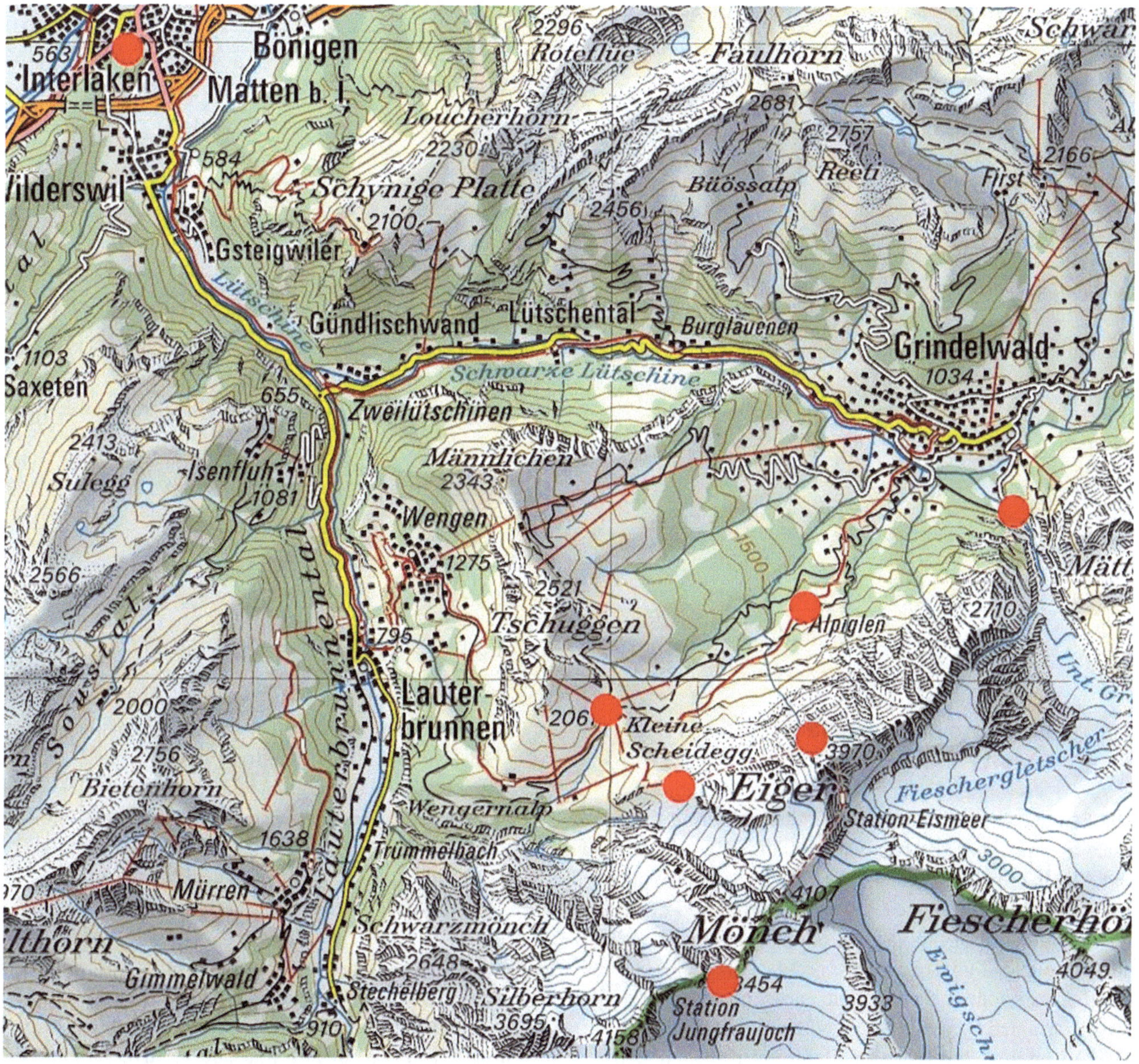


**Fig. 1** Location of the absolute stations of the calibration line

The five new absolute stations (see Fig. 1) were chosen in the vicinity of the original relative sites and are easily accessible by car or by train. Four of these sites are located directly at stations of the Jungfrau railway line. The station in Grindelwald was moved from the village center – which is in a sliding zone – to a stable cavern in the rocks. All the stations are suitable for FG5-measurements and easily accessible by car or train during the whole year. The location of the points has been chosen so that their long-term preservation should be guaranteed.

The coordinates and heights of all the stations were determined with an accuracy of about 10 cm by means of GPS measurements or by connections to surrounding points of cadastral or railway surveying. The points in Interlaken and Grindelwald are connected to the National Levelling network. Besides of the absolute stations, some eccenters for relative measurements form part of the network. Whenever possible, the still existing points of the measurements of 1980 have been selected. All points are documented in the national database of geodetic reference markers. Table 1 gives an overview about the characteristics of the stations.

**Table 1** Overview about the stations with approximate heights and gravity values

| Name | Height (m) | Gravity (mGal) | Δg (mGal) | Σ Δg (mGal) | Number of eccentric relative stations |
|---|---|---|---|---|---|
| Interlaken | 570 | 980,506 | 0 | 0 | 4 |
| Grindelwald | 1,010 | 980,379 | 127 | 127 | 3 |
| Alpiglen | 1,616 | 980,283 | 96 | 223 | 1 |
| Kleine Scheidegg | 2,063 | 980,205 | 78 | 301 | 3 |
| Eigergletscher | 2,320 | 980,143 | 62 | 363 | 2 |
| Eigerwand | 2,865 | 979,994 | 149 | 512 | 2 |
| Jungfraujoch | 3,456 | 979,900 | 94 | 606 | 4 |
| Jungfraujoch (Sphinx) | 3,578 | 979,860 | 40 | 646 | 1 (only relative) |

**Table 2** Absolute measurements on the calibration line with number of accepted sets/drops, gravity at the reference level, measured VGG, gravity reduced to the marker and the standard uncertainty u (k = 1) of gravity at the marker

| Station | sets/drops | g @ 130 cm (mGal) | VGG (mGal/m) | g @ Marker (mGal) | u @ Marker (μGal) |
|---|---|---|---|---|---|
| Interlaken | Not measured yet | | | | |
| Grindelwald | 34/3,205 | 980,378.785 | 0.1448 | 980,378.973 | 2.0 |
| Alpiglen | 34/3,293 | 980,283.019 | 0.2924 | 980,283.399 | 2.1 |
| Kl. Scheidegg | 20/1,982 | 980,204.261 | 0.2859 | 980,204.633 | 2.5 |
| Eigergletscher | 26/2,553 | 980,142.708 | 0.3091 | 980,143.110 | 2.4 |
| Eigerwand | 20/1,960 | 979,993.698 | 0.3623 | 979,994.169 | 2.5 |
| Jungfraujoch | 20/1,890 | 979,899.783 | 0.3972 | 979,900.299 | 2.5 |

## 4 Absolute Measurements and Vertical Gravity Gradients

The first absolute measurements on the calibration line took place in September 2010 on the two stations in Grindelwald and Alpiglen with the FG5 #209 of METAS. Approximately 24 sets with 100 drops each were observed and corrected for the influence of the variable components such as Earth tides, ocean loading, atmospheric pressure and polar motion. On the occasion of the absolute measurements the VGG was determined by observations with a Scintrex CG-3 M of METAS on three levels (about 40, 80 and 130 cm above the marker).

The four stations at Kleine Scheidegg, Eigergletscher, Eigerwand and Jungfraujoch were observed in October 2013 with the same instrument upgraded to a FG5-X. The measuring scenario was approximately the same as in 2010 with 20–26 sets and the determination of the VGGs. The measurements in Interlaken could not be realised yet until now. The VGG was determined on these stations by measurements with the ZLS Burris instrument #B-78 of the Bavarian Academy of Sciences and Humanities (BAdW).

The obtained standard uncertainties of the gravity values at the reference height of the instrument (130 cm above the marker) are all around 2.0 μGal, which is significantly better than the 4 μGal (k = 1) claimed by METAS in the CMC (Calibration and Measurement Capapability) of the MRA. The standard deviation of the VGGs is better than 1.2 μGal/m for all stations. This results in a total uncertainty of the gravity value at the ground markers of better than 2.5 μGal. The results are summarized in Table 2 and the details can be found in the global database for absolute measurements AGrav (agrav.bkg.bund.de, Wilmes et al. 2009).

## 5 Relative Measurements

In parallel to the absolute measurements, a first campaign with relative gravimeters took place in October 2013. The following three instruments were used: A Scintrex CG-5 of swisstopo/ETHZ on all absolute and relative stations, the ZLS Burris B-78 from BAdW on the absolute stations and a few selected eccenters and a LCR type G meter (G-87) with a digital feedback system of DGFI who measured always in parallel with the ZLS instrument. Theoretically, the whole calibration line can be measured in 1 day, but in order to increase accuracy and reliability, it is better to invest at least 2 days. In our case we did the measurements in 3 days and each station has been visited at least on 2 days with each instrument.

After the field measurements, several calculation runs were performed and a variety of results was produced. In a first step, daily solutions for each instrument with one linear drift per day were calculated. This allowed a first rough check of consistency and the detection of problematic parts. In a next step, complete solutions with all relative measurements per instrument were produced. Finally, a combined solution

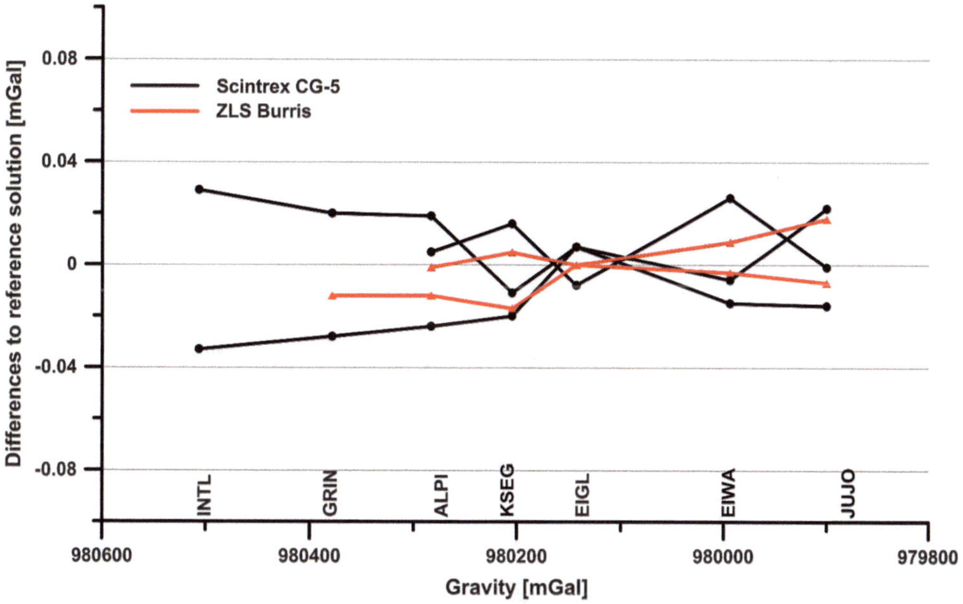


**Fig. 2** Comparison of the daily solutions of the CG-5 and the ZLS with the complete solution of the corresponding instrument

of all relative measurements together with the absolute measurements with an unknown scale factor per instrument was computed.

## 6 Comparisons of the Measurements of 2013

The first comparison was between the daily solutions of each instrument in relation to a complete solution of each instrument. This gives us the residuals of Fig. 2, where for simplicity reasons, only the results for 1 point per site (usually the absolute station) is shown. The CG-5 shows a repeatability in the order of ±0.03 mGal. This value seems to be a little higher than other measurements with this instrument but is explainable by some remaining hysteresis effects (Reudink et al. 2014) and some difficulties in estimating the drift parameters out of the data of one single day only. In the combined solution of the CG-5 measurements the residuals are usually smaller than 0.015 mGal and the drift parameters for each day get more consistent.

The residuals of the ZLS Burris daily solutions are smaller (±0.015 mGal) than the ones of the CG-5. But for the ZLS as well, it is valid, that the residuals of the observations become better in the combined solution (almost all below 0.01 mGal) and the drift parameters are determined more reliably. It has to be noted that the ZLS did not measure in Interlaken and only 1 day of the Grindelwald measurements could be used due to strong external perturbations on 1 day. The results of the LCR daily solutions are not displayed in Fig. 2 but their residuals are as well in agreement of better than 0.03 mGal. Due to the same perturbations as with the ZLS, it was impossible to evaluate the LCR measurements from Grindelwald with sufficient accuracy.

In a next step, the results of the complete solutions for each instrument were compared among each other and plotted with respect to the absolute measurements. For the LCR and ZLS two solutions from swisstopo and BAdW are available with independent software and different approaches for drift modeling. These two solutions agree in general better than 20 μGal. The result is shown in Fig. 3. The large residuals of the LCR measurements are clearly visible. They go from −97 μGal in Alpiglen to +125 μGal at Jungfraujoch and show a rather linear behavior with respect to the absolute gravity value. This indicates, that the scale factor of this instrument has to be improved. The results of the other two instruments (CG-5 and ZLS) are in rather good agreement of about ±20 μGal with the absolute measurement. An obvious scale is not visible. But there is one significant exception at the station of Kleine Scheidegg: There, the ZLS measurements show an offset of about −160 μGal towards the absolute and the CG-5 measurements, although their daily solutions are in good agreement (see Fig. 2). An error in the point identification can be excluded as a possible source of the differences and the reason remains basically unknown until now. One possible explanation could eventually be found in a problem of the calibration table of the instrument. In this table, transformation factors are given for every 50 counter units (CU) of the instrument with an accuracy of 15 μGal. If we translate these values into local scale factors, we get the values shown in Fig. 4 and we see that the scale at Kleine Scheidegg (point KS in Fig. 4)

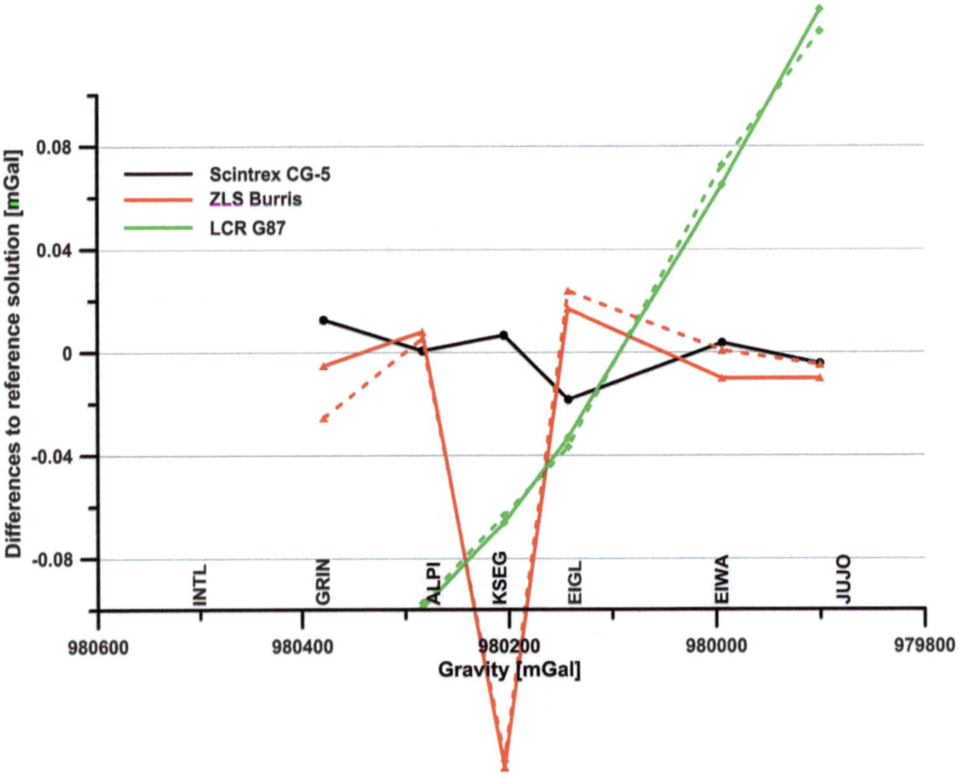


**Fig. 3** Comparison of the solutions of each instrument. Differences to the absolute measurements. *Dashed lines* indicate alternative evaluations by BAdW

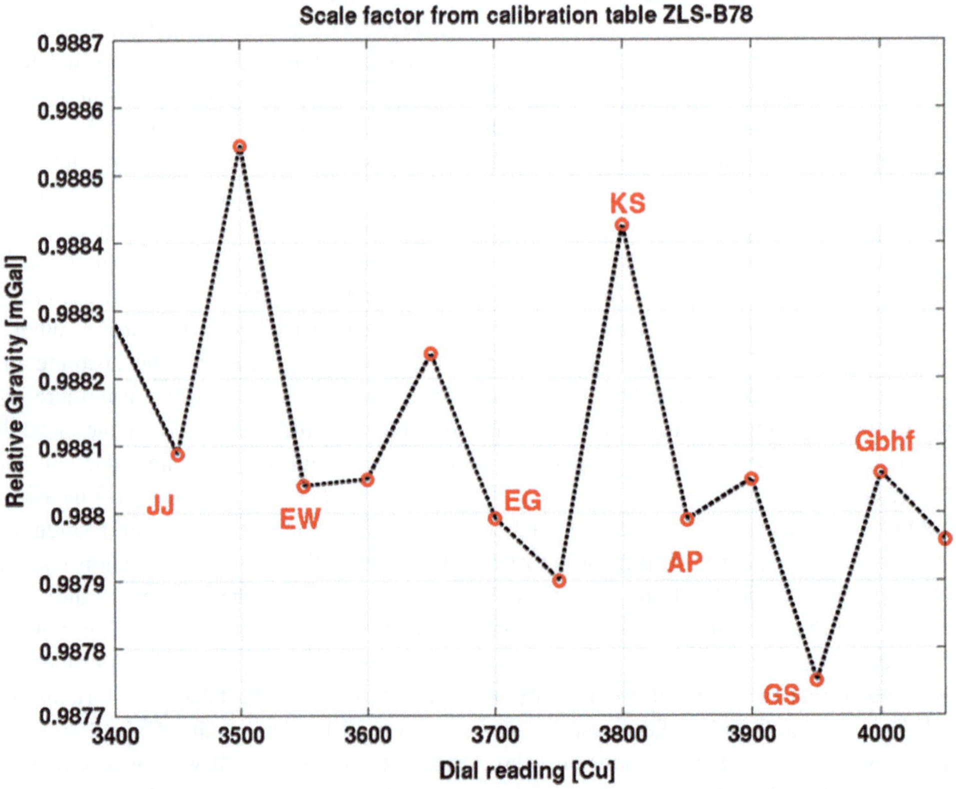


**Fig. 4** Local scale factors for the ZLS-B78 instrument with the extreme value at Kleine Scheidegg (point KS)

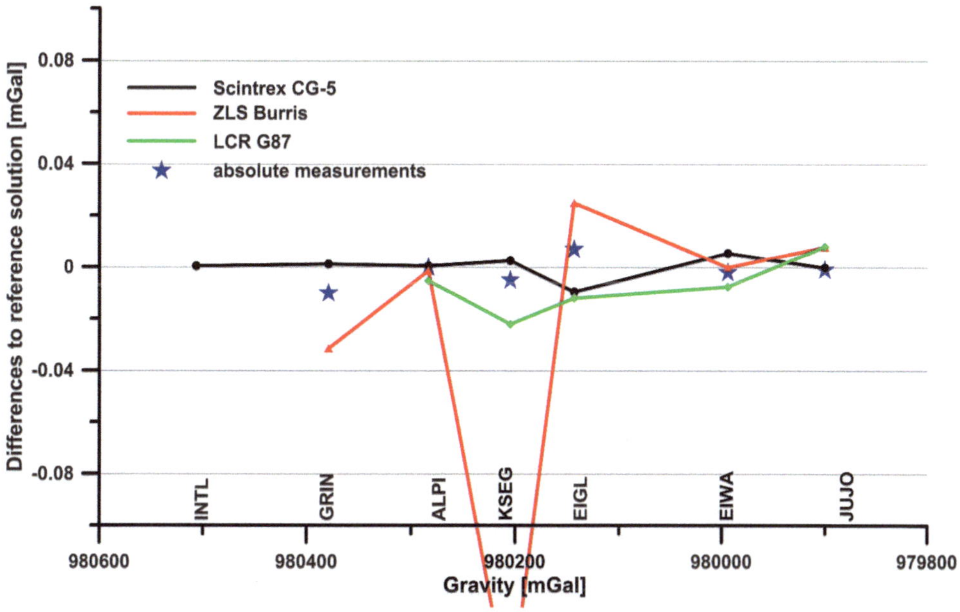


**Fig. 5** Comparison of the results of the relative measurements of each instrument and the absolute measurements after applying a constant scale factor for each relative gravimeter. Differences to the overall-adjustment

shows an extreme value. Whereas on most other points the factor is around 0.9880, on KS the factor is 0.9884. This scale difference would result in a gravity difference to the neighboring stations in the order of 100 μGal. If this is really the reason for the different result at KS has to be investigated more in other campaigns with points where the same counter positions as on KS are used.

Finally, we computed an overall-adjustment of all absolute and relative measurements together and estimated one linear scale factor for each relative instrument. The weighting of the absolute measurements was done according to the obtained accuracies as documented in Table 2. For the relative measurements, generally the accuracy as declared in the instrument specifications was introduced. In case of increased noise during one set of measurements, it was downweighted accordingly. The measurements of the ZLS at KSEG have been eliminated for the combined solution. The found scale factors are $-14 \pm 10$ ppm for the CG-5, $-74 \pm 23$ ppm for the ZLS and $+568 \pm 61$ ppm for the LCR. Whereas for CG-5 and ZLS these factors are almost negligible, for the LCR it is highly significant and has to be considered as we have seen already in Fig. 3. In this adjustment, the absolute measurements got some residuals as well (see Fig. 5). They are in the range from $-10$ μGal (Grindelwald) to $+7$ μGal (Eigergletscher). This is significantly larger than the accuracies we obtained (see Table 2) and indicate some systematic differences between absolute and relative measurements. The mean accuracy of one relative measurement were obtained as 8.6 μGal for the CG-5, 9.7 μGal for the ZLS and 20.1 μGal for the LCR. These values are approximately in accordance with the values given by the instrument manufacturers. In Fig. 5 the results of the measurements of each instrument after applying the linear scale factor are plotted against the results of the overall adjustment. On most stations, the agreement of the solutions is better than 20 μGal. Besides of the already discussed problems at Kleine Scheidegg, we see some larger discrepancies for the ZLS measurements in Grindelwald (external perturbations) and Eigergletscher of up to 40 μGal.

## 7 Comparison with the Measurements of 1980

The results of the 1980 relative measurements with three LCR meters are published in Klingelé and Kahle 1987 as gravity differences between neighboring stations. The estimated accuracy for these differences is in the order of 10–40 μGal. In 2013, all of these old stations who still could be identified have been re-measured with the CG-5 and are included in the campaign adjustment. The results of 1980 and 2013 can now be compared on the common points (see Fig. 6). The differences range from $-0.21$ mGal in Grindelwald to $+0.31$ mGal at Jungfraujoch. They show – except for the station in Interlaken, which is known to be located in an unstable region – a very linear behavior and a scale factor of 1,116 ppm was estimated. After the correction of this scale, the remaining residuals of the 1980 LCR measurements are

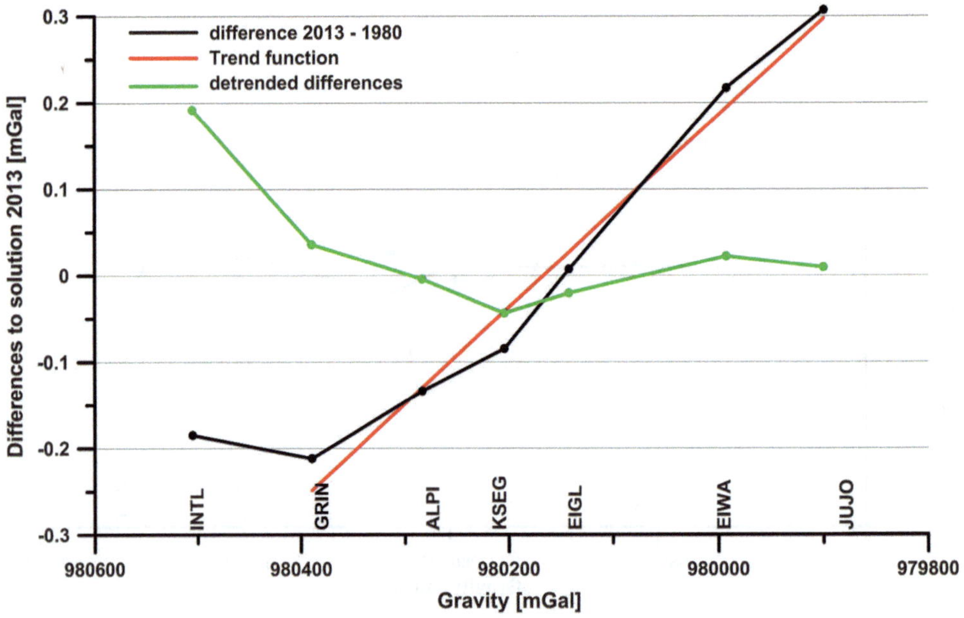


**Fig. 6** Comparison of the results of 2013 with the results of 1980

in accordance with the CG-5 measurements of better than ±0.04 mGal. This is even better than the accuracy given for the measurements in 1980. An exception, of course, is the instable station in Interlaken. But the calculated scale of more than 1,000 ppm for the 1980 measurements remains astonishingly large, even more if we consider that these relative measurements were in accordance of 0.057 mGal with the absolute measurements of that time (Klingelé and Kahle 1987).

## 8 Conclusions and Outlook

We established a traceable gravimetric calibration line in the Swiss Alps which fulfils the CIPM MRA. It consists of seven absolute stations with a gravity range of more than 600 mGal and allow the calibration of gravimeters with a limited measuring range or with a non-linear scale factor. The short distances between the stations allow – in terms of time and money – an economic check of relative gravimeters. The line includes as well some eccentric relative stations for easier access and to allow the connection to the former network of 1980. All the stations are easily accessible by car or train during the whole year. A first campaign with three different relative instruments revealed some problems in scale and their calibration factors and showed the necessity to have such a calibration line. All the site descriptions and measurements are openly documented and available. Interested institutions are invited to contact us and to visit the line with their own instruments in this very attractive region.

The next work to be done is the absolute measurement in Interlaken, which until now could not have been completed. It is foreseen to repeat the absolute measurements every 5–10 years. The interval of relative measurements is planned to be 1–2 years with as many gravimeters of different institutions as possible.

## References

BIPM (2003) Mutual recognition of national measurement standards and of calibration and measurement certificates issued by national metrology institutes. Comité international des poids et mesures (CIPM), Paris. http://www.bipm.org/en/cipm-mra

Francis O, Baumann H, Volarik T, Rothleitner C, Klein G, Seil M, Dando N et al (2013) The European comparison of absolute gravimeters 2011 (ECAG-2011) in Walferdange, Luxembourg: results and recommendations. Metrologia 50(3):257–268

Hunziker E (1959) Gravimetermessungen in den Jahren 1953–1957. Astronomisch-geodätische Arbeiten in der Schweiz, vol 25. Swiss Geodetic Commission

Jiang Z, Pálinkáš V, Arias FE, Liard J, Merlet S, Wilmes H, Vitushkin L et al (2012) The 8th international comparison of absolute gravimeters 2009: the first key comparison (CCM. G-K1) in the field of absolute gravimetry. Metrologia 49(6):666–684

Klingelé E, Kahle H-G (1987) The Swiss gravimetric calibration line from Interlaken to Jungfraujoch. BGI Bull Inform 60:64–77

Marson I, Kahle H-G, Chaperon F, Mueller S, Alasia F (1981) Absolute gravity measurements in Switzerland: definition of a base network for geodynamic investigations and for the Swiss fundamental gravity net. Bull Géod 55:203–217

Niebauer TM, Sasagawa GS, Faller JE, Hilt R, Klopping F (1995) A new generation of absolute gravimeters. Metrologia 32:159–180

Reudink R, Klees R, Francis O, Kusche J, Schlesinger R, Shabanloui A, Sneeuw N, Timmen L (2014) High tilt susceptibility of the scintrex CG-5 relative gravimeters. J Geod 88:617–622

Scintrex Limited (2012) CG-5 Scintrex autograv system operation manual, part #867700 Revision 8. Scintrex Limited, Ontario
Timmen L, Flury J, Peters T, Gitlein O (2006) A new absolute gravity base in the German Alps. Contributions to geophysics & geodesy, vol 36 (special issue), 2nd International Gravity Field Workshop Smolenice 2006
Wilmes H, Wziontek H, Falk R, Bonvalot S (2009) AGrav—the new international absolute gravity database of BGI and BKG and its benefit for the global geodynamics project (GGP). J Geodynam 48:305–309

# Airborne Gravimetry for Geoid and GOCE

R. Forsberg, A.V. Olesen, E. Nielsen, and I. Einarsson

**Abstract**

DTU-Space has since 1996 carried out large area airborne surveys over both polar, tropical and temperate regions, especially for geoid determination and global geopotential models. Recently we have started flying two gravimeters (LCR and Chekan-AM) side by side for increased reliability and redundancy. Typical gravity results are at the 2 mGal rms level, translating into 5–10 cm accuracy in geoid. However, in rough mountainous areas results can be more noisy, mainly due to long-period mountain waves and turbulence. In the paper we outline results of surveys and recent geoid determinations in Antarctica and Tanzania based on DTU-Space aerogravity and GOCE. In both cases the airborne data validate GOCE to very high degrees, and confirms the synergy of airborne gravity and GOCE. For Antarctica, the deep interior Antarctic survey (continued in 2013 from a remote field camp), shows that it is possible efficiently to cover even the most remote regions on the planet with good aerogravity. With the recent termination of the GOCE mission, it is therefore timely to initiate a coordinated, preferably international, airborne gravity effort to cover the polar gap south of 83° S; such a survey can in principle logistically be done in a single season.

**Keywords**

Airborne gravity • Geoid • GOCE • Gravimeter

## 1 Introduction

Airborne gravity measurements have since the early 1990s developed into a reliable production system to accurately measure the gravity field of the earth from aircraft. With the recent GOCE long-wavelength global gravity field models, airborne gravity field mapping has become even more important to determine the medium-wavelength features of the gravity field, and thus a precise geoid. A good consistency between airborne data and GOCE is essential for the proper combination of such data, and will be addressed in this paper through comparison of recent major DTU Space aerogravity results to GOCE.

The development of airborne gravimetry has mainly been driven in the commercial domain by the need for accurate and high-resolution gravity anomaly mapping for oil, gas and mineral exploration, and accuracies below 1 mGal are now reported for state-of-the art systems (Williams and MacQueen 2001; Elieff and Ferguson 2008). In the government and academic domain, long-range GPS-based aerogravity for regional geophysics was pioneered by both US and Russian researchers (Brozena 1992), and later implemented in smaller aircraft by several groups (Bell et al. 1992; Klingele et al. 1995), and lately adopted for large-scale geoid mapping e.g. in the US GRAV-D project. The DTU Space aerogravity applications for geoid determina-

R. Forsberg (✉) • A.V. Olesen • E. Nielsen • I. Einarsson
National Space Institute, Technical University of Denmark (DTU-Space), Elektrovej 327, 2800 Lyngby, Denmark
e-mail: rf@space.dtu.dk


S. Jin, R. Barzaghi (eds.), *IGFS 2014*, International Association of Geodesy Symposia 144,
DOI 10.1007/1345_2015_47

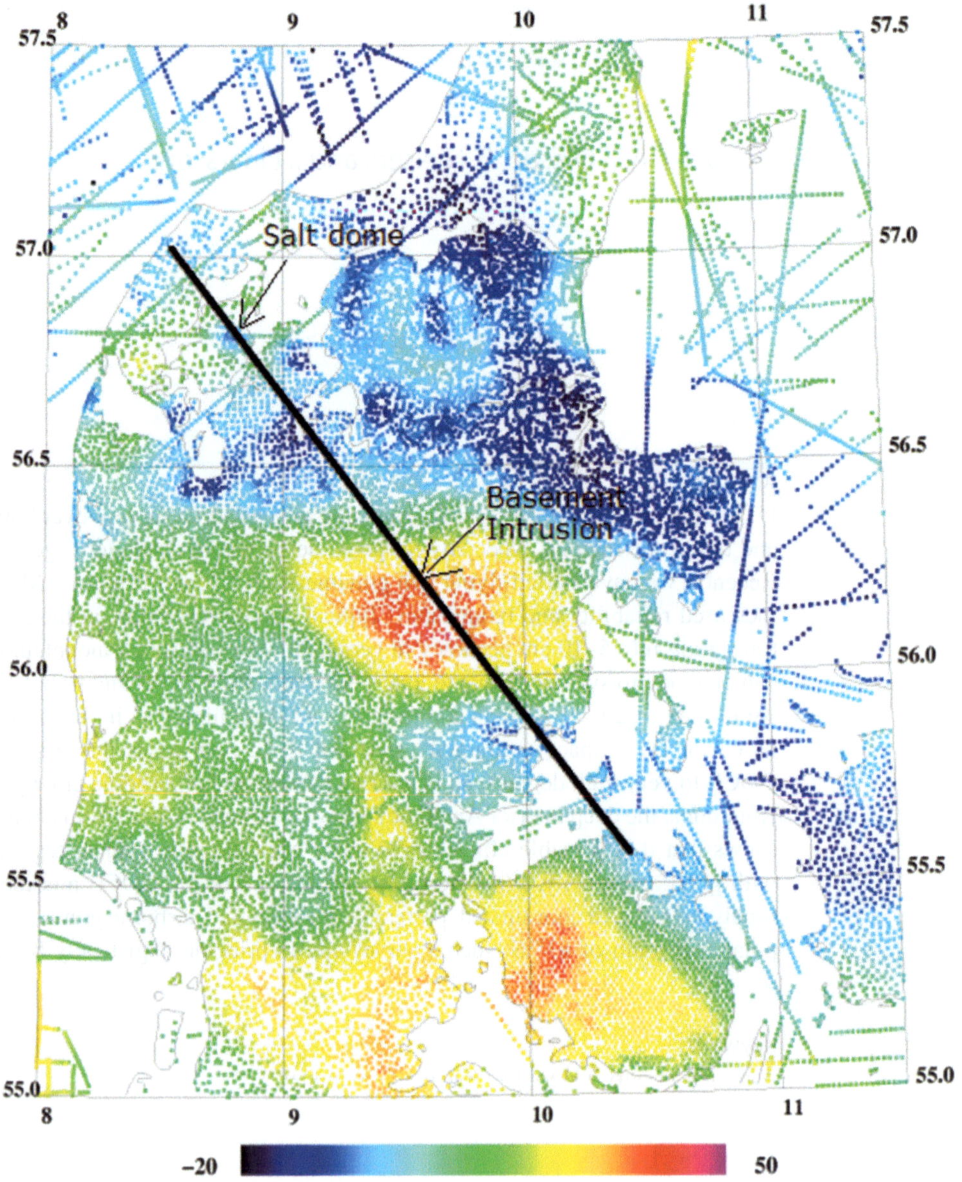


**Fig. 1** Flight line of the first Chekan-AM flight test, with surface gravity free-air anomalies (mGal)

tion were developed in the mid 1990s, as part of the EU AGMASCO (Airborne Geoid Mapping System for Coastal Oceanography), see Forsberg et al. (1996). The system setup and experience developed in the AGMASCO project have since been used extensively for small aircraft, long range surveys in many different regions of the world (e.g., Olesen et al. 2000; Forsberg et al. 2007; Forsberg et al. 2011; Forsberg et al. 2012). The DTU Space surveys have been based on the Lacoste and Romberg (LCR) S-type gravimeter, an air damped beam type instrument described in details by Valliant (1991), but recently a Chekan-AM gravimeter (Fig. 1) was added as a side-by-side auxillary gravity sensor. The Chekan gravimeter principle is based on a dual quartz flexible pendulum element system with fluid damping, and is mounted on a more modern GPS-stabilized inertial platform, for details see Krasnov et al. (2011). In the paper we will therefore initially show some novel results of joint LCR-Chekan flights, and then outline some large-scale GOCE comparisons to major campaigns in Tanzania and Antarctica.

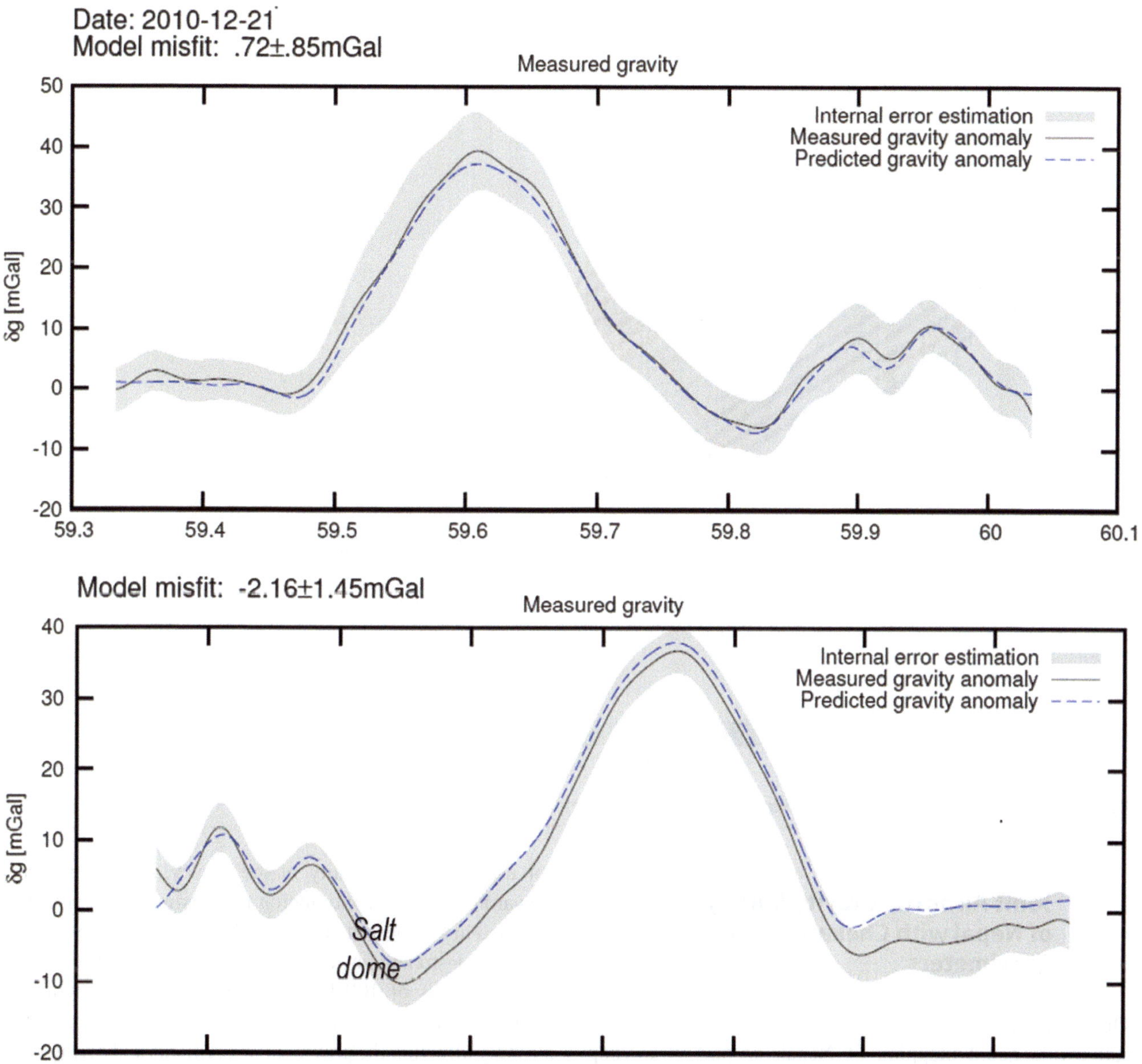


**Fig. 2** Results of the SE–NW and NW–SE flights, compared to upward continued ground gravity

## 2 First Flight Test of Chekan AM in Denmark

A flight test of Chekan-AM was carried out in December 21, 2010 using a Beech King Air 200 aircraft, belonging to COWI company, Denmark. The flight test was repeated forward and backward at two different low levels (less than 1,000 m) along a SE–NW oriented line across Denmark, with dense ground control gravity data from the Danish national gravity data base (Fig. 1). The profile crossed two major anomalies: the "Silkeborg high", a large anomaly due to assumed basement intrusions below thick sediment sequences, and the narrow "Mors low" anomaly, due to a major shallow salt dome. The profile results compared to the upward continue ground thruth are shown in Fig. 2.

The comparison of the two flights show a difference between the airborne free-air anomalies and the upward continued surface gravimetry data of $0.72 \pm 0.85$ mGal for the outbound (northward) flight, and $-2.16 \pm 1.45$ mGal for the return flight, a highly satisfactory result, given the unavoidable errors in the upward continuation FFT process (Schwarz et al. 1990), especially leakage from the less covered marine areas. The test therefore demonstrates that the Chekan-AM instrument is capable of generating results at the 1 mGal accuracy level under good flight conditions.

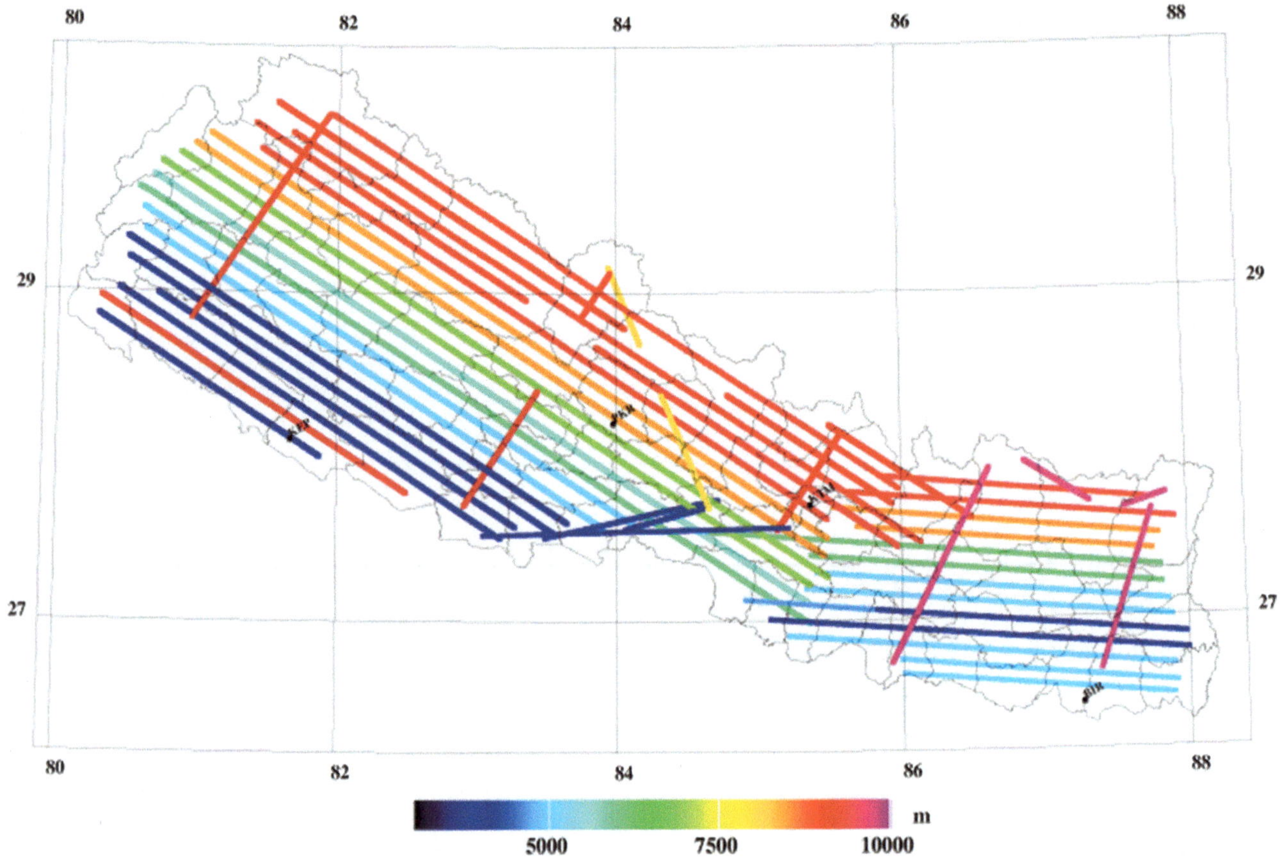


**Fig. 3** Flight line elevations (meter) of the Nepal airborne gravity survey, for the processed L&R data

## 3 Joint Airborne Gravity Survey of Nepal with Chekan-AM and L&R Gravimeters

The Nepal aerogravity flight campaign was carried out for nationwide geoid mapping of Nepal, as part of a cooperation with the Nepal Survey Department, supported by the US National Geospatial-Intelligence Agency. The campaign was carried out in the period December 4–17, 2010, using the same aircraft as in the Danish test. This aircraft, equipped with a pressurized cabin, was flown to Nepal from Denmark, as no local suitable aircraft was available for the challenging flights over the highest mountains on the earth.

The flight tracks flown over Nepal were spaced at approximately 6 nautical miles, cf. Fig. 3. Because of the highly varying topography, the individual flight heights were varying from 4 km for the southernmost lines to 10 km for the northern lines, as required by clearance of topography. Cross-lines were all flown at high altitude. For this reason cross-over errors of survey lines can not readily be used for estimating the quality of the survey, but an integrated upward/downward continuation process must be applied, as outlined in Forsberg et al. (2012). Especially for the northern flights, jet streams at altitude with wind speeds occasionally in excess of 100 knots provided a major operational challenge, generating major turbulence and mountain waves, strongly affecting the gravity measurements and giving occasional gaps.

While the processed L&R data covered the entire country, except for the minor gaps seen in Fig. 3, the Chekan-AM data showed more problems, with malfunctions on a couple lines (due to some power problems), but especially some cases where the sensor was saturated in turbulence, and hit the hard stops of the sensor. Figure 4 shows the overall processed Chekan-AM data, and Fig. 5 examples of flight line data for L&R and the Chekan-AM data. A good agreement between LCR and Chekan is seen on one line, but the other line shows clearly the effect of out-of-scale Chekan measurements. These effects can be detected from asymmetry ($m_1$–$m_2$) in the dual measurements of the Chekan sensor, and some recovery is possible.

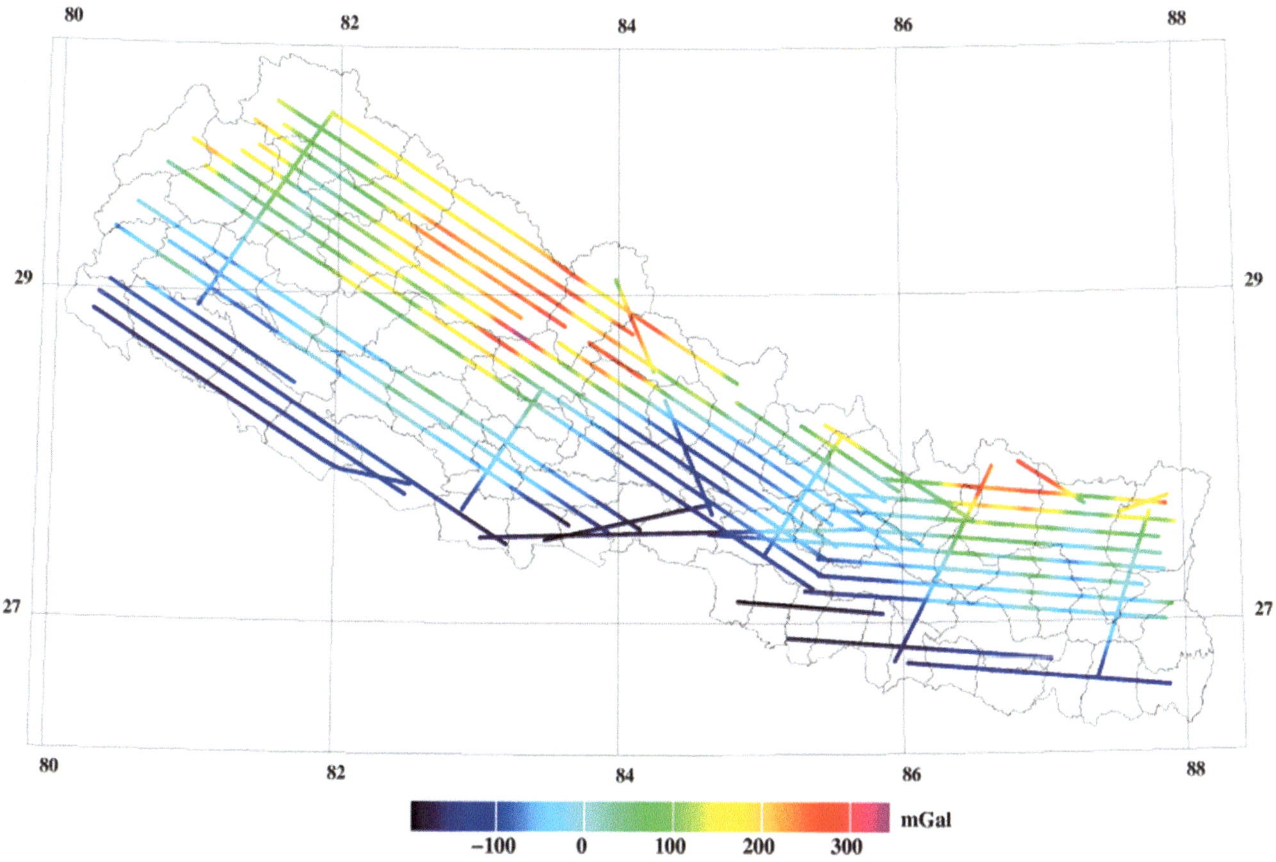


**Fig. 4** Chekan-AM processed data (free-air anomaly data at altitude)

Overall, however, the turbulence tolerance is less for the Chekan than the L&R, but when conditions are less extreme the two instruments supplement each other, and give an overall more redundant survey system. The estimated r.m.s. accuracy of the separately processed LCR and Chekan data was 4.6 and 5.1 mGal respectively, and the combined data set 3.9 mGal, when continued to a common level of 6,000 m using a linear gravity gradient approximation and least squares collocation (Forsberg et al. 2011). This is a highly satisfactory results given the extremely rough conditions and the large gravity field variability.

The merged airborne gravity data were used for an improved geoid determination of Nepal, together with new GOCE satellite data, existing surface gravimetry data, and terrain information from SRTM. The GOCE data were used as a reference field, and combined with the airborne and surface data using remove-restore methods with modified Stokes' kernels; for details of the geoid determination and downward continuation by least-squares collocation, and the used spherical FFT methods, see Forsberg et al. (2011) and Forsberg and Olesen (2010). Figure 6 shows the overall distribution of airborne and surface gravimetry data, Fig. 7 the airborne data and the GOCE "Direct" model (Pail et al. 2011) Release 4, and Fig. 8 the resulting geoid. In the Kathmandu Valley, this new geoid showed an r.m.s. agreement of GPS-levelling data at the 7 cm level.

## 4 GOCE Comparison of New Surveys in Tanzania and Antarctica

Two major surveys carried out in East Antarctica 2011 and 2013, and Tanzania 2012–2013, were used to quantify the comparison to GOCE. Both regions were covered with no or only sparse terrestrial data prior to the aerogravity surveys. The data from these surveys were based primarily on the LCR data, with limited Tanzania fill-in from Chekan-AM. Both surveys were tied to absolute ground gravity, and the r.m.s. gravity error for the Antarctica survey estimated at 3.0–3.6 mGal r.m.s., and 2.3 mGal for Tanzania, with *no* cross-

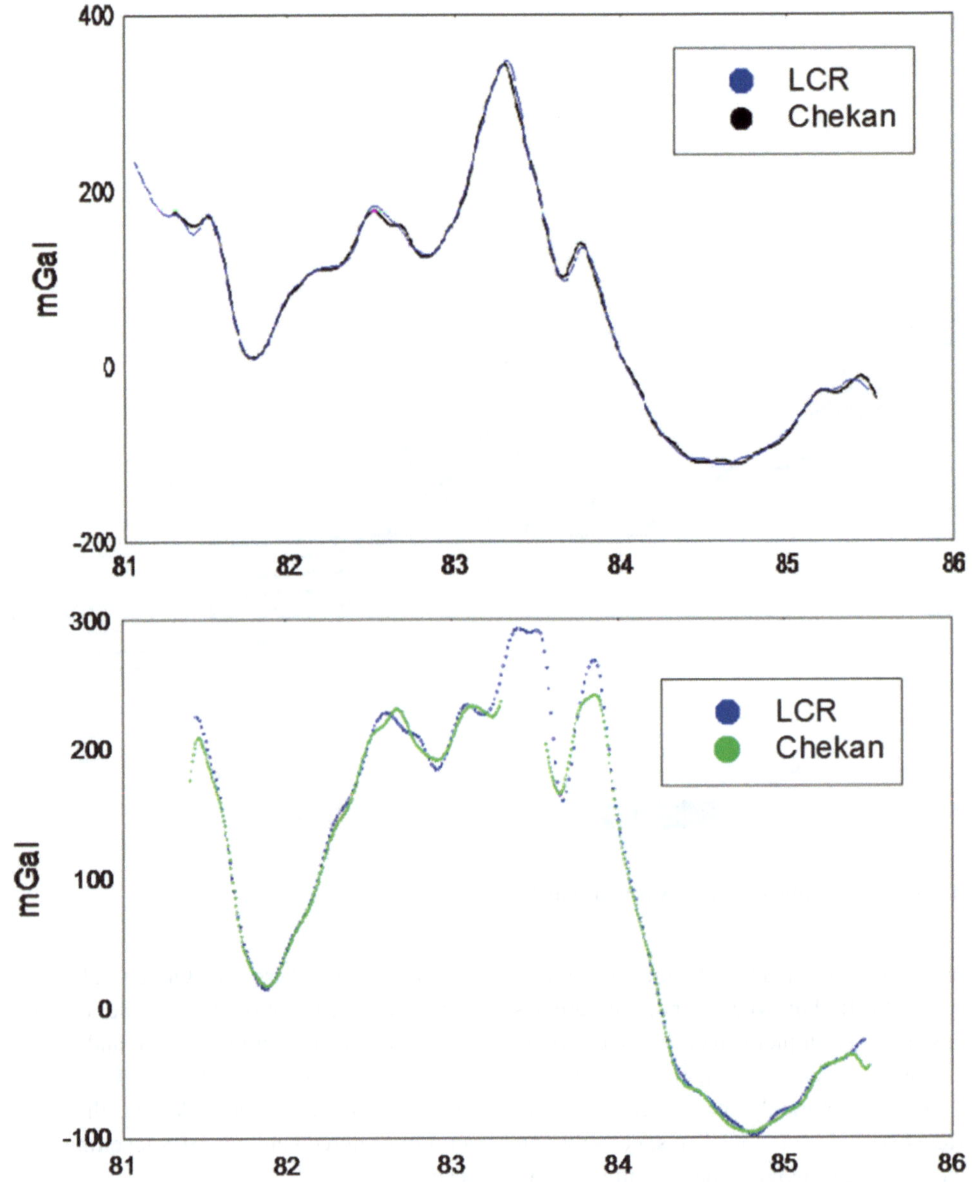


**Fig. 5** Example of two lines (line 20, upper, and line 24, lower) in the high mountains, with LCR and Chekan-AM data processed independently, similar filtering parameters. Line 24 shows examples of Chekan saturation. x-axis is the longitude

over adjustment applied. This r.m.s. error variation reflects both difference in aircraft and environmetal conditions; the 2011 Antarctica survey was flown with a long-range vintage 1942 DC3 modernized aircraft (Basler BT-67), the 2013 survey with a Twin Otter without autopilot, and the Tanzania survey with a modern "glass-cockpit" Cessna Caravan. The 2013 Antarctica survey was done primary from a remote field camp (FD83), with fuel airdropped from a Russian jet by ALCI logistics company.

Figures 9 and 10 show the flown tracks in Antarctica and Tanzania respectively, on top of the GOCE RL-4 "Direct" gravity field model results. The comparison of the airborne data and GOCE as a function of the maximal degree used in the GOCE field synthesis is shown in Table 1. It is seen that an excellent fit is obtained, with the major regional anomalies seen in the airborne data, also reflected in the GOCE data. The GOCE data seems to contain gravity field information even up to degree 240–260, and the overall bias agreement between airborne data and GOCE is excellent at the 1 mGal level (this small bias is likely coming from non-linear sensor effects in LCR airborne gravimetry during turbulence). Such a minor bias if the difference will not play a major role when

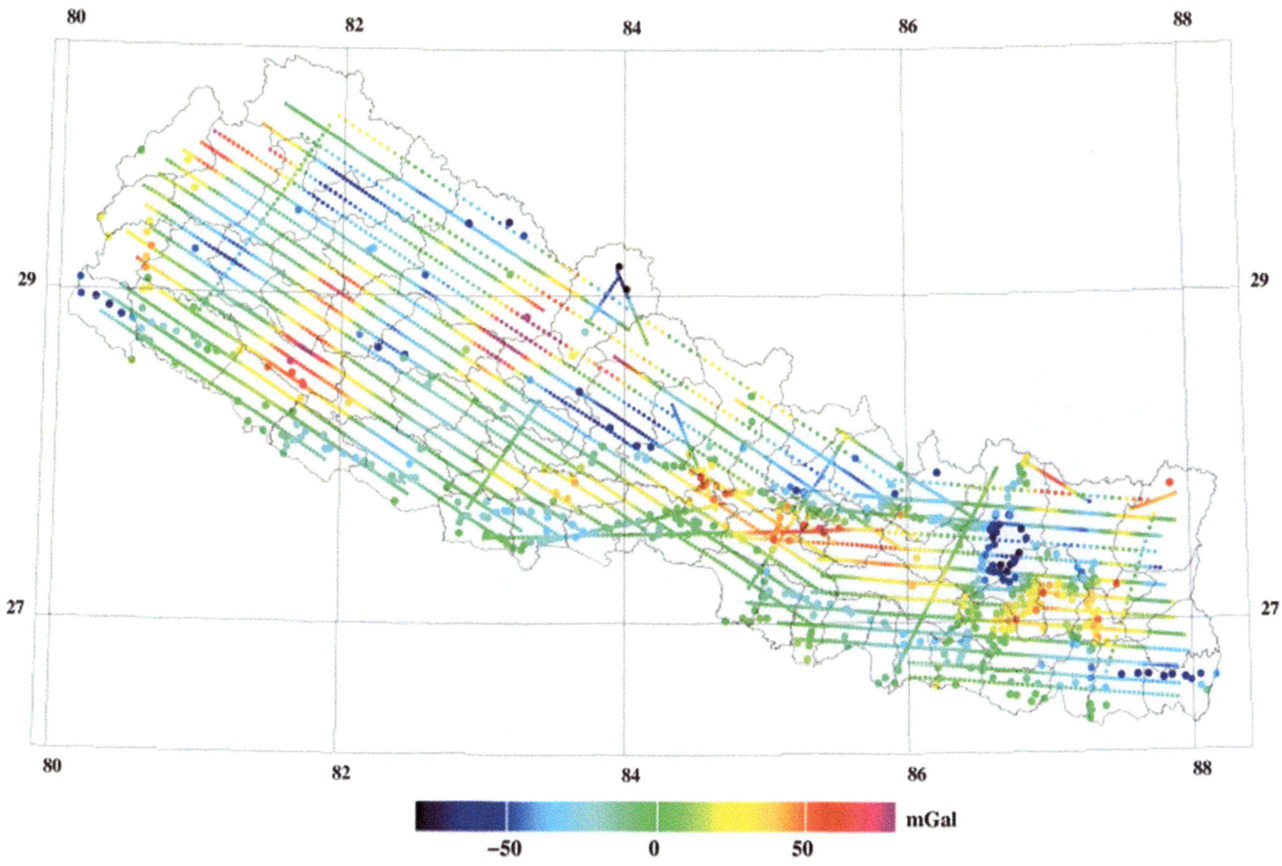


**Fig. 6** Combined data set of airborne and surface gravimetry data for Nepal. The airborne data thinned to 30 s for the geoid determination process; the apparent change in ground spacing reflect the strong winds and the direction of flights

airborne gravimetry and GOCE are combined properly in the spectral domain for geoid determination, where long-wavelength gravity field information will be determined primarily from GOCE.

## 5 Conclusions

It is seen how airborne gravimetry can give useful results at the few-mGal accuracy level, and thus complement GOCE nicely for geoid determination. The first flights results of a new acquired Chekan-AM gravimeter shows 1 mGal accuracy compared to ground control for a repeated line with good flight conditions in Denmark, and for a major airborne geoid survey of Nepal, with major challenges due to a rugged gravity field and turbulent flight conditions, a combined accuracy of around 3.9 mGal r.m.s. were estimated. The computed geoid showed an accuracy of 7 cm r.m.s. when compared to levelling in the central Kathmandu Valley, the only place where sufficiently accurate GPS-levelling data were available.

The error comparison of GOCE to large scale airborne surveys were done for recent surveys in East Antarctica and Tanzania, showing a good fit of GOCE R4 data to the airborne surveys, both in terms of detectability and coherence of free-air anomaly features, and quantitatively showing how the R4 GOCE solution seems to contain useful information even the maximal expansion harmonic degree of 260. Airborne data is therefore an efficient way to cover the two polar gaps of GOCE with gravity field data, so that future GOCE-based gravity field models can indeed by truly global. The Arctic gap is already filled with sufficient aerogravity and surface data for this purpose, primarily coming from the Arctic Gravity Project and later UNCLOS surveys, while the Antarctic polar gap is essentially uncovered, except for a few scattered flight lines, mainly from recent NASA IceBridge activities. Based on the experience of the 2013 airborne gravity campaign, operating out of the FD83 remote field

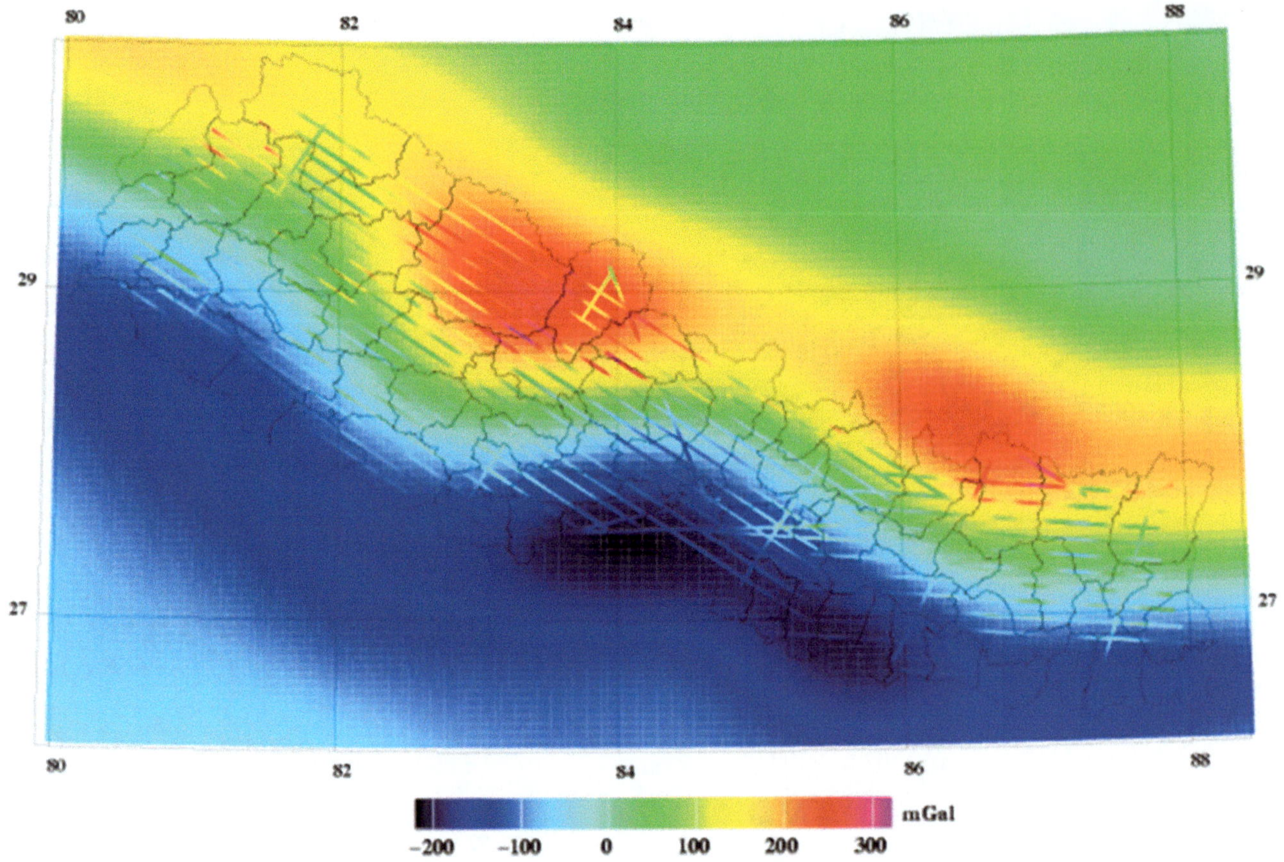


**Fig. 7** Combined airborne gravity data set, overlaid on the GOCE gravity, and plotted with a common colour scale. An excellent visual agreement is apparent

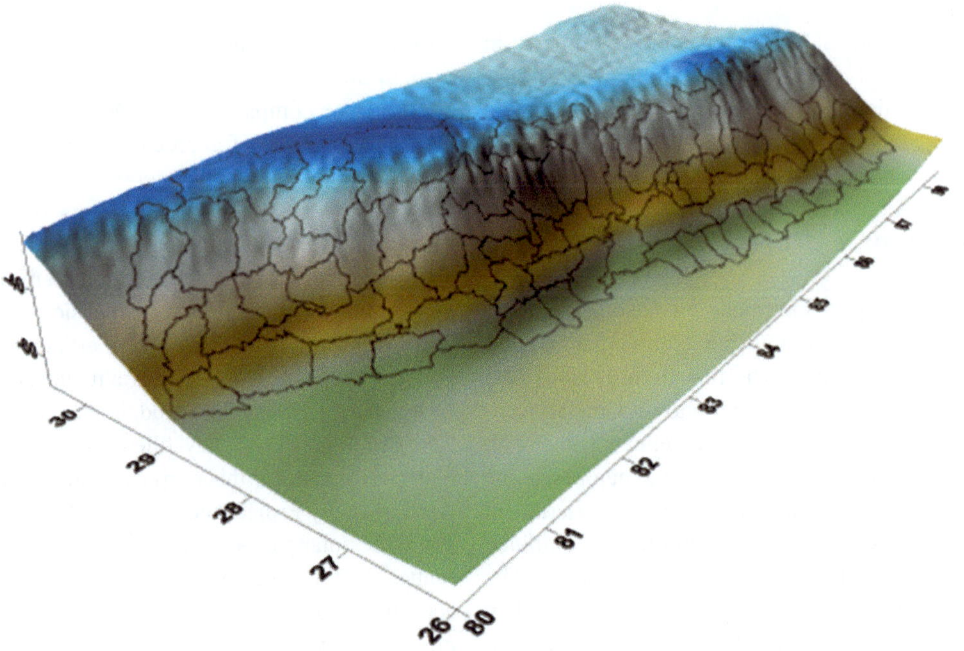


**Fig. 8** Geoid model of Nepal from airborne gravity data and GOCE. Labels are latitude, longitude, and height (range −70 to −30 m)

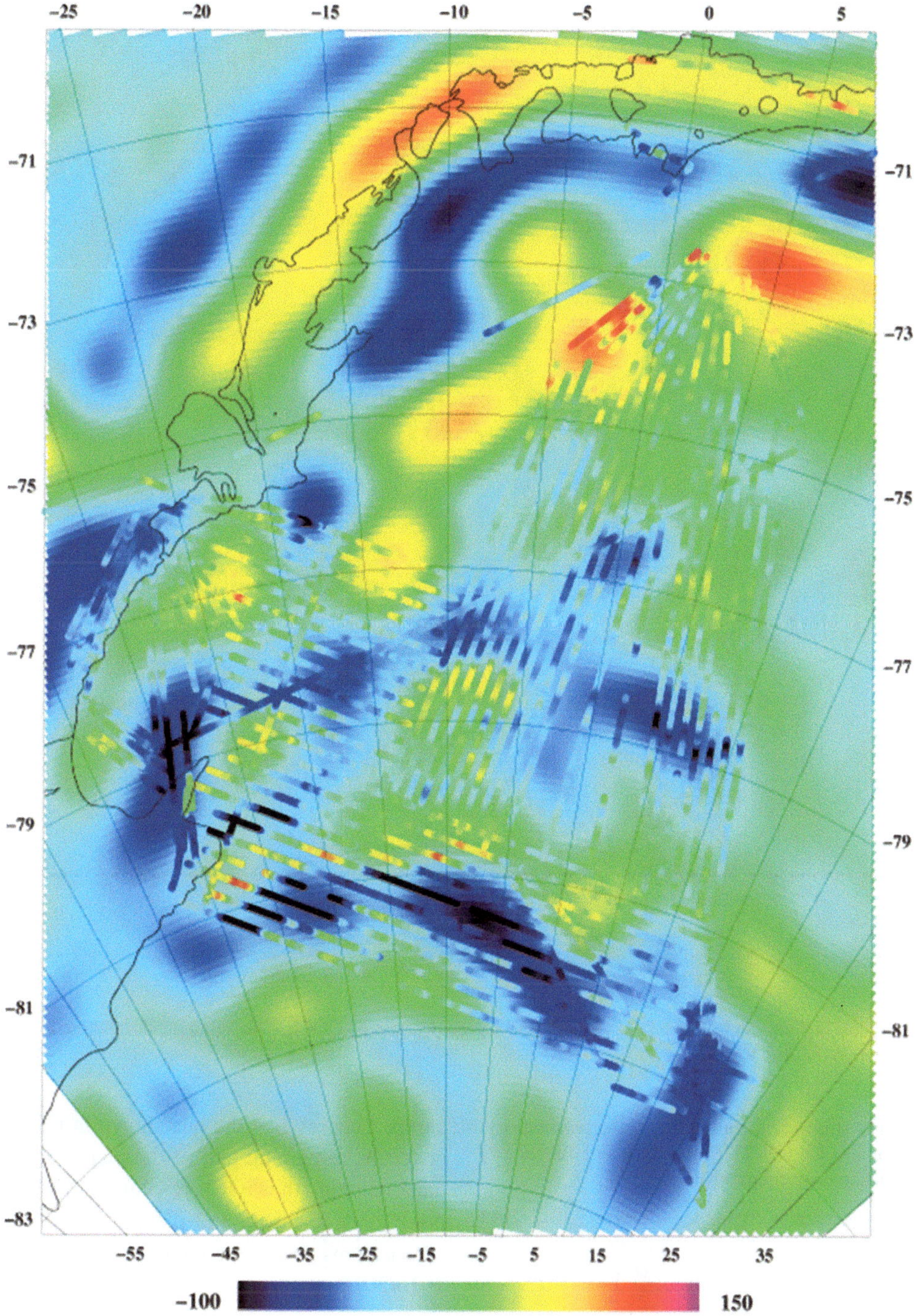


**Fig. 9** Free-air anomalies from the 2011 (NNE-SSW trending lines) and 2013 airborne gravity surveys in East Antarctica, overlain on GOCE RL4. Unit mGal. The central negative anomalies reflect mainly major ice stream subglacial valleys

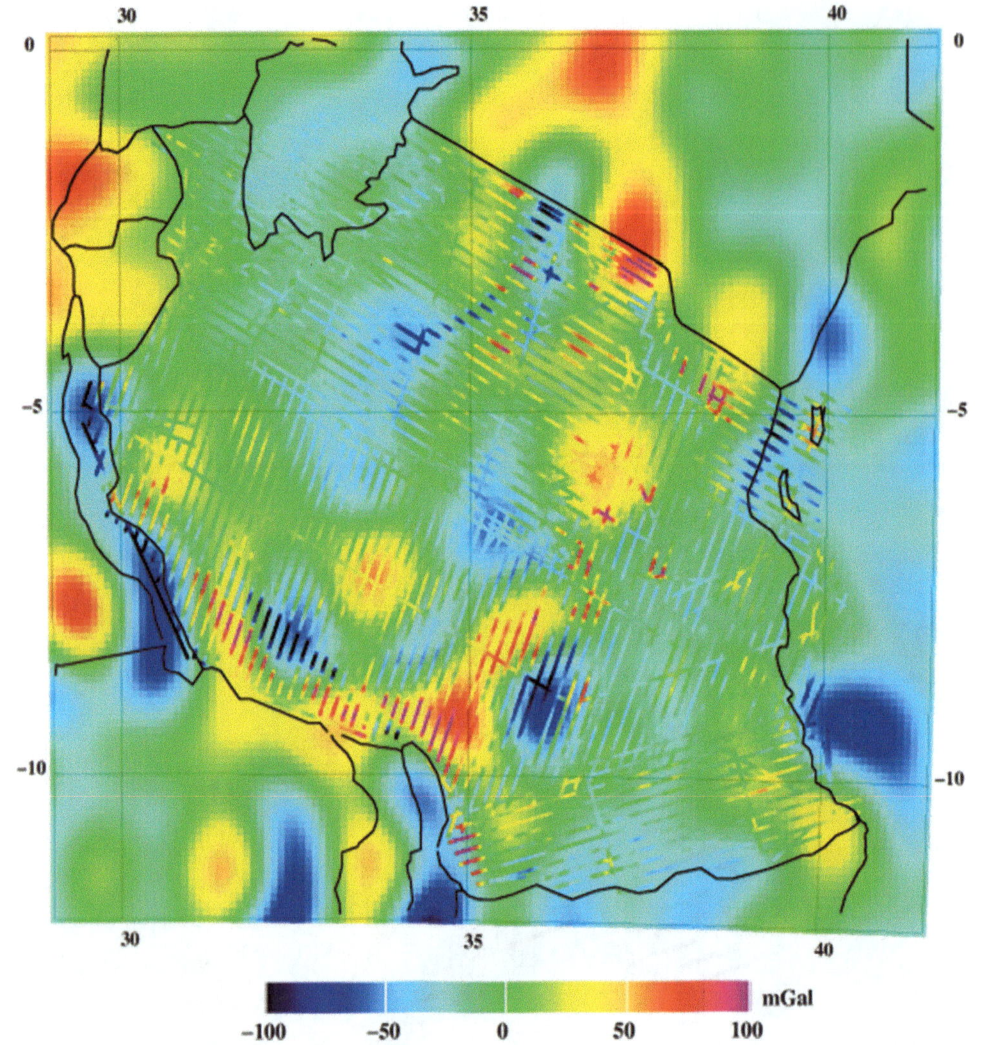


**Fig. 10** Free-air anomalies in Tanzania from the 2012–13 airborne survey overlain on GOCE. The "strange" linear structure in the SW part of the survey is real, and related to the Rift Valley

**Table 1** Comparison of atmosphere-corrected airborne gravimetry in Antarctica and Tanzania to the GOCE RL-4 model, as a function of maximal GOCE expansion degree; unit mGal

| | Antarctica 2011 | | Tanzania 2012–2013 | |
|---|---|---|---|---|
| Max degree of GOCE expansion | Mean | Std.dev. | Mean | Std.dev. |
| *Data itself* | *4.0* | *37.7* | *−4.7* | *33.5* |
| 180 | 0.1 | 26.6 | 1.6 | 28.4 |
| 200 | 0.3 | 24.6 | 1.3 | 27.4 |
| 220 | 1.2 | 22.5 | 1.3 | 26.1 |
| 240 | 1.2 | 22.1 | 1.3 | 25.1 |
| 260 | 1.1 | 22.2 | 1.3 | 24.5 |

camp, a survey like the one shown in Fig. 11, could readily cover the remaining part of the planet with airborne gravity data matching the GOCE resolution. Such a survey could be done in one season, and plans are currently underway for preparations of an international effort to cover this gap by 2015/2016.

**Acknowledgements** The survey costs for Antarctica were sponsored by NGA, ESA and DTU Space, and for Tanzania by the Department of Survey, Tanzania, the World Bank and NGA. The Antarctica operations where carried out in cooperation with British Antarctic Survey, the Norwegian Polar Institute, and the University of Texas at Austin.

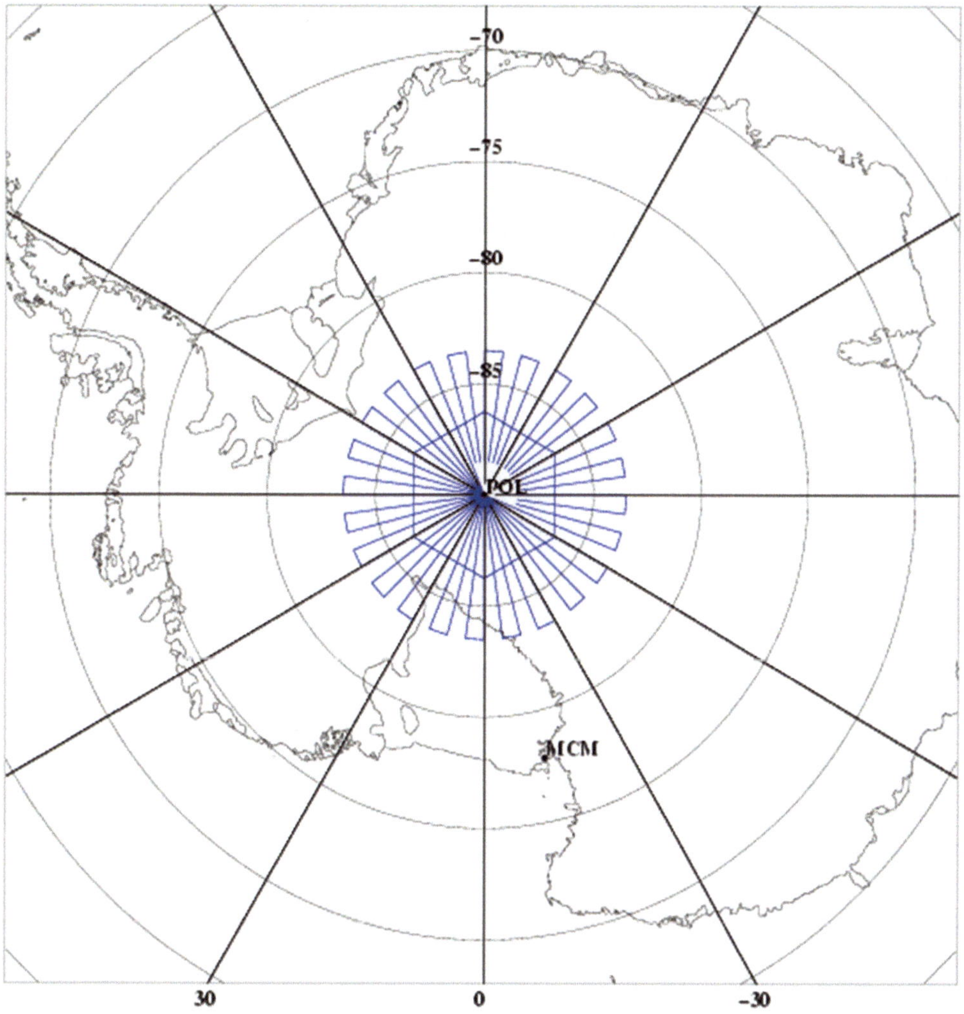


**Fig. 11** Schematic flight tracks for an airborne survey to cover the southern GOCE polar gap by 2015/2016

## References


Bell RE, Coakley BJ, Blankenship DD, Hodge SM, Brozena JM, Jarvis J (1992) Airborne gravity from a light aircraft: CASERTZ 1990–91. In: Yoshida Y (ed) Recent progress in antarctic earth science. Terrapub, Tokyo, pp 571–577

Brozena JM (1992) The greenland aerogeophysics project: airborne gravity, topographic and magnetic mapping of an entire continent. In: Colombo O (ed) From mars to greenland, proc. IAG symposia, vol 110. Springer Verlag, Heidelberg, pp 203–214

Elieff S, Ferguson S (2008) Establishing the "air truth" from 10 years of airborne gravimeter data. First Break, 26, November 2008

Forsberg R, Brozena J (1996) Airborne geoid measurements in the Arctic ocean. In: Proceedings international symposium on gravity, Geoid and marine geodesy, Tokyo, Sept. 1996. IAG series, vol 117. Springer Verlag, Heidelberg, pp 139–147

Forsberg R, Olesen AV (2010) Airborne gravity field determination. In: Xu G (ed) Sciences of geodesy – I, Advances and future directions. Springer Verlag, Heidelberg, pp 83–104. ISBN 978-3-642-11741-1

Forsberg R, Hehl K, Bastos L, Giskehaug A, Meyer U (1996) Development of an airborne geoid mapping system for coastal oceanograoh (AGMASCO). In: Proceedings international symposium on gravity, Geoid and marine geodesy, Tokyo, Sept. 1996. IAG series, vol 117. Springer Verlag, Heidelberg, pp. 163–170.

Forsberg, R, Olesen A, Munkhtsetseg D, Amarzaya A (2007): Downward continuation and geoid determination in Mongolia from airborne and surface gravimetry and SRTM topography. Proceedings of the 1st Symposium of the International Gravity Field Service, Harita Dergisi, 73(18), 259–264, Ankara.

Forsberg R, Olesen AV, Einarsson I, Manandhar N, Shreshta K (2011) Geoid of Nepal from airborne gravity survey. Proceedings IAG Symposia, vol 139. Springer Verlag, Heidelberg, Melbourne, pp 521–528

Forsberg R, Olesen A, Alshasi A, Gidskehaug A, Ses S, Kadir M, Majid K, Benny P (2012) Airborne gravimetry survey for the marine area of the United Arab Emirates. Marine Geodesy 35(3):221–232. doi:10.1080/01490419.2012.672874

Klingele E, Halliday M, Cocard M, Kahle H-G (1995) Airborne gravimetric survey of Switzerland. Vermessung Photogrammetrie Kulturtechnik 4:248–253

Krasnov AA, Nesenyuk LP, Peshekhonov VG, Sokolov AV, Elinson LS (2011) Integrated marine gravimetric system. Development and operation results. Gyrosc Navig 2(2):75–81

Olesen AV (2002) Improved airborne scalar gravimetry for regional gravity field mapping and geoid determination. Ph.d. dissertation,

National Survey and Cadastre of Denmark Technical Report 24, p 123
Olesen AV, Forsberg R, Keller K, Gidskehaug A (2000) Airborne gravity survey of the Lincoln Sea and Wandel Sea, North Greenland. Phys Chem Earth A 25(1):25–29
Pail R, Bruinsma S, Migliaccio F, Foerste C, Goiginger H, Schuh W-D, Hoeck E, Reguzzoni M, Brockmann JM, Abrikosov O, Veicherts M, Fecher T, Mayrhofer R, Krasbutter I, Sanso F, Tscherning CC (2011) First GOCE gravity field models derived by three different approaches. J Geod 85(11):819–843
Schwarz K-P, Sideris MG, Forsberg R (1990) Use of FFT methods in physical geodesy. Geophys J Int 100:485–514
Valliant HD (1991) The LaCoste and Romberg air/sea gravity meter: an overview. Geophysical exploration at sea, 2nd edn, vol 1, CRC Press, Boca Raton, Florida, pp 141–176
Williams S, MacQueen JD (2001) Development of a versatile, commercially proven, and cost-effective airborne gravity system. Leading Edge 60:651–654

# Testing Airborne Gravity Data in the Large-Scale Area of Italy and Adjacent Seas

Riccardo Barzaghi, Alberta Albertella, Daniela Carrion, Franz Barthelmes, Svetozar Petrovic, and Mirko Scheinert

**Abstract**

In 2012 the GEOHALO flight mission was carried out using the new German research aircraft HALO. The surveyed zone covers the Central-South part of Italy, roughly from latitude 36°N to 44°N. In this area, seven main tracks NW to SE were surveyed having a spacing of about 40 km and an altitude of 3,500 m, complemented by an eighth track in an altitude of 10,000 m. Four perpendicular cross tracks were also added.

Amongst the geodetic-geophysical equipment GEOHALO carried two gravimeters. In this paper we will focus on the GFZ instrument, a CHEKAN-AM gravimeter. The present investigation aims at defining the spectral properties and the level of precision of the observed gravity data. Comparisons with gravity anomalies predicted from Italian ground data are presented. The gravity field in the surveyed area as derived from these ground data is propagated to the aerogravimetry survey points and compared to the observed gravity anomalies. Upward continuation is performed using the remove-restore approach and collocation. High-resolution global geopotential models are compared with the observed data as well. The statistics of the gravity residuals show that the survey data fit the predicted gravity at 2–3 mGal standard deviation level which proves that a good standard has been reached. A trackwise analysis is also performed to check for possible local discrepancies between observed and predicted gravity.

**Keywords**

Aerogravimetry • Collocation • Remove-restore • Global geopotential models

## 1 Introduction

In June 2012 an aerogravimetry test took place in the framework of the GEOHALO campaign that was carried out over the Italian Peninsula. It was a multidisciplinary geoscientific mission realized by an international team from universities and research institutions and managed by one of the co-authors (Scheinert 2013). In this campaign, the German High Altitude and Long Range Research Aircraft (HALO) was used, a modified G550 business jet manufactured by Gulfstream Aerospace Corporation. One major aim of the GEOHALO mission was to prove the feasibility of a high speed and high altitude geoscientific campaign which allows to survey a wide region in a reasonable length

R. Barzaghi (✉) • A. Albertella • D. Carrion
DICA – Politecnico di Milano, Piazza Leonardo da Vinci, 32, 20133 Milan, Italy
e-mail: Riccardo.barzaghi@polimi.it; alberta.albertella@polimi.it; daniela.carrion@polimi.it

F. Barthelmes • S. Petrovic
Helmholtz Centre Potsdam German Research Centre for Geosciences (GFZ), Telegrafenberg, 14473 Potsdam, Germany
e-mail: bar@gfz-potsdam.de; sp@gfz-potsdam.de

M. Scheinert
Institut für Planetare Geodäsie, Technische Universität Dresden, 01062 Dresden, Germany
e-mail: Mirko.Scheinert@tu-dresden.de



S. Jin, R. Barzaghi (eds.), *IGFS 2014*, International Association of Geodesy Symposia 144,
DOI 10.1007/1345_2015_45

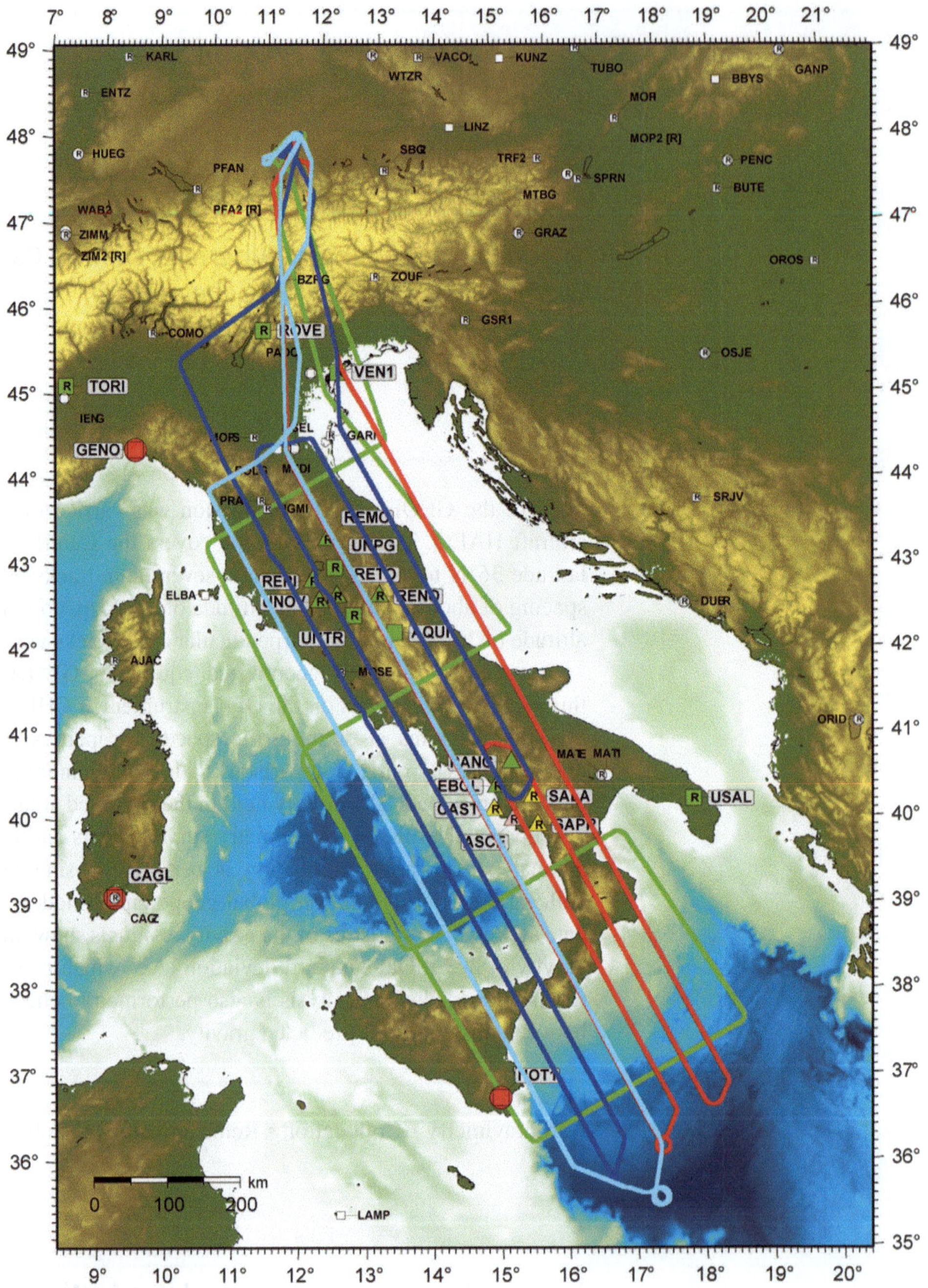


**Fig. 1** The GEOHALO flight pattern (the *colors* denote the four different mission days in the sequence *green*, *red*, *blue* and *cyan*. *Red circles* and further *markers* denote GNSS ground stations (EUREF, IGS and local sites) where potentially data are available as reference to process the flight trajectory.)

of time. This was done in order to gain experience for future missions to improve e.g. satellite only gravity models in remote areas with poor gravity data coverage. The mission flights were realized on 4 days, complemented by two preceding test flights to accomplish the overall mission certification. The mission flights comprise a total length of approx. 16,500 km with seven long north–south directed profiles and four cross lines flown in an altitude of 3,500 m, while an eighth profile was flown at 10,000 m. The flight velocity amounted to mean value of approx. 425 km/h. Gravity was observed using two gravimeters, a CHEKAN-AM gravimeter and a KSS-32M gravimeter. The flight profiles are shown in Fig. 1.

A preliminary analysis of the CHEKAN-AM airborne gravity data was done at GFZ. Comparisons with upward continued ground based gravity data were performed at Politecnico di Milano. Furthermore, global geopotential models (namely EIGEN-6C4 and EGM2008) were considered, too, in order to gain a more general comparison. In the next sections, the data filtering and the comparisons with the different gravity models are described in order to assess the precision of the collected aerogravimetry data.

## 2 The GEOHALO Trackwise Data Processing

The principles of airborne gravimetry are discussed e.g. by Forsberg and Olesen (2010). The recovery of gravimeter readings from the recorded raw data along the trajectory of the aircraft is based on the mathematical model and calibration constants provided by the manufacturer of the instrument (Krasnov et al. 2011). In order to calculate the gravity values (at flight altitude) observations have been corrected for the Eötvös effect (Jekeli 2001, Eq. 10.95, p. 334) and the vertical component of the kinematic acceleration. The last term is usually computed by a numerical double differentiation of the GNSS positions of the flight trajectory. Due to the higher speed of the HALO aircraft, this procedure did not yield satisfactory results. Therefore, the kinematic acceleration was computed by differentiating velocities derived from the GNSS processing. This was done using a new software developed at GFZ (Kaifei 2014), which includes the possibility to compute velocities from Doppler observations and to perform robust estimates. Since all acceleration components (sampled at 10 Hz) contain high-frequency noise they have to be low-pass filtered applying the same filter characteristics. A cosine type transfer function in the frequency domain was realized by applying direct and inverse Fast Fourier Transform (FFT). The results presented here were obtained by using a cut-off wavelength of 220 s (i.e. 110 s half-wavelength) and a transition region from 110 to 330 s (Fig. 2).

At an aircraft speed of 120 m/s this corresponds to a spatial resolution of about 12 km (half wavelength). Consequently, all components (gravimeter recordings, GNSS derived vertical accelerations, Eötvös corrections, positions) were filtered in this same way.

In order to convert the aerogravimeter readings to absolute gravity values an accessible ground reference has to be established. For this, a local reference point was marked at the apron of the HALO hangar and linked to gravity reference points of the German Gravity Network in the vicinity of the airfield by a local gravity survey. During the mission days, while the aircraft was still standing close to the local reference point, readings by the airborne gravimeters were taken before take-off and after landing and returning to the hangar. However, some jumps in the airborne recordings (in the order of some mGal), most probably due to landing, are still under investigation. Therefore, a careful determination of the drift parameters of the gravimeter is still to be done. In the present comparisons a constant offset has been fitted separately for each profile using a global model. For campaigns like this, comprising relatively long tracks (compared to the spatial resolution of recent global satellite-only models), such a trackwise fit can be regarded as a reasonable alternative. Further aspects have still to be investigated, among others the impact of different low-pass filters. Hence, the results presented in this paper have to be regarded as preliminary ones.

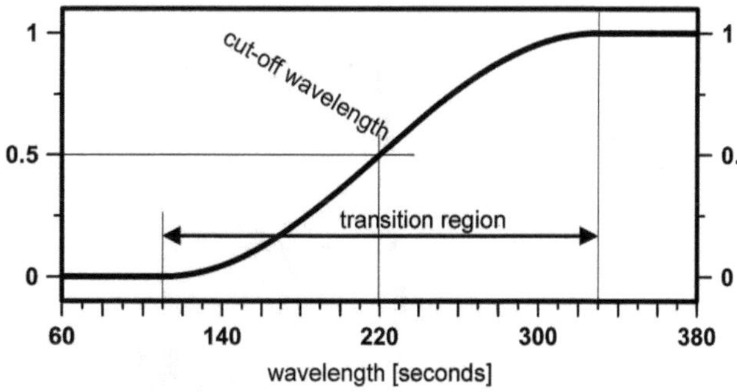


**Fig. 2** Transfer function of the used cosine type low-pass filter

## 3 Gravity Prediction Using the Italian Database

The Italian gravity database contains 410,397 gravity values in the area

$$5^\circ \leq \lambda \leq 20^\circ,\ 35^\circ \leq \varphi \leq 48^\circ$$

with a mean density of 20″. The data are not homogeneously distributed in this area. As one can see in Fig. 3, the gravity coverage is quite poor in the Alpine region and in some areas of the Apennines where the data coverage is unsatisfactory (the lowest data density is in the Alpine region, in the area $6^\circ \leq \lambda \leq 7^\circ$, $44^\circ \leq \varphi \leq 46^\circ$, with a mean density of 1 point/30 km$^2$). This can be reflected by some poor results of the predictions at the aerogravimetry tracks crossing these regions (e.g. the Alpine zone).

The standard remove-restore procedure was adopted to estimate the gravity anomalies at the aerogravimetry points. The low-frequency component of the gravity signal was accounted for by the GPM98CR model (up to degree 720, Wenzel 1998) which was used to compute the last Italian geoid (Barzaghi et al. 2007). The high-frequency gravity component was evaluated using the SRTM derived DTM at a resolution of $3'' \times 3''$ completed by the $1' \times 1'$ NOAA bathymetry (Barzaghi et al. 2007). The reference DTM used in RTC evaluation has been obtained filtering this detailed DTM with a moving average window having a $10'$ radius (this window size has been defined based on a statistical analysis on the residuals for different smoothing versions of the detailed DTM).

The residual gravity values, i.e. the observed data minus the global model effect and the residual terrain effect component (RTC), were gridded on a regular $2' \times 2'$ grid on the

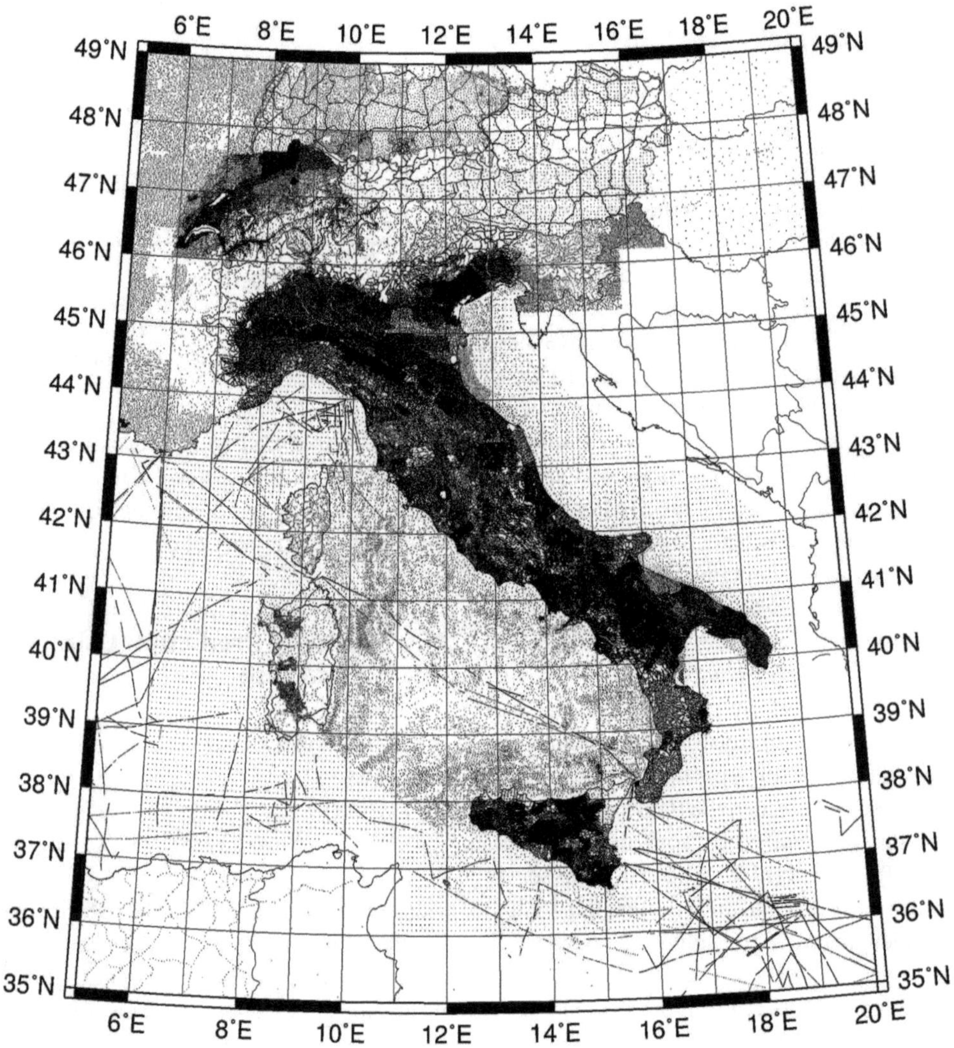


**Fig. 3** The Italian gravity database

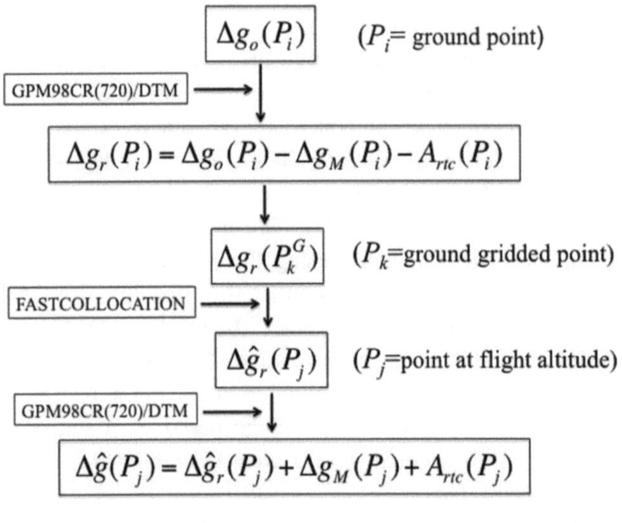


**Fig. 4** The gravity prediction procedure. $\Delta g_M$ global model gravity anomaly, $A_{RTC}$ RTC gravity component

ellipsoid and used to predict the residual gravity component at each aerogravimetric point by fast collocation (Bottoni and Barzaghi 1993). In the restore step, the model and the RTC components were added to the predicted residual gravity values in the same prediction points to estimate upward continued gravity anomalies.

This procedure is schematically summarized in Fig. 4.

## 4 The Comparison Between Airborne and Predicted Gravity Values

The estimated gravity values derived following the procedure of Sect. 3 were compared with the observed aerogravimetry data. Furthermore, the global models EGM2008 (Pavlis et al. 2012) and EIGEN-6C4 (Förste et al. 2014), both to its full resolution of degree and order 2,190, were considered in the comparisons with the observed values. The overall statistics

**Table 1** Statistics of the differences between observed airborne gravity and predicted gravity at flight altitude

| Model/database | RMS of gravity anomaly differences (observation – predicted) (mGal) |
|---|---|
| Italian database | 3.64 |
| EGM 2008 | 2.94 |
| EIGEN-6C4 | 2.87 |

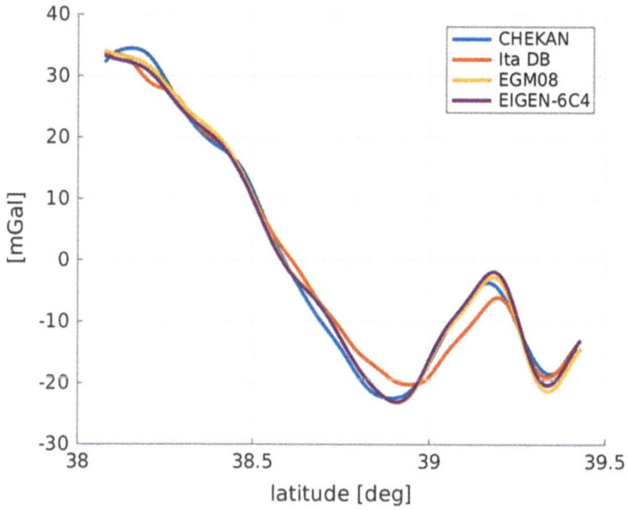


**Fig. 5** Medium frequency discrepancies between observed and Italian database prediction

of airborne gravity minus predicted gravity (62,717 data points) are given in Table 1. Since the aerogravimetry data are biased, as explained in Sect. 2, only standard deviations are considered.

These statistics are slightly better than those obtained in a similar test performed in the central and southern part of Italy (Barzaghi et al. 2009). Subsequently, trackwise analyses were performed. They show that the main features of the gravity signal as observed in the aerogravity data are similarly reproduced by the different tested models. However, over some tracks misalignements are present at medium frequencies with respect to the Italian database prediction (see Fig. 5). They are probably due to the GPM98CR model that was used in predicting the values (they are not present in the EGM2008 and EIGEN-6C4 predictions).

High frequency discrepancies can also be seen in some areas, particularly in the Central Apennines area (see Fig. 6). This might reflect some smoothing in the GEOHALO values which possibly do not account for some RTC effects of this particular area, an already known phenomenon described e.g. by Verdun et al. (2003) (the same holds true for the EGM2008 and EIGEN-6C4 models that have a resolution comparable with the along-track resolution of the GEOHALO data).

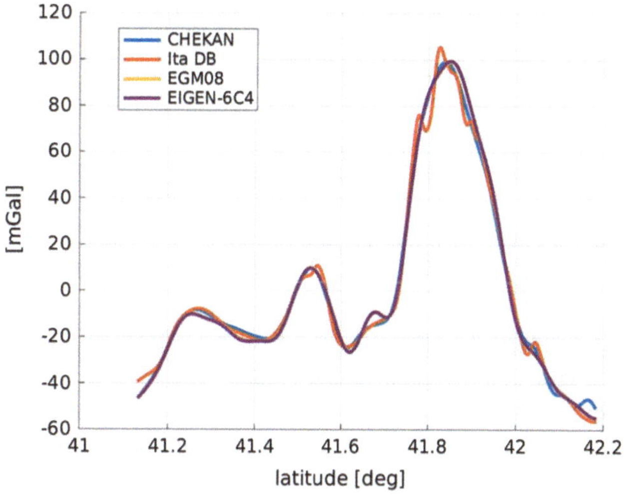


**Fig. 6** High-frequency discrepancies between observed and Italian database prediction over the Central Apennines

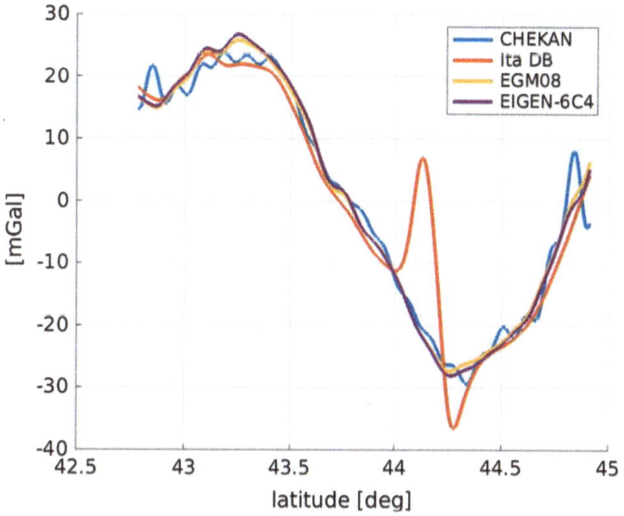


**Fig. 7** Medium frequency discrepancies due to biased data in the Italian database

Furthermore, medium frequency differences were also observed in the tracks covering the Northern area of the Adriatic Sea (see Fig. 7, at latitude between 44° and 44.5°).

They cannot reasonably be explained in terms of real gravity field features. By checking the Italian database it was found that there exist a small area containing some biased data. Since these irregular values are coherent and close to each other, the adopted outlier rejection procedure which was used to check the Italian gravity database failed. This, in turn, demonstrates the benefit of the aerogravity data helping to assess the quality and homogeneity of the ground based data. Most probably, the discrepancies at the beginning and the end of this profile can be explained by oscillations caused by a propagation of the strong vertical accelerations from outside of the shown track due to the filter properties. This profile is an example why in the final processing the accepted data span is usually shortened accordingly.

## 5 Conclusions

The GEOHALO flight mission proved that reliable gravity data can be acquired using a high-speed jet aircraft. The statistics of the residuals show that there is a satisfactory agreement with the predictions based on EIGEN-6C4 and EGM2008 global geopotential models. The same holds for the predictions based on the Italian gravity database. However, in this case the results are more controversial. Some high frequency discrepancies can reflect a smoothing in the aerogravity values (and in the signal predicted by global models). As a matter of fact, these discrepancies occur in the central part of the Apennines in the Maiella and the Gran Sasso massif areas. Therefore, it can be assumed that part of the terrain effect is not properly reproduced in the aerogravity signal. On the contrary, some other high frequency differences can be explained by biased data in the Italian database. They were highlighted by these comparisons. This gives the clear implication that the GEOHALO gravity data can profitably be used in checking and improving ground gravity databases.

## References

Barzaghi R, Borghi A, Carrion D, Sona G (2007) Refining the estimate of the Italian quasi-geoid. Boll Geod Sci 3:145–159

Barzaghi R, Borghi A, Keller K, Forsberg R, Giori I, Loretti I, Olesen AV, Stenseng L (2009) Airborne gravity tests in the Italian area to improve the geoid model of Italy. Geophys Prospect 57:625–632

Bottoni G, Barzaghi R (1993) Fast collocation. Bull Geod 67(2):119–126

Forsberg R, Olesen AV (2010) Airborne gravity field determination. In: Xu G (ed) Sciences of geodesy – I. Springer, Berlin/Heidelberg, pp 83–104. doi:10.1007/978-3-642-11741-13

Förste C, Bruinsma S, Marty J-C, Flechtner F, Abrikosov O, Dahle C, Lemoine H-M, Neumayer KH, Biancale R, Barthelmes F, König R (2014) EIGEN-6C4 – the latest combined global gravity field model including GOCE data up to degree and order 1949 of GFZ Potsdam and GRGS Toulouse. Presented at the 3rd International Gravity Field Service (IGFS) General Assembly, June 30 – July 6, 2014, Shanghai, China

Jekeli C (2001) Inertial navigation systems with geodetic applications. De Gruyter, Berlin

Kaifei HE (2014) GNSS kinematic position and velocity determination for airborne gravimetry. PhD thesis, Technical University Berlin, Germany

Krasnov A, Sokolov A, Usov S (2011) Modern equipment and methods for gravity investigation in hard-to-reach regions. Gyrosc Navig 2(3):178–183. doi:10.1134/S2075108711030072

Pavlis N, Holmes SA, Kenyon SC, Factor JK (2012) The development and evaluation of the earth gravitational model 2008 (EGM2008). JGR Solid Earth 117(B4):1–38

Scheinert M (2013) The geodetic-geophysical flight mission GEOHALO to acquire measurements of the gravity and magnetic fields, of GNSS remote sensing and of laser altimetry over the Mediterranean. Geophys Res Abstr 15:EGU2013-5160-1

Verdun J, Klingelé E, Bayer R, Cocard M, Geiger A, Kahle H-G (2003) The alpine Swiss-French airborne gravity survey. Geophys J Int 152(1):8–19

Wenzel G (1998) Ultra high degree geopotential models GPM98A, B and C to degree 1800. Proceedings Joint Meeting of the International Gravity Commission and International Geoid Commission, OGS, Trieste

# The Effect of Helium Emissions by a Superconducting Gravimeter on the Rubidium Frequency Standards of Absolute Gravimeters

Jaakko Mäkinen, Heikki Virtanen, Mirjam Bilker-Koivula, Hannu Ruotsalainen, Jyri Näränen, and Arttu Raja-Halli

**Abstract**

Recently, large offsets in the frequency of rubidium oscillators of absolute gravimeters have been reported, due to contamination by helium from a superconducting gravimeter that shares the same laboratory space. Such offsets can cause large errors in measured gravity if they go unnoticed and are not corrected for in the data processing. We give an account of the calibration histories of the rubidium oscillators of the FG5-221 (2003–2012) and of the FG5X-221 (2013–2014) of the FGI from the viewpoint of helium contamination. These gravimeters have been stationed (when not in field work) in the Metsähovi gravity laboratory, about 5 m from the superconducting gravimeter GWR TT020, in a separate room. The drift of the oscillator of the FG5-221 during the 10 years has been larger than specifications but reasonably linear. Evidence for possible helium effects is indirect: when the gravimeter is away from the laboratory some relaxation of the frequency offset is seen. This could be an indication that contamination by helium contributed to the large drift. The drift of the oscillator of the FG5X-221 has been much larger and correlates with known episodic helium emissions. During the installation of a new superconducting gravimeter GWR Dual OSG 073 in January-February 2014 and the ensuing large helium emissions we monitored the response of the FG5X-221 oscillator continuously. In 5 days, the clock frequency increased by $4.7 \times 10^{-9}$. Such an offset, if not corrected for in the data processing would cause an offset of $-9.2$ μGal in observed gravity.

**Keywords**

Absolute gravimeter • Helium contamination • Rubidium oscillator

## 1 Introduction

Absolute gravimeters (AGs) nowadays mostly use a rubidium oscillator as the standard of frequency (see e.g. Niebauer et al. 1986, 1995). The specifications of commercial rubidium oscillators typically quote monthly relative aging rates of $5 \times 10^{-11}$ or less (corresponding to 0.1 μGal in observed gravity). Calibration at a few months' interval and the interpolation of frequency offsets between the calibrations could then be expected to be sufficient to maintain the frequency contribution to the standard uncertainty of measured gravity at or below 0.1 μGal.

The sensitivity of rubidium frequency standards to contamination by helium has been well known to their constructors (e.g. Goldberg et al. 1985) and to time standard specialists from the very beginning. This sensitivity also includes the normal atmospheric helium content (about 0.52 Pa in partial pressure) which contributes to the initial frequency drift after manufacture. However, until the recent observations reported by TM Niebauer (e-mail information on September 21, 2012) and the paper by van Westrum et al.

J. Mäkinen (✉) • H. Virtanen • M. Bilker-Koivula • H. Ruotsalainen • J. Näränen • A. Raja-Halli
Finnish Geospatial Research Institute FGI, Geodeetinrinne 2, 02430 Masala, Finland
e-mail: Jaakko.Makinen@nls.fi



S. Jin, R. Barzaghi (eds.), *IGFS 2014*, International Association of Geodesy Symposia 144,
DOI 10.1007/1345_2015_205

(2014) the users of AGs (including the authors of this paper) have been largely unaware of the potential risk of helium contamination. The contamination increases the frequency and causes an apparent decrease in measured gravity, if the frequency offset is not corrected for.

Over the years numerous AGs have been exposed to helium levels well above the normal atmospheric content. The pertinent helium source is a superconducting gravimeter (SG): many gravity laboratories operate both SGs and AGs because of the complementary characteristics of the two instruments. Even when the instruments are not permanently co-located, SG laboratories are frequently visited by AGs, to perform SG calibration (Hinderer et al. 1991) and/or SG drift control. While latest-generation SGs normally operate with zero loss of liquid helium (http://www.gwrinstruments.com/index.html), this was not the case with previous generations. And in any case, malfunctions of the refrigeration system, services, and refills of the liquid helium of the SG lead to emissions. Thus it is important to obtain information about oscillator behaviour in relevant helium surroundings.

Van Westrum et al. (2014) describe an incident where two rubidium oscillators from FG5 gravimeters were accidentally exposed to large concentrations of helium in connection with the refill of liquid helium in a SG. After 6 weeks the frequency offsets of the oscillators were still $+5 \times 10^{-9}$ and $+16 \times 10^{-9}$, respectively. Van Westrum et al. (2014) showed that when the oscillators were continuously powered, the offsets then relaxed exponentially towards zero with a time constant of about 100 days. The "power on/off" parameter is important: the borosilicate glass of the rubidium cell is more porous when warm, and the cell is more susceptible to helium diffusion. It is then more affected by the helium concentration in the surroundings (van Westrum et al. 2014) and probably also more able to shed its helium contamination once the concentration is lowered.

They then performed a controlled experiment by immersing the oscillators in 100 % helium. They found offsets of similar magnitudes as above, and again a time constant of about 100 days.

We do not have a controlled experiment. Instead, we will give an account of the calibration histories of the Datum LPRO rubidium oscillator (2003–2012) of the FG5-221 (Niebauer et al. 1995) and of the Symmetricom SA22C oscillator (2013–2014) of the FG5X-221 (Niebauer et al. 2011), hereafter oscillator#1 and oscillator#2, respectively. These absolute gravimeters of the FGI have been stationed (when not in field work) in the Metsähovi gravity laboratory, about 5 m from the superconducting gravimeter GWR TT020, in a separate room.

For the physics of the helium contamination the reader is referred e.g. to van Westrum et al. (2014).

We express the frequency offset as the dimensionless ratio of the offset to the nominal frequency, rather than in units of frequency (Hz). The nominal frequency of the rubidium oscillators is 10 MHz; then 0.01 Hz corresponds to $1 \times 10^{-9}$. An offset of $+1 \times 10^{-9}$, if not accounted for in the data processing, causes an error of $-1.96$ μGal in measured gravity.

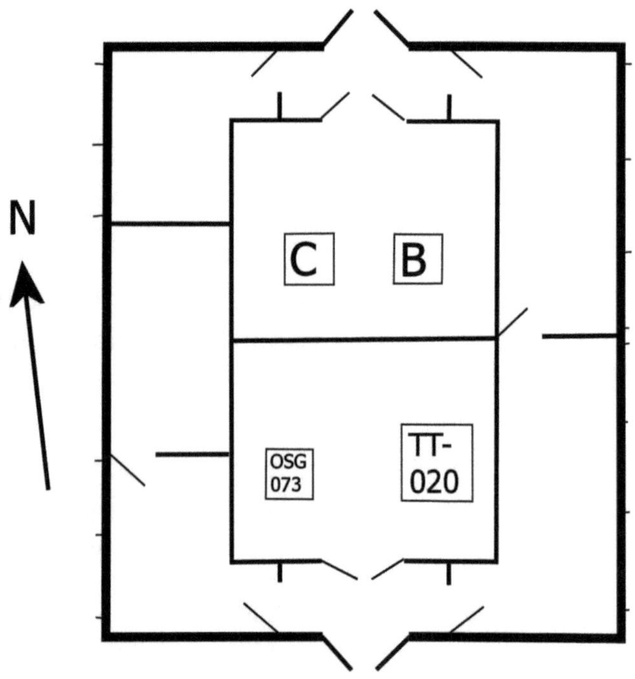


**Fig. 1** Plan of the Metsähovi gravity laboratory. The building is 11.26 m long in east–west direction. The positions of the SGs GWR TT020 and GWR Dual OSG 073, and of the absolute-gravity piers B and C are shown. The electronics box of the AG with the rubidium oscillator stands between B and C. The rooms around the SG and AG laboratory rooms are used as offices, storage, and workshops. They have a ventilation system but the laboratories do not

## 2 Experimental Conditions and Methods

### 2.1 The Gravity Laboratory

The Metsähovi gravity laboratory is shown in Figs. 1 and 2. It was completed in 1994 and is unchanged ever since. It is a relatively lightweight wooden structure with a concrete floor over granite bedrock. The external walls and the roof structure consist of wooden boards and mineral wool insulation, the separating walls and ceilings likewise. The roofing is made of bitumen felt. The main obstacle to helium escaping the building is probably the polyethene sheet that lines the external walls and the roof structure along the inner side of the insulation. This is standard building practice in Finland; its purpose is to prevent the normal water vapour content in the interior air from penetrating the insulation and

**Fig. 2** Metsähovi gravity laboratory seen from east

condensing there. We have not tried to model the transport of the helium inside and out of the building.

## 2.2 The Rubidium Oscillators

Nearly all calibrations have been made by us at Metsähovi, using the 5 MHz frequency from the hydrogen maser of the nearby Metsähovi Radio Observatory of the Aalto University as a reference. The standard uncertainty of these calibrations is better than $1 \times 10^{-11}$, typically around $1 \times 10^{-12}$. We also use six calibrations performed during field campaigns, referred to the hydrogen masers or GPS-disciplined rubidium standards of host laboratories. In addition, first calibrations for the two oscillators were delivered by the manufacturer of the two gravimeters, Micro-g LaCoste Inc. As the hydrogen masers are long-term monitored through GPS, all our calibrations are traceable to caesium frequency standards.

We compiled the location history of the gravimeters (the in/out of the gravity laboratory) from the manually kept logbooks of the gravimeters, where such entries are made with the precision of 1 h. The power on/off aspect was not taken into account.

The FG5-221 first entered the Metsähovi gravity laboratory on March 3, 2003. It was shipped back to the factory December 19, 2012 to be upgraded to FG5X-221. During the 10 years altogether 55 individual calibrations of oscillator#1 were made. The FG5X-221 arrived at the laboratory on May 8, 2013, and 18 individual calibrations of oscillator#2 had been made by January 26, 2014. In view of the impending large releases of helium we then started continuous frequency monitoring. This monitoring continues even today whenever the program of AG allows it. Calibration data until June 9, 2014 is used here.

## 2.3 The Helium Source: The Superconducting Gravimeters

The superconducting gravimeter GWR TT020 has been operating in the gravity laboratory (Figs. 1 and 2) since August 1994 (Virtanen 2006). It normally consumes about 0.5 L of liquid helium per day. If its refrigeration system stops, say because of a malfunction or a power cut, the helium consumption increases about eightfold to 4 L/day. An interruption in the refrigeration causes an increase in the dewar temperature and can thus be detected in the record of the auxiliary parameters of the SG.

A refill of liquid helium is needed approximately every 10 months and is accompanied by an emission of 15–50 L. The amount can be estimated from the difference between the quantity drawn from the transport container and the quantity added to the dewar. In addition, helium gas is used for service tasks. The quantity released at one time can correspond to a couple of days of normal consumption of liquid helium, and we neglect it here.

In January–February 2014 the new superconducting gravimeter GWR Dual OSG 073 was installed in the gravity

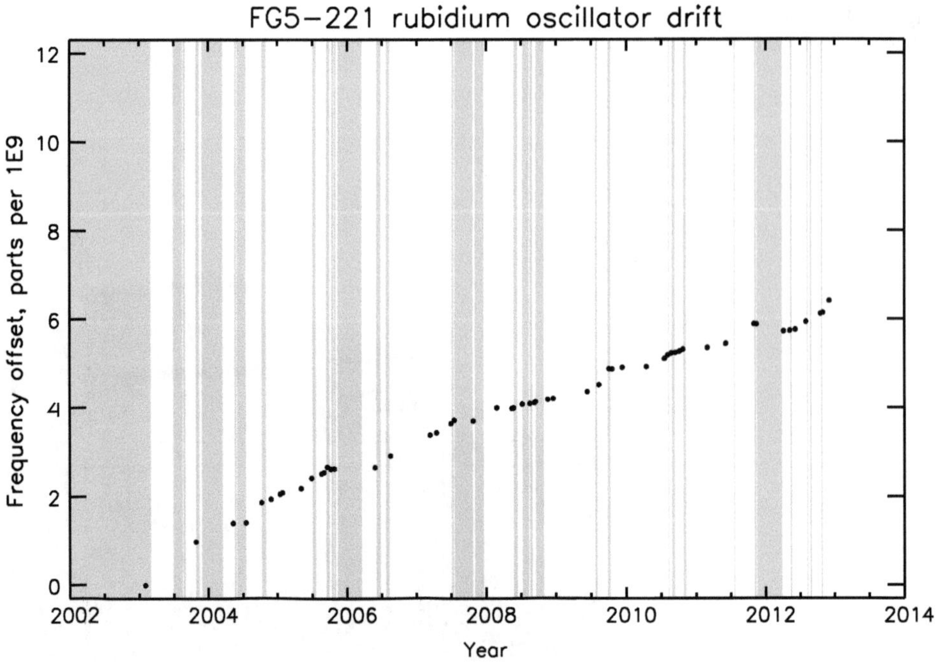


**Fig. 3** Calibration history of the rubidium oscillator of the FG5-221 (oscillator#1). The *dots* show the relative frequency offset in units of $1 \times 10^{-9}$, which corresponds to 0.01 Hz in the frequency of the oscillator (10 MHz) and an error of $-1.96$ μGal in measured gravity if the offset is not taken into account in the data processing. Uncertainties are not shown; standard uncertainty bars would mostly have the vertical extension of the dots themselves. *Shaded areas* indicate when the oscillator has been away from the gravity laboratory

laboratory. The initial fill-up released helium but the normal operation is with zero loss.

For this study the information about episodic helium emissions was compiled only for the operation of oscillator#2, the more sensitive of the two.

## 3 Results

Figure 3 shows the frequency history of oscillator#1. The total drift over 10 years is about $+6.5 \times 10^{-9}$. This is 5–15 times more than the average annual drift of the 4 rubidium oscillators shown by van Westrum et al. (2014) over similar time spans. The specifications of oscillator#1 quote aging less than $2.5 \times 10^{-11}$/month and less than $1 \times 10^{-9}$ over 10 years. However, there are potential drivers for drift other than aging. For instance, in the AG application the oscillator is switched on and off frequently, 20–50 times per year. Retrace problems could therefore contribute to the drift. Evidence for a possible helium contribution to the drift comes mainly from the drift characteristics during the gravimeter's absence from the laboratory: the drift decreases, or becomes zero, or even reverses to negative (Fig. 3).

The history of oscillator#2 and helium emissions during it are presented in Fig. 4. The aging specification of oscillator#2 is less than $4 \times 10^{-11}$/month. But in the 8 months (240 days) even before the major helium events the total drift is more than tenfold, $4 \times 10^{-9}$. The oscillator reacts to episodic helium events. The major events near day 400 cause a rise a $4.7 \times 10^{-9}$ in the frequency, about the same size as van Westrum et al. (2014) found from a 10-min total helium immersion with the oscillator switched off.

The drift reverses during absence from the laboratory. The decay of 14 % during 32 days of absence (after day 300 in the plot) corresponds to a time constant of about 220 days, and the decay of 9 % during 11 days (around day 500 in the plot) to a time constant of about 120 days. Both were field campaigns but the former included long travels with the oscillator switched off most of the time, and in the latter campaign with short trips the oscillator was powered almost all the time. This agrees with van Westrum et al. (2014) who found a time constant of 100 days when the oscillator was powered continuously.

After the major helium events near day 390 (Fig. 4) the frequency of oscillator#2 starts relaxing despite the continuous background (0.5 L/day) helium emission. But apparently not towards zero: assuming that the time constant even here is 100 days, the steady-state frequency offset for the background emission would be as high as $+7.9 \times 10^{-9}$.

The frequency and helium events around day 390 in Fig. 4 are plotted in Fig. 5 at the same vertical scale but with more resolution in time. First the refrigeration of the TT020 stopped because of a malfunction. The OSG 073 was filled on January 28, 2014 (the first emission peak), after which the transport container was left simmering in the SG laboratory. We use for it the typical loss estimate of 1% of container volume or 3 L/day. The TT020 was refilled on January 31 (the second peak), after which the container was wheeled out and refrigeration of the TT020 was re-started. The small bump in frequency after day 400 is known to be a minor helium emission in the course of operator training for OSG 073. Other than this, estimates of the emissions related to the installation of the OSG 073 are not well-constrained. From

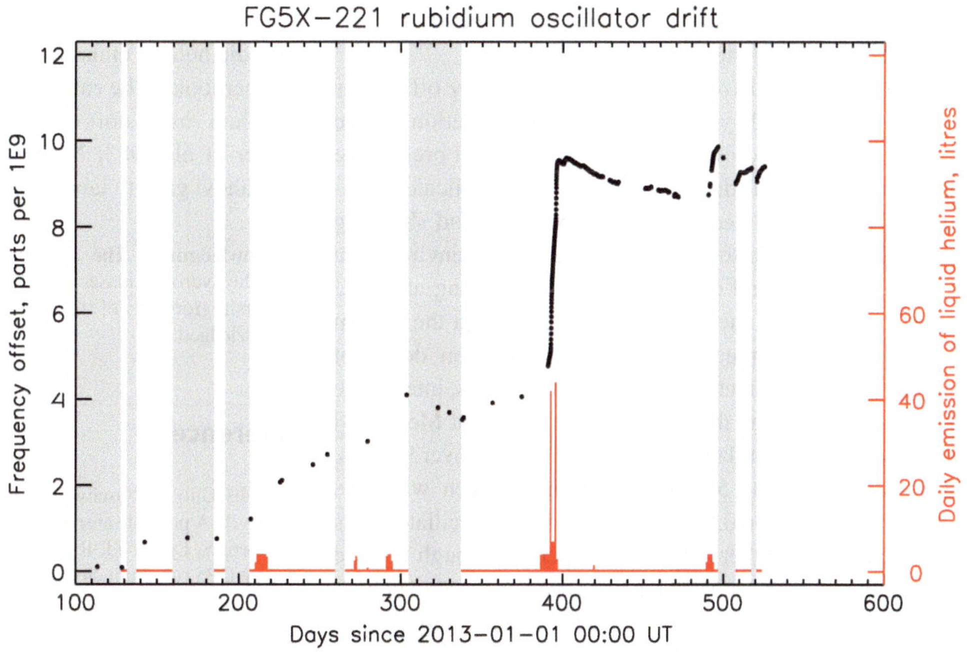


**Fig. 4** Calibration history of the rubidium oscillator of the FG5X-221 (oscillator#2). The *dots* (*left-hand scale*) show the relative frequency offset in units of $1 \times 10^{-9}$, which corresponds to 0.01 Hz in the frequency of the oscillator (10 MHz) and an error of $-1.96$ μGal in measured gravity if the offset is not taken into account in the data processing. Uncertainties are not shown; standard uncertainty bars would mostly have the vertical extension of the dots themselves. Continuous frequency monitoring is presented as 1-h averages. The bar in the *bottom* (*right-hand scale*) shows the estimated release of liquid helium per day. *Shaded areas* indicate when the oscillator has been away from the gravity laboratory

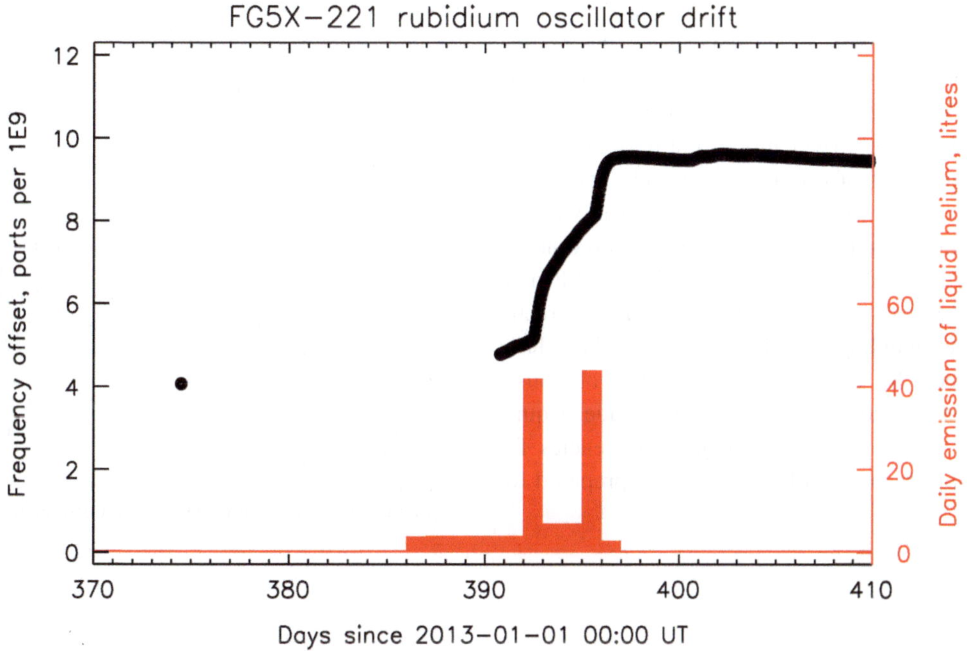


**Fig. 5** Details from Fig. 4 around day 390. For explanations see Fig. 4. To avoid saturating the vertical scale all helium events are averaged over one calendar day (UTC). Comments in text

the frequency plot it appears possible that not all emissions between the two peaks have been accounted for.

## 4 Discussion

In our case study of two rubidium oscillators we have found large effects due to helium contamination in one of them (#2), similar to those by van Westrum et al. (2014). Oscillator#1 appears to be influenced by helium, too, but considerably less.

Van Westrum et al. (2014) provide general considerations and instructions how to avoid and mitigate helium effects in rubidium oscillators of AGs. But what are the consequences of the helium effect for the gravity data taken with the two AGs discussed here? For this discussion it is important to understand that even very large offsets of the oscillators (from helium contamination or for whatever reason) have no effect on the gravity results, as long as the offsets are known, and corrected for in the processing.

For the FG5-221 (oscillator#1) the frequency offset was originally estimated by piecewise linear interpolation in time between the two closest calibrations (Fig. 3). At present we know that the "real drift" is probably a complicated mixture of exponential decays of both absorbing and shedding helium, in response to a time-variable helium environment in the laboratory and out of it. However, looking at Fig. 3 even in the light of the present knowledge about the helium issue, the simple piecewise linear interpolation does not appear such a poor method. To test it in extremis, interpolate between the first and the last calibration made at Metsähovi (the second and last data points in Fig. 3), i.e., over 9 years. This reproduces the 52 calibrations in between with rms error $0.3 \times 10^{-9}$, or 0.6 μGal in gravity. If oscillator#1 is representative of others of the same period (though this we do not know), many AGs which have been exposed to helium may have escaped without major harm to the gravity results, even where the operators have not taken the exposure into account in any way in their calibration schedules.

The helium risk was already known when the FG5X-221 (oscillator#2) arrived at the laboratory. During the measurements described here we did not have access to a GPS-disciplined oscillator (which does not have an offset). Instead, the monitoring of oscillator#2 in field campaigns was planned with care, with calibrations immediately before leaving and after returning, and additional calibrations of opportunity in the field to constrain the decay. When in the laboratory, we use the hydrogen maser signal directly as AG reference, doubling it to 10 MHz. Thus there have been no ill effects for the gravity results. But obviously the large sudden helium effects would have made it quite difficult for us to estimate the oscillator frequencies for the processing of gravity data had we been unprepared for them.

In the follow-up research we shall compile a record of episodic helium emissions for oscillator#1 as well, and take a closer look at the calibration history of the Efratom FRK-L rubidium oscillator of the JILAg-5 absolute gravimeter (Faller et al. 1983; Niebauer et al. 1986), resident in the Metsähovi gravity laboratory 1995–2003.

**Acknowledgment** The opportunity to use the reference frequency from the hydrogen maser of the Metsähovi Radio Observatory of Aalto University (formerly of Helsinki University of Technology) is gratefully acknowledged.

## References

Faller JE, Guo YG, Gschwind J, Niebauer TM, Rinker RL, Xue J (1983) The JILA portable absolute gravity apparatus. BGI Bull Inf 53:87–97

Goldberg S, Lynch TJ, Riley WJ (1985) Further test results for prototype GPS rubidium clock. In: Proceedings of the 17th annual precise time and time interval (PTTI) application and planning meeting, Dec 1985. NASA, Washington DC, pp 145–155. http://tycho.usno.navy.mil/ptti/1985papers/Vol%2017_14.pdf

Hinderer J, Florsch N, Mäkinen J, Legros H, Faller JE (1991) On the calibration of a superconducting gravimeter using absolute gravity measurements. Geophys J Int 106:491–497

Niebauer TM, Hoskins JK, Faller JE (1986) Absolute gravity: a reconnaissance tool for studying vertical crustal motion. J Geophys Res 91:9145–9149

Niebauer TM, Sasagawa GS, Faller JE, Hilt R, Klopping F (1995) A new generation of absolute gravimeters. Metrologia 32:159–180

Niebauer M, Billson R, Ellis B, van Westrum D, Klopping F (2011) Simultaneous gravity and gradient measurements from a recoil-compensated absolute gravimeter. Metrologia 48:154–163

van Westrum D, Bianchi T, Billson R, Ellis B, Niebauer TM, Rohner H (2014) The effect of helium contamination on rubidium clock references in absolute gravity meters. In: Peshekhonov VG (ed) Proceedings of the IAG symposium on terrestrial gravimetry: static and mobile measurements (TG-SMM2013), Saint Petersburg, Russia, 17–20 Sept 2013. State Research Center of Russia Elektropribor, St Petersburg, pp 125–130

Virtanen H (2006) Studies of earth dynamics with the superconducting gravimeter, vol 133. Finnish Geodetic Institute, Kirkkonummi, 130 p

# Part II

# Global Geopotential Models and Vertical Datum Unification

# Wavelet Multi-Resolution Analysis of Recent GOCE/GRACE GGMs

A.C. Peidou and G.S. Vergos

**Abstract**

The realization of the GRACE/GOCE missions offer new opportunities for gravity field approximation with higher accuracy at the medium wavebands, while wavelets (WL) provide powerful gravity field analysis tools in the space/frequency domain. This work focuses on the spectral analysis of GOCE, GOCE/GRACE and combined Global Geopotential Models (GGMs) through wavelet decomposition, filtering and reconstruction to improve their performance in the higher bands of the spectrum. The GGMs evaluated refer to the latest DIR-R4/R5, TIM-R4/R5 and GOCO03s models, which are compared with local GPS/Leveling geoid heights and gravity anomalies, while EGM2008 is used as a reference. Through a WL-based multi-resolution analysis, gravity anomalies and geoid heights are analyzed to derive their approximation and detail coefficients for various levels of decomposition, which correspond to different spatial scales. The content and signal power of each level of decomposition is analyzed to conclude on the amount and quality of signal power that GOCE/GRACE GGMs represent compared to EGM2008, especially up to the targeted waveband of 100–150 km. Filtering is investigated as well to remove high-frequency information from the low resolution GOCE models and adjust the WL reconstruction. The model synthesis that follows, through WL coefficient reconstruction, aims at the generation of new synthesized GGMs, where both GOCE and EGM2008 information is used, the latter serving to model the omission error in the GOCE GGMs. The synthesized GOCE GGMs offer an improvement of more than 30 cm compared to the original GOCE GGMs, while they provide a 1–2 cm improvement compared to EGM2008. In terms of the validation with gravity anomalies, a 5 mGal improvement was found, w.r.t. to the original GOCE GGMs, while w.r.t. EGM2008 there was no improvement. Finally, it was concluded that the GOCE GGMs show improved, between 5–22%, correlation with the land topography compared to EGM2008 for spatial scales between 176–704 km.

**Keywords**

Correlation • Filtering • Gravity field • Multi-resolution analysis • Spectrum coherency • Validation • Wavelets

## 1 Introduction

Monitoring and understanding of the Earth's gravity field parameters at various spatial scales has been the focus of many studies during the past decades. The satellite missions of GOCE and GRACE have provided new insights

A.C. Peidou • G.S. Vergos (✉)
Department of Geodesy and Surveying, Aristotle University of Thessaloniki, University Box 440, 54124 Thessaloniki, Greece
e-mail: vergos@topo.auth.gr


S. Jin, R. Barzaghi (eds.), *IGFS 2014*, International Association of Geodesy Symposia 144,
DOI 10.1007/1345_2015_44

to gravity field monitoring and interpretation. Moreover, significant results related to the time variation and evolution of the gravity field have emerged, the latter being a result of mass/water redistribution in system Earth as well as a response to geodynamic phenomena e.g., mega-earthquakes (Fuchs et al. 2013). GOCE has offered invaluable data on sea-level change, ocean currents and circulation and ice dynamics (Knudsen et al. 2011; Menna et al. 2014). Especially as far as gravity field and geoid approximation is concerned, GOCE has offered improved, compared to EGM2008, representations of the medium band of the spectrum, i.e., degree and order between 160 and 220, by as much as 4–5 cm over Europe (Gruber et al. 2011; Hirt et al. 2011; Vergos et al 2014).

This work focuses on the evaluation of the spectral content of GOCE/GRACE-based GGMs, both satellite only and combined ones, by assessing their accuracy in terms of both gravity anomalies and geoid heights. Gravity anomaly evaluation is carried out through local gravity measurements covering the entire European continent. Moreover, an extensive network of collocated GPS/Leveling benchmarks, covering continental Greece, are used for the geoid height evaluation.

Contrary to the usual evaluation in terms of the Global Geopotential Model (GGM) combination with EGM2008 (see e.g., Gruber et al. 2011; Vergos et al. 2014), in this work we employ wavelet (WL) decomposition as a multi-resolution analysis (MRA) tool. MRA describes the (infinite) sequence of closed linear subspaces of the space of square integrable functions $V_j \subset L^2(\Re)$ and can be applied with or without data. The transition from a MRA model $\{V_j\}$ to an associated wavelet basis function is what constitutes the decomposition scheme for every MRA subspace. Although WL MRA is relatively new as an analysis tool compared to Fourier analysis, wavelets have been developed in order to overcome the deficiencies of the Fourier transform (Mallat 1989). The advent of the WL transform and WL modelling in geosciences brought flexibility in the analysis process for over a decade, since it allows the decomposition of the signal under study to specific spatial scales that correspond to the levels of Wl decomposition. This is especially important since it allows the study of the properties of each individual level (corresponding to specific spatial scales) without influencing the rest. Within that frame, WL have been employed for local/regional determination of the Earth's gravity field (Panet et al. 2011), the identification of large-scale geoid undulations and their relation to mantle processes (Hayn et al. 2012), the solution of the Altimetry-Gravimetry Boundary Value Problem (Grebenitcharsky and Sideris 2005) and lately to the analysis of GOCE satellite gravity gradiometry data (Grebenitcharsky and Moore 2014). In this work WL transform and MRA are used to analyse both gravity anomalies and geoid heights in approximation and detail coefficients for various levels of decomposition. Given the initial resolution of the available data, the decomposition levels can be translated to spatial scales, hence WL MRA allows the study of GOCE/GRACE GGM contribution to various parts of the gravity field spectrum. To improve the performance of GOCE/GRACE GGMs, as to their spectral content in the higher bands of the spectrum, they are combined through wavelet decomposition, filtering and reconstruction with EGM2008. Hence EGM2008 is used to model the omission error in the low-degree GOCE/GRACE GGMs. Both the original and synthesized GOCE/GRACE fields are evaluated with in situ gravity anomalies and GPS/Leveling observations on benchmarks (BMs).

Coherence and correlation are powerful tools for testing the relation between land topography and gravity anomalies ($\Delta g$) for different spectral bands. They are both employed over the Amazon area and Europe, in combination with WL decomposition, to determine the coherency and correlation of GOCE/GRACE GGMs for different bands of the spectrum.

# 2 Methodology, GGMs and Data Availability

## 2.1 GOCE/GRACE GGM Analysis

Wavelets are base functions with localization properties in both space (time) and scale (frequency) domains and allow the synchronous analysis of GGM data at different levels/scales. Therefore wavelet signal processing can be a multiresolution analysis (MRA) at various levels of decomposition (Chui 1992). The two-dimensional wavelet transform gives coefficients that correspond to different spatial resolutions, related to the signal scales (Grebenitcharsky and Moore 2014). According to the WL decomposition algorithm, each scale (level $L_n$) of the signal is analyzed in an approximation coefficient ($A_n$), which carries the main (large scale) information and three detail coefficients (horizontal, vertical and diagonal $(H, V, D)_i|_{i=1,2,\ldots,n}$ that carry the high-frequency (short-scale) information of the signal (Mallat 1989, 1999). Through the synthesis process, various GGMs can be combined, since each level can be represented by a different GGM, based on the GGM performance at each specific level of analysis. Of course, if after the WL MRA the WL coefficients remain unaltered during the reconstruction process, the the original signal will be reconstructed perfectly due to the orthogonality of the WL base functions.

The synthesis is defined as the algebraic sum of the detail coefficients of each level used $(H, V, D)_i|_{i=1,2,\ldots,n}$ and the approximation coefficient of the last level ($A_n$) as:

$$\begin{aligned} Synthesis = {} & \mathrm{A}_n + (H, V, D)_\mathrm{n} + (H, V, D)_{\mathrm{n}-1} \\ & + \cdots + (H, V, D)_1 \end{aligned} \tag{1}$$

Therefore, for the available GOCE/GRACE GGMs, their spectral content at each level is analyzed and compared among each other and with EGM2008. The goal is to construct combined GGMs where both GOCE/GRACE and EGM2008 information will be used, so that the gravity field signal will be represented with higher accuracy. The choice of the GGM that will be used at each level depends on its resolution and the gravity field content with respect to EGM2008. Then, different GGMs can be combined during the synthesis process in order to determine a combined/hybrid GGM. In that sense, the first levels of the EGM2008 decomposed signal (small spatial scales) can be used to model the omission error in the GOCE GGMs. Likewise, the decomposed GOCE GGM signal can provide gravity information for those levels that correspond to the spatial scales targeted by the GOCE mission (larger than 100 km). Hence, during the synthesis process, various levels from different GGMs can be combined, in order to provide improved representations of the Earth's gravity field. This aims at reducing the omission error in GOCE GGMs and augmenting EGM2008 with gravity information from the GOCE mission.

Moreover, when GOCE/GRACE GGMs are analyzed, the gravity signal of the first levels (high-frequencies) is dominated by noise since these spatial scales are not mapped by the GOCE mission. This is especially profound at the limits of the spatial scales targeted by GOCE, i.e., around 100–120 km. For those levels that the noise is either dominant or contaminates the gravity field signal, increasing the SNR (signal-to-noise-ratio) demands a digital or spatial filter implementation. In this work, both Gaussian and box-car filters have been used to remove noise. The former is considered very effective for filtering in the space domain and the latter being a smoothing mathematical function, which uses a rectangular window in the frequency domain. After this synthesis and filtering process, the new combined GGMs can be evaluated w.r.t. some external information, e.g., GPS/Leveling geoid heights and gravity anomalies, to quantify the improvement reached.

Another valuable tool in the evaluation of the GOCE GGMs is in terms of the relationship between the GGM-derived gravity information and land topography. Therefore, a further investigation in terms of the spectral coherence and correlation was realized. Spectral coherence is a measure of the relation between two signals or data sets and if applied to the analysis of gravity field data it can give insight to how well they relate to the Earth's topography. The basic idea behind this evaluation, performed for each level of WL decomposition, is that if a GGM has higher coherence with the topography for a specific level, i.e., for a specific spectral band, then it represents better compared to the other GGMs the Earth's gravity field. In that way, the possible improvement by GOCE can be envisaged at specific targeted wavebands. Coherence is defined as (Bendat and Piersol 2010):

$$\mathrm{C_{gH}} = \frac{\left|\mathrm{G_{gH}}\right|^2}{\mathrm{G_{gg}G_{HH}}} \tag{2}$$

where, $G_{gH}$ denotes the cross-spectral density between the gravity and topography signals $g$ and $H$, and $G_{gg}$ and $G_{HH}$ the auto-spectral density. Another useful measure to be employed is the correlation matrix that can show whether and how strongly two signals are related. In our case, we construct the correlation matrix by estimating the correlation coefficients between the various levels of decomposition of both gravity and topography. Then the correlation coefficients are estimated as (Bendat and Piersol 2010):

$$R_{gH}^{ij} = \frac{C_{gH}^{ij}}{\sqrt{C_{gg}^{ii} C_{HH}^{jj}}} \tag{3}$$

In Eq. (3), $R_{gH}^{ij}$ denotes the correlation between level $i$ of gravity data from the GGMs and level $j$ of the topography, $C_{gH}^{ij}$ is the cross-covariance matrix of the two datasets and $C_{gg}^{ii}$ and $C_{HH}^{jj}$ the covariance matrices. In that way, the correlation matrix has as diagonal elements the coefficients for the same levels of decomposition. Finally, it should be noted that we have used the "awkward" term gravity to describe the GGM contribution in the sense that the gravity signal $g$ can be any functional related to the Earth's gravity field. In our study, as far as coherence and correlation are concerned, we focus on GGM-derived gravity anomalies over the Amazon area and over Europe.

## 2.2 Available Data and External Validation

The present study focuses on the GGM evaluation with external data, over the European Continent, within the region bounded between $30° \leq \varphi \leq 60°$ and $-10° \leq \lambda \leq 30°$. For the investigation of the correlation and coherence between the topography and GGM-induced gravity, the study focuses on two regions. The first one is the aforementioned European area while the second one focuses over the Amazon region, bounded between $-10° \leq \varphi \leq 20°$ and $275° \leq \lambda \leq 335°$. The main reason for focusing in two areas is their different topographic features and gravity field representation. The former stems from the Amazon area being mostly flat, while Europe has highly varying terrain. The latter refers to the fact that Amazon is a poorly surveyed area in terms of terrestrial gravity data, contrary to Europe. Therefore, GOCE GGMs should have improved performance over the Amazon, in terms of the correlation and coherency with land topography compared to EGM2008.

The GOCE/GRACE GGMs evaluated refer to the latest DIR and TIM models employing the fourth (R4) and fifth (R5) release, i.e., an effective data volume of 26.5 months of GOCE observations (R4) and the low orbit data (R5). TIM-R4 (Pail et al. 2011) presents a spherical harmonics expansion of the Earth's potential to a maximum degree $n_{max} = 250$ while TIM-R5 reaches a maximum degree of 280, both employing the time-wise approach and being pure GOCE models. DIR-R4 (Bruinsma et al. 2010, 2013), presents a spherical harmonics expansion of the Earth's potential to a maximum degree 260 while DIR-R5 reached a maximum degree 300, employing the direct approach and in addition to the GOCE observations, 9 years of GRACE data (10 for DIR-R5) and SLR have also been used. Apart from these GGMs, GOCO03s is used as well (Mayer-Gurr et al. 2012), which is based on both GOCE and GRACE data with a maximum degree of expansion to 250. In all cases, the performance of GOCE/GRACE GGMs is evaluated against EGM2008 which is used as reference (Pavlis et al. 2012). From these models, gravity anomalies and geoid heights have been determined, for all areas under study, at grid points with a spatial resolution of $3' \times 3'$ (5.5 km).

As far as the external GPS/Leveling and gravity data are concerned, the former refer to a set of 1542 collocated GPS and Leveling observations over the Hellenic trigonometric network (Vergos et al. 2014) in mainland Greece. The gravity data, refer to the gravity anomaly field derived in the frame of the World Gravity Map project (WGM2012) covering the entire European area (Balmino et al. 2012). WGM2012 is largely based on EGM2008 with the addition of a worldwide $1' \times 1'$ grid of ETOPO1-induced gravity disturbances. Therefore, the improvements brought by GOCE to the gravity anomaly comparison are expected to be marginal. Finally, the coherence and correlation between land topography and the Earth's gravity field are computed for the Amazon and Europe area, where ETOPO1 (Amante and Eakins 2009) was used for the topographic information. All computations were carried out in the Tide Free (TF) system, while GRS80 has been used as a normal field in the evaluation of the geoid zero-degree term.

**Table 1** Statistics of the differences between GPS/levelling and geoid heights from the GGMs before (normal font) and after (italics) the WL MRA synthesis

| | Min | Max | Mean | Std |
|---|---|---|---|---|
| EGM08 | −0.853 | 0.104 | −0.372 | ±0.134 |
| GOCO03s | −1.735 | 1.110 | −0.359 | ±0.464 |
| *GOCO03s* ($L_5, L_6, L_7$) | *−1.083* | *0.453* | *−0.387* | *±0.259* |
| *GOCO03s* ($L_6, L_7$) | *−0.855* | *0.093* | *−0.378* | *±0.124* |
| TIM-R4 | −1.597 | 1.155 | −0.358 | ±0.450 |
| *TIM-R4* ($L_5, L_6, L_7$) | *−1.151* | *0.399* | *−0.381* | *±0.239* |
| *TIM-R4* ($L_6, L_7$) | *−0.838* | *0.053* | *−0.383* | *±0.122* |
| TIM-R5 | −1.569 | 1.123 | −0.394 | ±0.469 |
| *TIM-R5* ($L_5, L_6, L_7$) | *−1.142* | *0.408* | *−0.393* | *±0.242* |
| *TIM-R5* ($L_6, L_7$) | *−0.831* | *0.047* | *−0.383* | *±0.121* |
| DIR-R4 | −1.540 | 1.105 | −0.366 | ±0.442 |
| *DIR-R4* ($L_5, L_6, L_7$) | *−1.048* | *0.401* | *−0.392* | *±0.223* |
| *DIR-R4* ($L_6, L_7$) | *−0.802* | *0.064* | *−0.394* | *±0.120* |
| DIR-R5 | −1.530 | 1.122 | −0.388 | ±0.454 |
| *DIR-R5* ($L_5, L_6, L_7$) | *−1.031* | *0.388* | *−0.389* | *±0.217* |
| *DIR-R5* ($L_6, L_7$) | *−0.811* | *0.032* | *−0.381* | *±0.118* |

*Units*: m

**Table 2** Statistics of the gravity anomaly differences between WGM2012 and the GGMs before (normal font) and after (italics) the WL MRA synthesis

| | Min | Max | Mean | Std |
|---|---|---|---|---|
| EGM08 | −49.66 | 128.50 | 0.31 | ±3.24 |
| GOCO03s | −204.97 | 272.23 | 0.11 | ±22.49 |
| *GOCO03s* ($L_5, L_6, L_7$) | *−89.09* | *129.87* | *0.32* | *±9.38* |
| *GOCO03s* ($L_6, L_7$) | *−51.19* | *123.95* | *0.31* | *±3.52* |
| TIM-R4 | −206.98 | 269.35 | 0.11 | ±22.14 |
| *TIM-R4* ($L_5, L_6, L_7$) | *−90.52* | *134.80* | *0.31* | *±8.85* |
| *TIM-R4* ($L_6, L_7$) | *−50.17* | *123.54* | *0.30* | *±3.48* |
| TIM-R5 | −196.90 | 272.70 | 0.11 | ±19.34 |
| *TIM-R5* ($L_5, L_6, L_7$) | *−59.78* | *129.77* | *0.49* | *±7.66* |
| *TIM-R5* ($L_6, L_7$) | *−40.53* | *124.22* | *0.43* | *±3.37* |
| DIR-R4 | −201.93 | 271.43 | 0.11 | ±21.93 |
| *DIR-R4* ($L_5, L_6, L_7$) | *−87.10* | *129.69* | *0.29* | *±8.47* |
| *DIR-R4* ($L_6, L_7$) | *−51.12* | *123.89* | *0.28* | *±3.44* |
| DIR-R5 | −203.66 | 270.65 | 0.11 | ±19.10 |
| *DIR-R5* ($L_5, L_6, L_7$) | *−65.17* | *129.42* | *0.54* | *±7.38* |
| *DIR-R5* ($L_6, L_7$) | *−41.02* | *125.50* | *0.41* | *±3.35* |

*Units*: mGal

## 3 GGM External Validation with MRA

Tables 1 and 2 summarize the statistics of the differences between the GPS/Levelling geoid heights, the WGM2010 gravity anomalies and the GOCE/GRACE GGMs. In these Tables, the normal faced lettering denotes the differences before any analysis of the GGM data has been performed, i.e., the original ones to the GGM $n_{max}$. EGM2008, which will be used as the reference GGM, provides a std at the 13.4 cm and 3.2 mGal, while the GOCE/GRACE GGMs reach the 44–46 cm and 21–22 mGal, respectively. The latter is expected due to the omission error in both geoid heights and gravity anomalies, due to the small $n_{max}$.

The goal now is to try and enhance this performance of the GOCE/GRACE GGMs through WL-based MRA. To that respect the derived geoid heights and gravity anomalies from all GGMs have been decomposed through a discrete wavelet transform (DWT) into 12 levels ($L_1$, $L_2$, ..., $L_{12}$), where each level was analyzed in an approximation coefficient and three detail coefficients. For all GGMs, $L_1$ corresponds to spatial scales between 5.5–11 km (smallest scales), $L_2$ to 11–22 km, $L_3$ to 22–44 km, $L_4$ to 44–88 km, $L_5$ to 88–176 km, $L_6$ to 176–352 km, $L_7$ to 352–704 km, $L_8$ to 704–1,408 km, $L_9$ to 1,408–2,816 km, $L_{10}$ to 2,816–5,632 km, $L_{11}$ to 5,632–11,264 km, and $L_{12}$ to 11,264–22,528 km (largest scales). Given the decomposition, the signals are then synthesized for the GOCE/GRACE GGMs, by replacing some of their levels with those of EGM2008, so that a new combined GGM would be generated, where both GOCE/GRACE and EGM2008 information is used. When the new GGM is denoted as, e.g., *GOCO03s* ($L_5$, $L_6$, $L_7$) this means that the first four levels come from EGM2008, levels 5–7 from GOCO03s and then levels 8–12 from EGM2008 again. $L_5$, $L_6$, $L_7$ span the spatial scales between 88 km and 704, corresponding to harmonic degrees ~28–225, therefore they represent the main waveband that GOCE should primarily provide its highest improvement.

For the WL-based MRA implementation, various mother WLs have been tested, from the simple Haar WL, to Coiflet and Daubechies (db), while for the latter two various orders have been investigated. It was finally decided that db10, i.e., the daubechies WL with ten vanishing moments, would be used for the WL analysis, since it provided the best results to the GPS/Leveling geoid heights. After the decomposition of the GGMs followed the reconstruction of their levels, by combining their detail coefficients, and then the synthesis, as outlined in the preceding section and Eq. (1). In Tables 1 and 2 we summarize the synthesized results from two scenarios, where GOCE/GRACE provides the information for $L_5$, $L_6$, $L_7$ and $L_6$, $L_7$. In the first case, the synthesized GGMs (herein denoted as SynthGOCO03s, SynthTIM-R4, SynthTIM-R5, SynthDIR-R4 and SynthDIR-R5) provide improved differences with the GPS/Leveling data by 20 cm in terms of the std compared to the original ones. SynthGOCO03s reduced the std to 26 cm, while SynthTIM-R4/SynthTIM-R5 and SynthDIR-R4/SynthDIR-R5 are at the 24 and 22 cm. Still, these are about 10 cm worse than EGM2008, signalling that some of the GOCE levels used for the synthesis are of lower quality and contain more noise than signal. When only $L_6$, $L_7$ from GOCE are used (corresponding to spatial scales 176–352 km and 352–704 km, respectively) then the situation improves drastically. SynthGOCO03s now reaches the 12.4 cm when compared to the GPS/Leveling geoid heights, while SynthTIM-R4 and SynthDIR-R4 provide even better results at the 12.2 and 12 cm. The R5 versions of the TIM and DIR models manage to improve the results of their R4 counterparts to 12.1 and 11.8 cm. These are 1–1.6 cm better than the performance of EGM2008, signalling the improvement offered by GOCE in the specific spectral range. Moreover, it shows that the WL-based MRA is an effective tool in order to analyse the GOCE GGMs and model their omission error in the shorter spatial scales.

The same holds for the external evaluation with gravity anomalies, since the Wavelet MRA Synthesis with $L_5$, $L_6$, $L_7$ improves the original ones by as much as 13–14 mGal. What is striking in the gravity anomaly differences is the significant improvement brought by TIM-R5 and DIR-R5. The std of the differences before any analysis is ~3 mGal lower than that of the R4 GGMs, while after the synthesis they give improved std by ~1 mGal. This signals the value of the low orbit GOCE data, which is especially evident in the gravity anomalies compared to geoid heights due to the larger spectral content of the former in the smaller scales. When the synthesis is based only on $L_6$, $L_7$ from GOCE the std of the differences is the same as that of EGM2008 (~3 mGal). It is not unexpected that the synthesized GOCE/GRACE GGMs do not manage to improve the results of EGM2008, since most of the WGM2012 gravity data have been used in the compilation of EGM2008.

From these results, it becomes apparent that indeed GOCE manages to improve the geoid and gravity field representation in the medium bands of the spectrum and that the followed methodology manages to provide reasonable results and can be employed in related studies where a synthesized geoid and/or gravity field is needed. A further test performed, was to examine the behavior of $L_5$ from the GOCE/GRACE GGMs, given that its inclusion in the synthesis worsens the results. $L_5$ corresponds to spatial scales between 88–176 km, therefore given that GOCE was to provide useful signal up to spatial scales of 100 it means that part of the signal in $L_5$ is beyond the mesuring waveband of GOCE (ESA 1999). To that respect, and in order to remove some of the noise present in $L_5$ of the decomposed GOCE/GRACE GGM signal, a simple Gaussian and a boxcar type of filter have been tested. Both will act as low-pass filters, where we intend to remove some of the high-frequencies in the waveband between 88–176 km. Various cut-off wavelengths between 90 and 140 km, with an increment of 5 km have been tested, and finally it was decided to keep the one corresponding to a spatial scale of 120 km. For all these tests, after the filtering, the signals were synthesized and comparisons with the GPS/Leveling and gravity data have been performed. The cut-off wavelength of 120 km was the one that offered the best results in the comparisons with the external data. For smaller wavelengths, the results deteriorate due to the presence of noise, while for larger wavelengths signal was removed along with noise deteriorating the results. Tables 3 and 4 presents the statistics of the differences between the filtered and then synthesized GOCE/GRACE GGMs (denoted as, e.g., *GOCO03s-f* where *f* stands for filtered) with the GPS/Leveling and gravity data. For GOCO03s, the std of the differences with the GPS/Leveling data reduces from 25.9 to

**Table 3** Statistics of the differences between GPS/levelling and geoid heights from the synthesized GGMs after filtering GOCE/GRACE $L_5$

| | Min | Max | Mean | Std |
|---|---|---|---|---|
| GOCO03s-f ($L_5$,$L_6$, $L_7$) | −0.870 | 0.206 | −0.377 | ±0.176 |
| *GOCO03s-f ($L_5$,$L_6$, $L_7$)* | −0.898 | 0.244 | −0.373 | ±0.192 |
| TIM-R4-f ($L_5$,$L_6$, $L_7$) | −0.927 | 0.279 | −0.377 | ±0.190 |
| *TIM-R4-f ($L_5$,$L_6$, $L_7$)* | −1.011 | 0.314 | −0.373 | ±0.213 |
| TIM-R5-f ($L_5$,$L_6$, $L_7$) | −0.907 | 0.202 | −0.384 | ±0.163 |
| *TIM-R5-f ($L_5$,$L_6$, $L_7$)* | −0.874 | 0.237 | −0.379 | ±0.180 |
| DIR-R4-f ($L_5$,$L_6$, $L_7$) | −0.925 | 0.279 | −0.378 | ±0.187 |
| *DIR-R4-f ($L_5$,$L_6$, $L_7$)* | −1.027 | 0.316 | −0.373 | ±0.212 |
| DIR-R5-f ($L_5$,$L_6$, $L_7$) | −0.875 | 0.189 | −0.381 | ±0.157 |
| *DIR-R5-f ($L_5$,$L_6$, $L_7$)* | −0.874 | 0.228 | −0.376 | ±0.177 |

*Units*: m
Normal lettering for Gauss and italics for boxcar filtering

**Table 4** Statistics of the differences between WGM2012 and gravity anomalies from the synthesized GGMs after filtering GOCE/GRACE $L_5$

| | Min | Max | Mean | Std |
|---|---|---|---|---|
| GOCO03s-f ($L_5$, $L_6$, $L_7$) | −72.77 | 127.91 | 0.31 | ±6.48 |
| *GOCO03s-f ($L_5$, $L_6$, $L_7$)* | −74.65 | 130.64 | 0.32 | ±6.94 |
| TIM-R4-f ($L_5$, $L_6$, $L_7$) | −74.88 | 132.14 | 0.29 | ±6.36 |
| *TIM-R4-f ($L_5$, $L_6$, $L_7$)* | −75.29 | 135.47 | 0.29 | ±6.81 |
| TIM-R5-f ($L_5$, $L_6$, $L_7$) | −51.01 | 130.51 | 1.09 | ±5.73 |
| *TIM-R5-f ($L_5$, $L_6$, $L_7$)* | −53.43 | 134.46 | 1.35 | ±6.19 |
| DIR-R4-f ($L_5$, $L_6$, $L_7$) | −73.01 | 130.56 | 0.29 | ±6.25 |
| *DIR-R4-f ($L_5$, $L_6$, $L_7$)* | −73.96 | 134.30 | 0.29 | ±6.72 |
| DIR-R5-f ($L_5$, $L_6$, $L_7$) | −53.83 | 129.66 | 1.09 | ±5.63 |
| *DIR-R5-f ($L_5$, $L_6$, $L_7$)* | −55.70 | 132.90 | 1.35 | ±6.14 |

*Units*: mGal
Normal lettering for Gauss and italics for boxcar filtering

17.6 cm and 19.2 cm for the Gaussian and boxcar filters. A smaller improvement is found for the TIM-R4 and DIR-R4 models, for which the std decreases by ~4–5 cm. Noticeably though, while the synthesis of TIM-R4 ($L_5$, $L_6$, $L_7$) is better than that of GOCO03s (23.9 cm compared to 25.9 cm), after the filtering with the Gaussian filter GOCO03s is better by 1.6 cm. This may be due to the fact that the latter is based on fewer GOCE data, hence the $L_5$ geoid signal is poorly modeled compared to TIM-R4. In that way, filtering benefits more the GGM with the least amount of information in that waveband, by removing the noise, while in TIM-R4 it removes not only the noise but some useful geoid signal that is present. With TIM-R4 and DIR-R4 probably a more elaborate filtering process is needed, so that the noise and signal can be better decomposed and separated. This is out of the scope of the present work and is actually the field of future investigation. For DIR-R4, after filtering $L_5$ the std drops to 18.7 and 21.2 cm, showing slightly better results than TIM-R4. Once again, more interesting results from the filtering are acquired for the R5 GGMs, since for TIM-R5 the std drops to 16.3 and 18.0 cm and for DIR-R5 to 15.7 and 17.7 cm, for the Gaussian and boxcar filter respectively. This is a 3 cm improvement compared to the R4 GGMs and is due to the fact that the R5 models contain the low orbit GOCE data, hence the SNR is larger at $L_5$. The same results are acquired for the external validation with gravity anomalies, where an improvement of 2–3 mGal is achieved, compared to the unfiltered synthesis, but this is still worse than the synthesis where only $L_6$ and $L_7$ are used. Concluding on the filtering process investigated, the results acquired are indeed improved compared to no filtering $L_5$ at all, but in no case did we achieve the results when using for the synthesis only $L_6$ and $L_7$. If some useful signal is to be derived from $L_5$, so that the results will be further improved, then more elaborate filtering and/or error modeling is needed to decorrelate signal from noise.

The final part of the GOCE/GRACE GGM evaluation was carried out over the Amazon area and Europe by investigating the spectral coherence and correlation between the GGM-derived gravity anomalies and land topography. To derive the coherence between the GGM gravity anomalies and topography/bathymetry, WL MRA has been used again so that both signals have been decomposed in twelve levels and then the signal of each level has been reconstructed. Then, employing Eq. (2) the spectral coherence has been evaluated for each level between the topography signal and gravity anomalies from EGM2008, TIM-R4/R5, DIR-R4/R5 and GOCO03s.

For Amazon, the same results with the external evaluation have been derived, with the higher coherence found in $L_6$ and $L_7$ for the GOCE GGMs and lower coherence for $L_5$ and $L_4$. These are depicted in Fig. 1, where the coherence for $L_3$, $L_5$, $L_6$ and $L_7$ is shown for the area over Amazon. As expected, the coherence for the GOCE/GRACE GGMs for $L_3$ is very low with practically no coherence up to 30–35 km and then only up to 20%, which is probably just noise. EGM2008 on the other hand has a more or less uniform coherence between 55 and 72%. In $L_5$ the situation starts to improve for the GOCE/GRACE GGMs, with higher coherence up to ~42%, while in all cases their coherence is below that of EGM2008. Notice that the filtering cut-off wavelength set in the previous test to 120 km coincides in the coherence plot with the point that the coherence starts to raise for the GOCE/GRACE GGMs. Therefore, in the second half of the GOCE/GRACE $L_5$ there seems indeed to be some useful signal that still remains to be accounted for within the present methodological scheme. The situation changes completely for $L_6$ and $L_7$, where the GOCE /GRACE GGMs are equivalent to EGM2008 and in most cases better than that. Between the GOCE/GRACE GGMs it is interesting to notice that for $L_5$, the release 5 versions of TIM and DIR have better coherency

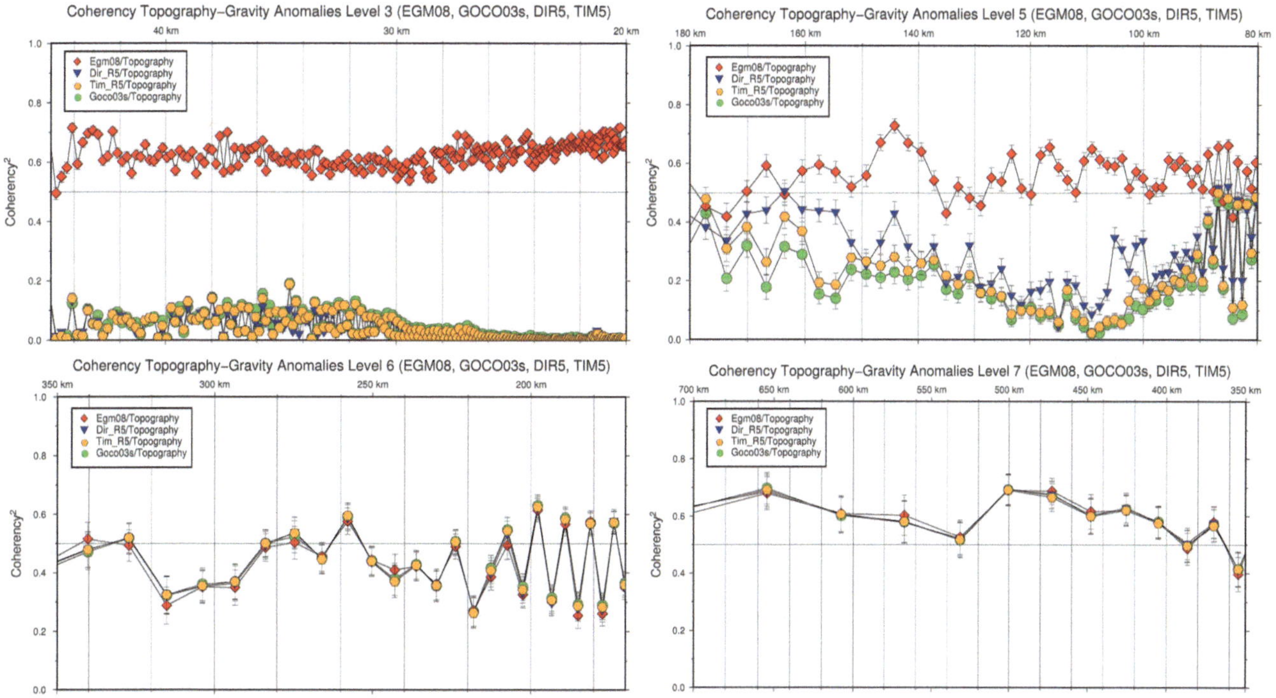


**Fig. 1** Spectral coherency for various levels of decomposition between topography and GGM-derived gravity anomalies over the Amazon area

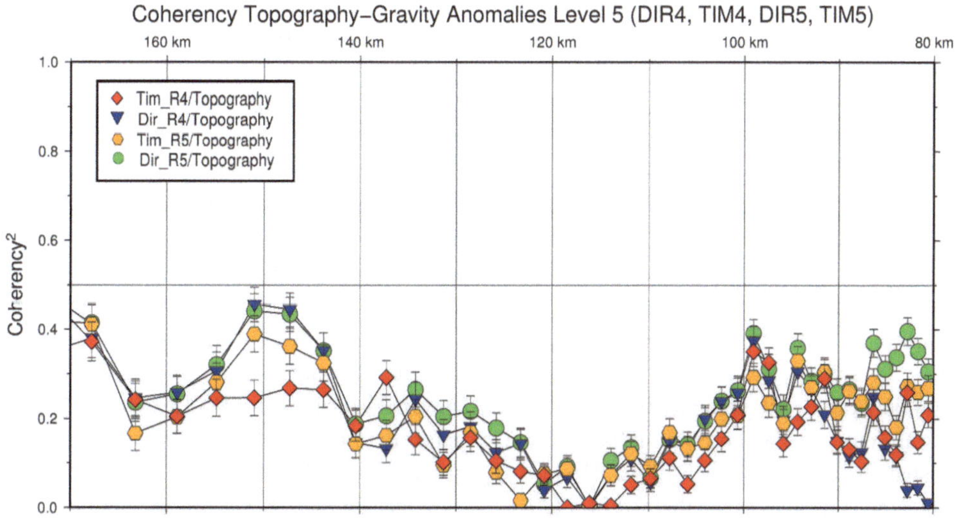


**Fig. 2** Spectral coherency for $L_5$ between topography and TIM-R4/R5 and DIR-R4/R5 GGM-derived gravity anomalies over the Amazon area

than GOCO, which is due to the more GOCE data used in their development. For larger wavelengths, GOCO performs equally well given that these lower harmonic degrees of the spectrum are sufficiently modeled by the release 5 of GOCE data along with the information form GRACE which was used in its development. Comparing the R4 and R5 versions of the TIM and DIR models, Fig. 2 shows their coherence for $L_5$ over Amazon. From this Figure it is readily seen how the addition of the low-orbit GOCE data manage to provide improved results by about 10–20%. This is especially evident at spatial scales between 140–160 km, where the difference between TIM-R4 and TIM-R5 is at the 20% level. Finally, DIR-R5 manages to provide improved results especially for scales between 80–100 km, where it is better by as much as 35%.

Over Europe the results are slightly different, given that EGM2008 is dominant. For $L_3$ EGM2008 has a mean coherence of 70% while it reaches the 85% level as well. For $L_5$ EGM2008 retains a high coherence between 50 and 80% while the GOCE GGMs have a maximum of 50% and a mean of $\sim$30%. This is due to fact that Europe is a well surveyed area in terms of terrestrial gravity data, hence EGM2008

**Table 5** Correlation between Topography and GGM-derived gravity anomalies over Amazon

| | EGM2008 (%) | GOCO03s (%) | TIM-R4 (%) | TIM-R5 (%) | DIR-R4 (%) | DIR-R5 (%) |
|---|---|---|---|---|---|---|
| $L_1$ | 28.80 | 0.10 | 2.10 | 1.80 | 0.30 | 0.50 |
| $L_2$ | 59.80 | 0.20 | 1.40 | 0.90 | 0.20 | 0.21 |
| $L_3$ | 71.90 | 3.30 | 5.50 | 8.58 | 2.50 | 8.27 |
| $L_4$ | 74.20 | 28.20 | 32.10 | 31.70 | 30.10 | 33.47 |
| $L_5$ | 65.50 | 60.80 | 65.20 | 63.90 | 63.00 | 63.90 |
| $L_6$ | 62.40 | 63.10 | 66.20 | 69.87 | 68.70 | 69.87 |
| $L_7$ | 67.70 | 70.20 | 71.50 | 74.10 | 70.10 | 74.07 |
| $L_8$ | 42.10 | 44.50 | 46.20 | 64.30 | 45.10 | 64.16 |
| $L_9$ | 23.00 | 29.20 | 25.50 | 26.00 | 26.50 | 29.80 |
| $L_{10}$ | 30.10 | 35.00 | 39.10 | 53.20 | 39.80 | 53.60 |
| $L_{11}$ | 51.80 | 52.10 | 46.80 | 56.10 | 52.30 | 64.20 |
| $L_{12}$ | 97.90 | 98.10 | 98.20 | 99.00 | 98.50 | 99.63 |

manages to represent very well the gravity field over this region. From this analysis it can be concluded that indeed in poorly surveyed areas, GOCE GGMs can be expected to contribute significantly in gravity field mapping.

Likewise, the correlation between the topographic signal and that of the GGM-derived gravity anomalies for the various levels of decomposition has been analyzed. Table 5 presents the correlation coefficients for EGM2008 and the GOCE/GRACE GGMs only for Amazon given the aforementioned discussion. The correlation results for Europe are again in favor of EGM2008, since most of the terrestrial gravity anomalies over the entire continent have been used in its development. A similar picture with the coherence analysis is found. For the very first levels ($L_1$ to $L_4$), EGM2008 dominates since the GOCE/GRACE GGMs present little or no correlation at all (up to 33% for $L_4$). From $L_5$ onwards the contribution of GOCE is evident, since the latter GGMs manage to perform equally well with EGM2008 (for $L_5$) and outperform it for $L_6$, $L_7$ and $L_8$. For $L_6$ and $L_7$ they present a higher correlation between 3–5% compared to EGM2008, which increases to 22% for $L_8$. The correlation found for $L_9$ (corresponding to spatial scales between 1,408–2,816 km) is somewhat puzzling, since it is lower than that for $L_8$ and $L_{10}$. Analyzing the topography and gravity anomaly signals for $L_9$, it was found that the former has two (positive) dominant features in the EW direction with a low over the Amazon basin, probably coming from the Andes to the East and a merge of the Brazilian and Guyana shields in the West. On the other hand, the gravity anomaly signal shows very little variation for these spatial scales over the area under study, with two predominant positive features in the NS direction and a low over the Amazon basin. This inconsistency can be due to the fact that the gravity signal from the Andes is more high frequency in nature and is represented in the lower levels (it is mostly seen in $L_4$–$L_7$) compared to the Brazilian and Guyana shield topography that dominate the area under study in the southern and northern parts respectively. In any case, the contribution of GOCE to the medium wavebands of the spectrum is once again evident, while if a higher resolution digital terrain model have been used, then the superiority of the GOCE/GRACE GGMs would have been more evident in the coherence and correlation analysis.

## 4 Conclusions

A detailed evaluation, employing WL-based MRA, has been carried out for the recent GOCE and GOCE/GRACE GGMs both in terms of geoid heights and gravity anomalies. From the external evaluation that referred to geoid heights, it was concluded that the combined GGMs improve the estimated geoid heights, compared to local GPS/Leveling data, since the std is reduced from ±0.45 to ±0.22 m for the DIR-R4/R5 and TIM-R4/R5 models. When only $L_6$ and $L_7$ have been used from the GOCE/GRACE GGMs, then the results improved further to ±0.118 cm for DIR-R5, being 1.6 cm better than EGM2008. Contrary to that, the evaluation with gravity anomalies revealed that in the best case scenario the GOCE/GRACE GGMs manage to reach the agreement of EGM2008, something attributed to the fact that most of the data used for the compilation of WGM2012 have been included in EGM2008. Hence this gravity information is not a totally independent dataset due to the high correlation with EGM2008.

When filtering $L_5$ of the GOCE/GRACE GGMs the results improved between 4–8 cm and 2–3 mGal, showing that indeed $L_5$ contains useful geoid/gravity signal that can contribute to the overall GGM performance. To that respect, more advanced filtering options will be investigated to separate the noise from the signal and improve the GGM performance. Overall, the proposed WL MRA methodology, for the analysis and synthesis of GOCE/GRACE GGMs, provides promising results since spatial scales in the GGMs that are modeled with lower accuracy can by successfully replaced with other sources of information which are of higher accuracy/quality. It is worth mentioning that with the presented WL-based analysis there is no mixing of the spatial scales of the gravity field signal. This is so in the sense that WLs can isolate specific portions of the gravity field signal through the analysis in various levels of decomposition, the latter corresponding to specific spatial scales. Then each level can be manipulated separately from the rest, allowing the study of the gravity field signal properties for the specific spatial scales. This cannot be done with, e.g., Fourier transform based methods, where the entire spectrum of the gravity field signal is studied in its entity. From our analysis, the main problem with this WL-based approach is that the dyadic nature of WLs allow the isolation of specific spatial scales for each level (given the resolution of the original signal). If one would like to study, e.g., the GOCE GGM signal for spatial scales between 60 and 140 km only, then this approach cannot be followed, since the specific range belongs to two different levels ($L_4$ and $L_5$) so each level should be studied separately.

From the analysis of the spectral coherency and correlation between topography and the GGM-derived gravity anomalies it was concluded that EGM2008 has significantly better results for the first few levels. This is expected since the GOCE/GRACE GGMs investigated are satellite-only ones. The contribution of the satellite missions in seen again in $L_5$, $L_6$, $L_7$ and $L_8$, where the GOCE/GRACE GGMs show improved result compared to EGM2008 by 3–22%.

## References


Amante C, Eakins BW (2009) ETOPO1 1 arc-minute global relief model: procedures, data sources and analysis. NOAA Technical Memorandum NESDIS NGDC-24. National Geophysical Data Center, NOAA. doi:10.7289/V5C8276M. Accessed Feb 2014

Balmino G, Vales N, Bonvalot S, Briais A (2012) Spherical harmonic modeling to ultra-high degree of Bouguer and isostatic anomalies. J Geod 86(7):499–520. doi: 10.1007/s00190-011-0533-4

Bendat JS, Piersol AG (2010) Random data: analysis and measurement procedures, 4th edn. Wiley, New York. ISBN 978-0-470-24877-5

Bruinsma SL et al (2010) GOCE gravity field recovery by means of the direct numerical method. Presented at the ESA Living Planet Symposium, Bergen, Norway, 27 June – 2 July

Bruinsma SL et al (2013) The new ESA satellite-only gravity field model via the direct approach. Geophys Res Lett 40(14):3607–3612. doi: 10.1002/grl.50716

Chui C (1992) An Introduction to Wavelets, 1st edn. Academic Press, San Diego. ISBN: 978–0121745844

ESA (1999) Gravity field and steady-state ocean circulation mission. ESA Publication Division, ESTEC, Noordwijk, Netherlands, SP-1223

Fuchs MJ, Bouman J, Broerse T, Visser P, Vermeersen B (2013) Observing coseismic gravity change from the Japan Tohoku-Oki 2011 earthquake with GOCE gravity gradiometry. J Geophys Res 118(10):5712–5721

Grebenitcharsky R, Moore P (2014) Application of wavelets for along-tracking multi-resolution analysis of GOCE SGG data. In: Marti U (ed) Gravity, geoid and height systems, International association of geodesy symposia, vol 141, Springer, Switzerland, pp 41–50. doi:

Grebenitcharsky RS, Sideris MG (2005) The compatibility conditions in altimetry–gravimetry boundary value problems. J Geod 78(10):626–636. doi: 10.1007/s00190-004-0429-7

Gruber TH, Visser PNAM, Ackermann CH, Hosse M (2011) Validation of GOCE gravity field models by means of orbit residuals and geoid comparisons. J Geod 85(11):845–860

Hayn M, Panet I, Diamen M, Holschneider M, Mandea M, Davaille A (2012) Wavelet-based directional analysis of the gravity field: evidence for large-scale undulations. Geophys J Int 189(3):1430–1456. doi: 10.1111/j.1365-246X.2012.05455.x

Hirt C, Gruber T, Featherstone WE (2011) Evaluation of the first GOCE static gravity field models using terrestrial gravity, vertical deflections and EGM2008 quasigeoid heights. J Geod 85(10):723–740

Knudsen P, Bingham R, Andersen OB, Rio M-H (2011) A global mean dynamic topography and ocean circulation estimation using a preliminary GOCE gravity model. J Geod 85(11):861–879

Mallat SG (1989) A theory for multiresolution signal decomposition: the wavelet representation. IEEE Trans Pattern Anal Mach Intel 11:674–693. doi: 10.1109/34.192463

Mallat S (1999) A wavelet tour of signal processing, 3rd edn. Academic Press, San Diego. ISBN: 978-0-12-374370-1

Mayer-Gurr T et al (2012) The new combined satellite only model GOCO03s. Presentation at GGHS 2012 IAG Symposia, Venice, October 2012

Menna M, Poulain P-M, Mauri E, Samppietro D, Panzetta F, Reguzzoni M, Sansò F (2014) Mean surface geostrophic circulation of the Mediterranean Sea estimated from GOCE geoid models and altimetric mean sea surface: initial validation and accuracy assessment. Boll Geofis Teor Appl 54(4):347–365. doi: 10.4430/bgta0104

Pail R et al (2011) First GOCE gravity field models derived by three different approaches. J Geod 85(11):819–843

Panet I, Kuroishi Y, Holschneider M (2011) Wavelet modelling of the gravity field by domain decomposition methods: an example over Japan. Geophys J Int 184(1):203–219. doi: 10.1111/j.1365-246X.2010.04840.x

Pavlis NK, Holmes SA, Kenyon SC, Factor JK (2012) The development and evaluation of the earth gravitational model 2008 (EGM2008). J Geophys Res 117:B04406. doi: 10.1029/2011JB008916

Vergos GS, Grigoriadis VN, Tziavos IN, Kotsakis C (2014) Evaluation of GOCE/GRACE global geopotential models over Greece with collocated GPS/Levelling observations and local gravity data. In: Marti U (ed) Gravity, geoid and height systems, International association of geodesy symposia, vol 141. Springer, Switzerland, pp 85–92. doi: 10.1007/978-3-319-10837-7_11

# Evaluation of GOCE-Based Global Geopotential Models Versus EGM2008 and GPS/Levelling Data in Northwest of Turkey

N.B. Avsar, B. Erol, and S.H. Kutoglu

**Abstract**

The Gravity field and steady-state Ocean Circulation Explorer (GOCE), reached the end of its mission in mid-October 2013, was a milestone in Earth's gravity field determination. Several Global Geopotential Models (GGMs) have been published based on the data collected during the roughly 4-year mission of GOCE. This study focuses on evaluation of the contribution of GOCE data for the representation of gravity field with significant improved accuracy in Turkish territory. The evaluation is based on the consecutive releases as well as the releases generated by different computation strategies of GOCE-based models. In the study, a total of 7 GOCE-based models (EIGEN_6C3stat, JYY_GOCE02S, ITG-GOCE02, GO_CONS_GCF_2_TIM_Release 1, 2, 3, and 4) were assessed and the results were compared with the performance of EGM2008. The accuracy of GGMs was analyzed using the reference Global Positioning System (GPS)/levelling network of the case study for Bursa located in the northwest of the Anatolian peninsula. In the analysis, 433 GPS/levelling benchmarks after removing detected blunders were used for evaluation of the global geoid models. The validation results show the superior performance of the high resolution global combined model EIGEN_6C3stat among the evaluated models. Its fit with GPS/levelling-derived geoid heights in the study area is at the level of 9.1 cm in terms of the standard deviation of the differences.

**Keywords**

EGM2008 • Geoid • Global Geopotential Model • GOCE • GPS/levelling

## 1 Introduction

The Gravity field and steady-state Ocean Circulation Explorer (GOCE) was launched in March 17, 2009 as part of European Space Agency (ESA)'s Living Planet Programme.

N.B. Avsar (✉) • S.H. Kutoglu
Department of Geomatics Engineering, Bulent Ecevit University, Incivez, 67100 Zonguldak, Turkey
e-mail: nb_avsar@beun.edu.tr; kutogluh@hotmail.com

B. Erol
Department of Geomatics Engineering, Istanbul Technical University, Ayazaga, Maslak, 34469 Istanbul, Turkey
e-mail: bihter@itu.edu.tr

One of the main objectives of GOCE has been specified to be the determination of the Earth's gravity field with an accuracy of 1 mGal at a spatial resolution better than 100 km via on-board a high-precision gravity gradiometer as well as a GPS receiver (Drinkwater et al. 2003). Such an advance in the knowledge of the Earth's gravity field and the developments on computation algorithms/technologies have led to improve the Global Geopotential Models (GGMs). Since the launch of the GOCE mission, a number of satellite-only and combined models computed using the missions data by the different strategies have been released (ICGEM Global Gravity Field Models 2014). These models are expected to contribute to the improvement of the accuracy of regional geoid modelling as well as global geoid.


S. Jin, R. Barzaghi (eds.), *IGFS 2014*, International Association of Geodesy Symposia 144,
DOI 10.1007/1345_2015_55

In precise regional geoid model determination, the long-wavelength components of the gravity field are derived from a GGM. In this sense, there are various studies as to the determination of a regional geoid with improved accuracy for Turkey depending on the GGMs generated from data of the dedicated satellite gravity missions; CHAMP (CHAllenging Minisatellite Payload), GRACE (Gravity Recovery And Climate Experiment), and GOCE (Erol 2012; Erol et al. 2009; Kiliçoğlu et al. 2009; Ustun and Abbak 2010). Accordingly, the principal aim of this study is to investigate and clarify the contribution of GOCE-based GGMs to improvement of regional geoid determination in Turkey.

Moreover, within this study, the best fitting global model for geodetic applications, such as transformation from ellipsoidal heights to orthometric heights, is evaluated as well. Based on the simple relation between $H$ (orthometric height), $h$ (ellipsoidal height) and $N$ (geoid height)

$$H = h - N \tag{1}$$

when $h$ is provided by Global Navigation Satellite System (GNSS) and $N$ exists from a reliable and precise geoid model, $H$ can then be obtained immediately. In other words, the accurate determination of orthometric heights via GNSS/levelling requires primarily a high accuracy geoid model (Erol and Erol 2012). However, a high accuracy geoid model in global coverage is not possible due to lack of accurate and dense data covering the entire Earth. In this case, accurate geoid heights can be generated from regionally improved GGMs which fit well to the statistical properties of the regional gravity field of the territory. Also, it should be noted that for the very small areas, the geoid heights of GNSS benchmarks are usually interpolated utilizing the relationship between GNSS-derived ellipsoidal heights and regional orthometric heights from spirit levelling measurements at the reference points of high-order geodetic networks.

The accuracy of the most recent regional gravimetric geoid model of Turkey, Turkey Hybrid Geoid Model (THG-09), is reported as $\pm 8.4$ cm (Kiliçoğlu et al. 2011). The THG-09 is computed using the Remove-Compute-Restore algorithm based on EGM08 global model.

In order to clarify the optimal GGM, which fits regional gravity field the best, a comparison and validation of the GGMs is carried out with independent (external) data sets, such as geoid heights obtained from GNSS/levelling, gravity anomalies derived from terrestrial observations, vertical deflections, satellite orbit data etc. (Amos and Featherstone 2003; Erol et al. 2009; Godah and Krynski 2011; Guimaraes et al. 2012; Hirt 2011; Kotsakis and Katsambalos 2010; Merry 2007). It should also be considered that the reliability of such evaluations depends on several factors such as external data qualities, applied testing methodology. Several studies have been performed by different research teams to assess the quality of GOCE-based GGMs in regional gravity field modeling, so far, e.g. Amjadiparvar et al. (2013) for Canada and USA, Godah and Krynski (2011, 2013) for Poland, Guimaraes et al. (2012) for Brazil, Janak and Pitonak (2011) for Central Europe, Sprlak et al. (2012) for Norway, Voigt et al. (2010) for Germany, etc.

In this study, a total of 7 GOCE-based recently released models were selected to validate their fit to the gravity field in Turkey, using high-accuracy local GPS/levelling data of 433 benchmarks. Their performances were also compared with the performance of the ultra-high resolution combined EGM2008 model. The assessments base on the differences between GGMs and GPS/levelling geoid heights at the bechmarks.

## 2 Data Sets

### 2.1 Global Geopotential Models

Global Geopotential Model is a set of spherical harmonic coefficients representing the Earth's gravity field in varying wavelengths of the spectrum. These models are also qualified with the Earth parameters of geocentric gravitational constant (*GM*) and reference radius (*R*). The functionals of the Earth's gravitational field can be computed from the spherical harmonic coefficients on and above the Earth's surface. For instance, using the potential theory the disturbing potential (*T*) are expressed by spherical harmonic series as follows (Hofmann-Wellenhof and Moritz 2005):

$$T = \frac{GM}{r}\sum_{n=2}^{n_{\max}}\left(\frac{R}{r}\right)^{n}\sum_{m=0}^{n}\left(\Delta\overline{C}_{nm}\cos m\lambda + \overline{S}_{nm}\sin m\lambda\right)\overline{P}_{nm}(\cos\theta) \tag{2}$$

where $r$, $\theta$, $\lambda$ are the spherical geocentric coordinates of the computation point: radial distance, co-latitude and longitude, respectively; $\overline{C}_{nm}$ and $\overline{S}_{nm}$ are the fully normalized spherical harmonic (Stokes) coefficients with $n$, $m$ being degree and order, respectively; $n_{max}$ is maximum degree of spherical harmonic expansion; $\overline{P}_{nm}(\cos\theta)$ are the fully normalized associated Legendre functions. It should be stated that $\Delta\overline{C}_{nm}$ is provided by subtracting from the zonal coefficients ($m = 0$) the corresponding Stokes' coefficients of a normal gravity field.

The spatial resolution of a GGM (in km) at the Earth's surface is computed by dividing ~20,000 km by the maximum degree of the GGM (Featherstone 2002). On the other hand, the primary limitations on the precision of the GGMs result from the spatial coverage and quality of the gravity data.

**Table 1** List of the Global Geopotential Models evaluated

| GGM | Year | $n_{max}$ | Data | Citation |
|---|---|---|---|---|
| EIGEN-6C3stat | 2014 | 1,949 | S (GOCE, GRACE, LAGEOS), G, A | Förste et al. (2013) |
| JYY_GOCE02S | 2013 | 230 | S (GOCE) | Yi et al. (2013) |
| ITG-Goce02 | 2013 | 240 | S (GOCE) | Schall et al. (2014) |
| GO_CONS_GCF_2_TIM_R4 | 2013 | 250 | S (GOCE) | Pail et al. (2011) |
| GO_CONS_GCF_2_TIM_R3 | 2011 | 250 | S (GOCE) | Pail et al. (2011) |
| GO_CONS_GCF_2_TIM_R2 | 2011 | 250 | S (GOCE) | Pail et al. (2011) |
| GO_CONS_GCF_2_TIM_R1 | 2010 | 224 | S (GOCE) | Pail et al. (2010) |
| EGM2008 | 2008 | 2,190 | S (GRACE), G, A | Pavlis et al. (2012) |

*S*: Satellite tracking data, *G*: Gravity data, *A*: Altimetry data, *LAGEOS*: LAser GEOdynamics Satellites

Especially, with the contribution of the dedicated satellite gravity missions, the quality and coverage of the Earth gravity data has been improved until a certain extent. GOCE mission has enabled to short-scale gravity recovery, as well.

In this study, a total of seven GOCE-based models have been evaluated, available at the International Centre for Global Earth Models (ICGEM). Additionally, their progresses with respect to the ultra-high degree model EGM2008 have been analyzed. The differences among the evaluated models are their data period, the types of data used and the processing strategies applied. An overview of the tested GGMs is given in Table 1.

In the frame of ESA's GOCE High Level Processing Facility (HPF) project, three different approaches have been applied to compute GOCE-based GGMs. They are denoted as direct (DIR) approach, time-wise (TIM) approach and space-wise (SPW) approach (Pail et al. 2011). Moreover, several models, such as EIGEN (European Improved Gravity model of the Earth by New techniques), have been also developed by other institutions.

The GOCE-only GGMs allow revealing GOCE's abilities as independent because they are based solely on the GOCE data. For example, the time-wise approach (TIM) considers the gravity gradients and orbit observations as time series measured along the GOCE satellite orbit and use no gravity field information with the exception of GOCE data, neither as reference model, nor for constraining the solution (Pail et al. 2011). The consecutive releases based on this approach: GO_CONS_GCF_2_TIM_R1, R2, R3, and R4 span 01/11/2009–11/01/2010, 01/11/2009–05/07/2010, 01/11/2009–17/04/2011, 01/11/2009–19/06/2012 data periods, respectively.

ITG-Goce02, computed from approximately 7.5 months of (from November 2009 to June 2010) GOCE data, was provided by applying short arc approach to gradiometer and orbit data. Thus, it is aimed to decorrelate the highly correlated GOCE gradiometer and orbit data (Schall et al. 2014). JYY_GOCE02S is a satellite-only model which was derived using approximately 34 months (from November 2009 to August 2012) of GOCE data, as well. Its processing details are presented in Yi (2012) and Yi et al. (2013).

EIGEN-6C3stat, computed by collaboration of GFZ (GeoForschungsZentrum) and GRGS (Groupe de Recherche de Geodesie Spatiale), is the newest release of EIGEN-6C (Förste et al. 2013). The model, which corresponds to ~10 km spatial resolution, includes the largest periods of GRACE and LAGEOS data as well as of GOCE data up to now. It was also augmented with the altimetry/surface gravity data.

EGM2008, the last model evaluated in this study, has been developed using combination of the ITG-GRACE03S model (57 months of GRACE data) with the gravity data obtained from terrestrial/airborne gravimetry and altimetry by NGA (United States National Geospatial Agency) (Pavlis et al. 2012). Although this combined model does not involve GOCE data, the study includes it as well because it is unique GGM capable of resolving the Earth's gravity field beyond degree of 2000 at the present time (ICGEM Global Gravity Field Models 2014). Furthermore, as previously stated, EGM2008 is the underlying model of THG-09.

## 2.2 GPS/Levelling Data

In the assessment of the GOCE-based GGMs for Turkey territory, the GPS/levelling geoid heights were used as independent control data. For this purpose, 433 GPS/levelling benchmarks of a C order GNSS network was selected in the northwest of Turkey (Fig. 1). This local Bursa geodetic network has been established in frame of BUSKI M3 Project (see BUSKI (2009) for further information on the project). In this context, the GPS and levelling data quality of the test network satisfy the criteria in articles on vertical control and geoid testing, which are put forth by the current Turkish Large Scale Map and Map Information Production Regulation. In this manner the Helmert orthometric heights of the benchmarks were determined by spirit levelling in reference to Turkish National Verti-

**Fig. 1** The distribution of 433 GPS/levelling benchmarks in the local Bursa network on topographic data (Unit: m) (For interpretation of colour in this figure legend, the reader is referred to the original version of this paper)

**Table 2** Description of the local GPS/levelling network

| | | |
|---|---|---|
| Network and topography | Area (km) | 50 × 70 |
| | Number of benchmarks | 433 |
| | Density, benchmark per km$^2$ | ~8 |
| GPS observations | GPS receiver type | Dual frequency |
| | Measurement method | Static and Rapid-Static |
| | Coordinate datum | ITRF96 |
| | Type of benchmarks | B orders of TUTGA-99A, C1, C2, and C3 orders |
| Levelling | Levelling method | Spirit |
| | Vertical datum | TUDKA-99 |

| Latitude | | Longitude | | Ellipsoidal height (WGS84) | | | Helmert orthometric height | | |
|---|---|---|---|---|---|---|---|---|---|
| Min | Max | Min | Max | Min | Average | Max | Min | Average | Max |
| 39° 55′ 16.2182″ | 40° 31′ 22.9152″ | 28° 23′ 33.2959″ | 29° 22′ 55.9823″ | 38.6063 m | 195.6172 m | 1,354.2653 m | 0.8664 m | 157.6212 m | 1,315.2489 m |

*TUTGA-99A* Turkish National Fundamental GPS Network 1999-A, *WGS84* World Geodetic System 1984

cal Control Network 1999 (TUDKA-99) and their absolute accuracy is approximately 2.0 cm. GPS coordinates of the benchmarks refer to the ITRF96 datum with the approximate accuracies of 1.5 and 2.0 cm in the horizontal and vertical directions, respectively. The principal characteristics of the test network and its data are presented in Tables 2.

## 2.3 Numerical Evaluation

In this study, the validation of the GOCE-based GGMs was carried out by comparison with the GPS/levelling derived geoid heights in the local Bursa test network.

The geoid height ($N_{GPS/levelling}$) for each benchmark of the test network was derived from

$$N_{GPS/levelling} = h_{GPS} - H_{levelling} \quad (3)$$

depending on $h$ measured by GPS and $H$ provided by levelling. In this equation, the error caused by neglecting the difference between plumb line and ellipsoidal normal is negligible (Fotopoulos 2003). Nevertheless, the computed GPS/levelling geoid heights contain some effects resulted from an aggregate of random and systematic errors in the measurements of ellipsoidal and orthometric heights (Amjadiparvar et al. 2013). So, some of points were assigned as blunders through a third-order polynomial surface and eliminated from the data sets. Hereby, 2 benchmarks were removed from the test network.

On the other hand, the geoid height ($N_{GGM}$) of each benchmark was computed using GGMs, as well. The computations were carried out through the ICGEM Calculation Service for each GGM (ICGEM Calculation Service 2014). By means of the well-known Bruns' formula, it can be written as follows (Hofmann-Wellenhof and Moritz 2005):

$$N = \frac{GM}{r\gamma}\sum_{n=2}^{n_{\max}}\left(\frac{R}{r}\right)^{n}\sum_{m=0}^{n}\left(\Delta\overline{C}_{nm}\cos m\lambda + \overline{S}_{nm}\sin m\lambda\right)\overline{P}_{nm}(\cos\theta) \quad (4)$$

where $\gamma$ is normal gravity. However, it should be underlined here that the set of spherical harmonic coefficients are not used directly to derive the geoid heights. Its causes and solution procedure were detailed in Rapp (1997) and Smith (1998). Due to inexact knowledge of the mass of the Earth and the potential of the geoid, a correction term should be applied to geoid height computations. This term, known as zero-degree term, is directly related to the difference between the geocentric gravitational constant (*GM*) of the geopotential model and the reference ellipsoid. Moreover, Smith (1998) has been pointed out that the GGM should include the permanent tide system of the model. Although the shape of the geoid changes accordingly to the type of permanent tide system (mean, zero, and tide-free), its potential does not change from one system to another.

In the computation of $N_{GGM}$, a regular $30'' \times 30''$ grid of geoid heights was employed with Geodetic Reference System 1980 (GRS80) as the reference system considering the Turkish National Reference Frame and tide-free system as the permanent tide system. It was also incorporated the zero-degree term into all the calculations of GGM geoid heights. Eventually, we applied Inverse Distance Weighting Interpolation for generating the geoid heights in the test area.

In this study, the comparison of two sets of geoid heights, $N_{GPS/levelling}$ and $N_{GGM}$, was done in absolute sense. That is, the differences of the geoid heights were achieved for each benchmark as follows:

$$\Delta N = N_{GPS/levelling} - N_{GGM} \quad (5)$$

Figure 2 depicts the differences of the geoid heights for the eight models in Bursa. The statistic of the differences is also reported in Table 3.

Considering the standard deviations and the biases for all the GOCE-only GGMs, ITG-Goce02 performs slightly better than the other models. We can also infer from the analysis of four different generations of the GOCE-only GGMs derived via the time-wise approach that TIM_R1 presents the smallest standard deviation although it has the oldest data period and the lowest resolution among the TIM models. It is highly related to low-commission errors of ITG-Goce02 and TIM_R1 models because of their lower maximum degrees. On the other hand, it should be noted that omission error of the GGMs is a determinant of their accuracies (Amjadiparvar et al. 2013). Accordingly, spectral behaviors of the GGMs should be investigated in order to disclose the errors recorded in their spherical harmonic coefficients (Erol 2012; Ustun and Abbak 2010). However, in this study we focus to exhibit the performances of the GOCE-based GGMs using local GPS/levelling data for detection of their fit of local gravity field. In this sense, all the validation results show the superior performance of the high resolution global combined model EIGEN-6C3stat. Its fit with $N_{GPS/levelling}$ in the test area is at the level of 9.1 cm in terms of the standard deviations. Moreover, the bias value of $\Delta N$ for this model is −3.5 cm. It means a little systematic drift between the GGM surface and the local vertical datum in Bursa region, which can be removed via de-trending.

The evaluation of the global models with GPS/levelling data, distributed along the entire country, are expected to provide more sensitive results while the spatial resolution of the control benchmarks and GGMs correspond each other. The tests were possible to carry out in a small test area because of the data limitation in data-base for this study.

GPS/levelling data represents the full spectrum of the gravity signal. This condition is related to its spectral resolution for a GGM (Ustun and Abbak 2010). Accordingly, the better performance of the combined models versus the GOCE-only models is not a surprise due to more signal content in the combined models. The results have also indicated that the combined GGMs perform realistic than the GOCE-only GGMs over the mountainous part in southeast of the test area (see Fig. 2). Apparently, the higher frequency contents of the combined GGMs by means of the terrestrial gravity data have given a better approximation for the terrain-dependent gravity field features.

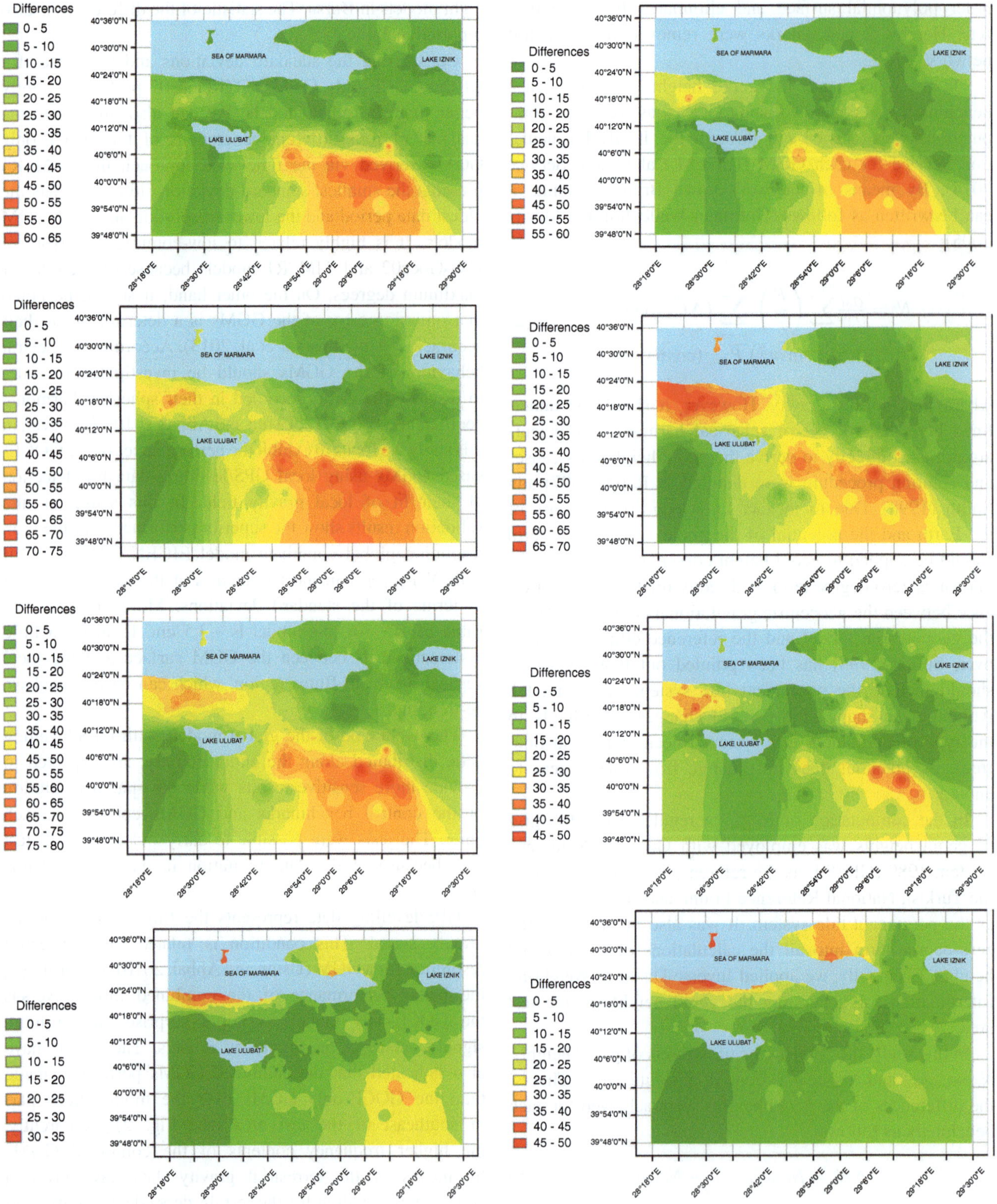

**Fig. 2** Geoid heights differences in the text area (absolute) (Unit: cm) (For interpretation of colour in these figures legend, the reader is referred to the original version of this paper)

**Table 3** Comparison and validation results of GOCE-based GGMs in the test area (Unit: cm)

| | Min | Max | Mean | Std dev | RMS |
|---|---|---|---|---|---|
| GO_CONS_GCF_2_TIM_R1 | −21.1 | 64.3 | 8.9 | 14.2 | 16.8 |
| GO_CONS_GCF_2_TIM_R2 | −27.5 | 60.0 | 7.2 | 15.1 | 16.7 |
| GO_CONS_GCF_2_TIM_R3 | −33.7 | 74.6 | 21.0 | 19.7 | 28.8 |
| GO_CONS_GCF_2_TIM_R4 | −30.9 | 66.6 | 22.8 | 20.3 | 30.5 |
| JYY_GOCE02S | −9.8 | 75.2 | 27.1 | 15.9 | 31.4 |
| ITG-Goce02 | −34.8 | 46.6 | 1.5 | 14.3 | 14.4 |
| EIGEN-6C3stat | −35.0 | 19.4 | −3.5 | 9.1 | 9.7 |
| EGM2008 | −49.2 | 19.7 | −9.9 | 11.4 | 15.1 |

## 3 Conclusions

In this study, the performance of the recent GOCE-based GGMs was examined using the available GPS/levelling data in the Bursa region of Turkey. We also included EGM2008 in the analysis to identify any improvement with respect to it. According the results, the combined model EIGEN-6C3stat fits best to the gravity field in the validation area. Also, it has better performance than EGM2008 in terms of statistical results of validations. Therefore, this model will probably be a suitable candidate as the reference model for an improved regional gravimetric geoid model of Turkey near future. Hence, it is expected to contribute to modernization of the local vertical datum in Turkey, as well. Moreover, it is believed that the results of this study will serve as reference for practical determination of orthometric heights in the test area using the GNSS technologies.

On the other hand, it is achieved an interesting finding that the global model performance drop down as the release of the satellite-only model develop as seen through the example of TIM_R1. The reason of this situation can be clarified when we have a closer inspection of the spectral characteristic of each release.

Consequently, more precise regional geoid model of Turkey is highly expected via GOCE-based combined GGMs based on the mission objective of GOCE that is the determination of the Earth's gravity field and geoid with high accuracy and maximum spatial resolution.

In the next stage of the study, the investigation of the omission errors and the spectral analysis of the GOCE-based GGMs are planned for more comprehensive evaluation of their performances in addition to their regional improvement with optimal combination of terrestrial gravity data in Turkey.

## References


Amjadiparvar B, Rangelova EV, Sideris MG (2013) North American height datums and their offsets: evaluation of the GOCE-based global geopotential models in Canada and the USA. J Appl Geodesy 7:191–203. doi:10.1515/jag-2012-0033

Amos MJ, Featherstone WE (2003) Comparisons of recent global geopotential models with terrestrial gravity field data over New Zealand and Australia. Geomatics Res Australas 79:1–20

BUSKI (2009) Bursa Water and Sewerage Administration General Directorate (BUSKI) M3 Project, Geodetic Report Vol. 1, Bursa (in Turkish)

Drinkwater MR, Floberghagen R, Haagmans R, Muzi D, Popescu A (2003) GOCE: ESA's first Earth Explorer Core mission. In: Beutler GB, Drinkwater MR, Rummel R, von Steiger R (eds) Earth gravity field from space – from sensors to Earth sciences. In the Space Sciences Series of ISSI, vol 18. Kluwer Academic Publishers, Dordrecht pp 419–432

Erol B (2012) Spectral evaluation of Earth geopotential models and an experiment on its regional improvement for geoid modelling. J Earth Syst Sci 121(3):823–835. doi:10.1007/s12040-012-0190-x

Erol B, Erol S (2012) GNSS in practical determination of regional heights. In: Jin S (ed) Global navigation satellite systems: signal, theory and applications. ISBN 978-953-307-843-4, doi:10.5772/28820

Erol B, Sideris MG, Celik RN (2009) Comparison of global geopotential models from the CHAMP and GRACE missions for regional geoid modelling in Turkey. Stud Geophys Geo 53(4):419–441. doi:10.1007/s11200-009-0032-8

Featherstone WE (2002) Expected contributions of dedicated satellite gravity field missions to regional geoid determination with some examples from Australia. J Geospat Eng 4(1):1–19

Förste C, Bruinsma S, Marty JC, Flechtner F, Abrikosov O, Dahle C, Lemoine JM, Neumayer KH, Biancale R, Barthelmes F, König R (2013) EIGEN-6C3stat -the newest high resolution global combined gravity field model based on the 4th release of the direct approach. http://icgem.gfz-potsdam.de/ICGEM/documents/Foerste-et-al-EIGEN-6C3stat.pdf. Accessed 2 Feb 2014

Fotopoulos G (2003) An analysis on the optimal combination of geoid, orthometric and ellipsoidal height data. Dissertation, University of Calgary

Godah W, Krynski J (2011) Validation of GOCE geopotential models over Poland using the EGM2008 and GPS/levelling data. Geoinf Issues 3(1):5–17. ISSN 1689-6440

Godah W, Krynski J (2013) Evaluation of recent GOCE geopotential models over the area of Poland. Acta Geodyn Geomater 10:379–386. doi:10.13168/AGG.2013.0037

Guimaraes G, Matos A, Blitzkow D (2012) An evaluation of recent GOCE geopotential models in Brazil. J Geodetic Sci 2(2):144–155. doi:10.2478/v10156-011-0033-8

Hirt C (2011) Assessment of EGM2008 over Germany using accurate quasigeoid heights from vertical deflections, GCG05 and GPS/levelling. Zeitschrift fuer Geodaesie, Geoinformation und Landmanagement (zfv) 136(3):138–149

Hofmann-Wellenhof B, Moritz H (2005) Physical Geodesy. Springer Wien, New York

ICGEM Calculation Service (2014) International Centre for Global Earth Models (ICGEM), Calculation Service. http://icgem.gfz-potsdam.de/ICGEM/. Accessed 27 Jan 2014

ICGEM Global Gravity Field Models (2014) International Centre for Global Earth Models (ICGEM), Global Gravity Field Models. http://icgem.gfz-potsdam.de/ICGEM/modelstab.html. Accessed 4 Feb 2014

Janak J, Pitonak M (2011) Comparison and testing of GOCE global gravity models in Central Europe. J Geodetic Sci 1(4):333–347. doi:10.2478/v10156-011-0010-2

Kiliçoğlu A, Direnç A, Simav M, Lenk O, Aktuğ B, Yildiz H (2009) Evaluation of the Earth Gravitational Model 2008 in Turkey. Newton's Bull, Special Issue: "External Quality Evaluation Reports of EGM08", vol 4, pp 164–171. ISSN 1810–8555

Kiliçoğlu A, Direnç A, Yildiz H, Bölme M, Aktuğ B, Simav M, Lenk O (2011) Regional gravimetric quasi-geoid model and transformation surface to national height system for Turkey (THG-09). Stud Geophys Geo 55(4):557–578. doi:10.1007/s11200-010-9023-z

Kotsakis C, Katsambalos K (2010) Quality analysis of global geopotential models at 1542 GPS/levelling benchmarks over the Hellenic Mainland. Surv Rev 42(318):327–344. doi:10.1179/003962610X12747001420500

Merry CL (2007) Evaluation of global geopotential models in determining the quasi-geoid for Southern Africa. Surv Rev 39(305):180–192. doi:10.1179/003962607X165159

Pail R, Goiginger H, Mayrhofer R, Schuh W, Brockmann JM, Krasbutter I, Hoeck E, Fecher T (2010) GOCE gravity field model derived from orbit and gradiometry data applying the time-wise method. Presented at the ESA Living Planet Symposium 2010, Bergen, June 27–July 2, Bergen

Pail R, Bruinsma SL, Migliaccio F, Foerste C, Goiginger H, Schuh WD, Hoeck E, Reguzzoni M, Brockmann JM, Abrikosov O, Veicherts M, Fecher T, Mayrhofer R, Krasbutter I, Sanso F, Tscherning CC (2011) First GOCE gravity field models derived by three different approaches. J Geod 85:819–843. doi:10.1007/s00190-011-0467-x

Pavlis NK, Holmes SA, Kenyon SC, Factor JK (2012) The development and evaluation of the Earth Gravitational Model 2008 (EGM2008). J Geophys Res Solid Earth 117(B4):4406

Rapp RH (1997) Use of potential coefficient models for geoid undulation determinations using a spherical harmonic representation of the height anomaly/geoid undulation difference. J Geod 71:282–289

Schall J, Eicker A, Kusche J (2014) The ITG-Goce02 gravity field model from GOCE orbit and gradiometer data based on the short arc approach. J Geod 88:403–409. doi:10.1007/s00190-014-0691-2

Smith DA (1998) There is no such thing as The EGM96 geoid: subtle points on the use of a global geopotential model. IGeS Bull 8:17–28. International Geoid Service, Milan

Sprlak M, Gerlach G, Pettersen BR (2012) Validation of GOCE global gravity field models using terrestrial gravity data in Norway. J Geodetic Sci 2(2):134–143. doi:10.2478/v10156-011-0030-y

Ustun A, Abbak RA (2010) On global and regional spectral evaluation of global geopotential models. J Geophys Eng 7:369–379. doi:10.1088/1742-2132/7/4/003

Voigt C, Rülke A, Denker H, Ihde J, Liebsch G (2010) Validation of GOCE products by terrestrial data sets in Germany. Geotechnologien Sci Rep 17:106–111. doi:10.2312/GFZ.gt.17.16

Yi W (2012) An alternative computation of a gravity field model from GOCE. Adv Space Res 50(3):371–384. ISSN 0273–1177, doi:10.1016/j.asr.2012.04.018

Yi W, Rummel R, Gruber T (2013) Gravity field contribution analysis of GOCE gravitational gradient components. Stud Geophys Geo 57(2):174–202. ISSN 1573–1626, doi:10.1007/s11200-011-1178-8

# Precise Modelling of the Static Gravity Field from GOCE Second Radial Derivatives of the Disturbing Potential Using the Method of Fundamental Solutions

Róbert Čunderlík

**Abstract**

The method of fundamental solutions (MFS) is used to derive the disturbing potential and gravity disturbances from the second derivatives observed by the GOCE satellite mission. Namely, the radial components $T_{rr}$ of the gravity disturbing tensor available from the EGG_TRF_2 product are processed to evaluate the unknown coefficients in the source points that are located directly on the real Earth's surface. MFS as a mesh-free boundary collocation technique uses the fundamental solution of the Laplace equation as its basis functions. Hence, the system matrix is created by the second radial derivatives of the fundamental solution that depend solely on 3D positions of the GOCE observations and the source points. Once the coefficients are evaluated, the disturbing potential and gravity disturbance can be computed in any point above the Earth's surface. This paper presents results of processing 20 datasets of the GOCE measurements, each for different 2-months period. To obtain "cm-level" precision, the source points are uniformly distributed over the Earth's surface with the high-resolution of 0.075° (5,760,002 points). For every dataset the radial components $T_{rr}$ as input data are filtered using the nonlinear diffusion filtering. The large-scale parallel computations are performed on the cluster with 1.2 TB of the distributed memory. A combination of numerical solutions obtained for different datasets/periods yields the final static gravity field model. Its comparison with the SH-based satellite only geopotential models like GOCO03S, GOCE-TIM5 or GOCE-DIR5 indicates its high accuracy. Standard deviation of differences evaluated at altitude 235 km above the reference ellipsoid is about 0.05 $m^2s^{-2}$ ($\sim$5 mm) in case of the disturbing potential, and 0.01 mGal for gravity disturbances.

**Keywords**

Global gravity field modelling • GOCE gravity gradients • Method of fundamental solutions • Point masses modelling

## 1 Introduction

The GOCE satellite mission has brought a significant improvement in modelling the gravity field, especially its medium-frequency part. The second derivatives of the geopotential precisely observed by GOCE provide unique information about the gravity field at altitude of its orbits. From the mathematical point of view, such information is sufficient to model the geopotential outside the Earth and gives opportunities in applications of alternative approaches for global gravity field modelling.

In this paper we present the method of fundamental solutions (MFS) to derive the disturbing potential and its first derivatives purely from the second derivatives observed by GOCE. MFS as a mesh-free boundary collocation

R. Čunderlík (✉)
Faculty of Civil Engineering, Department of Mathematics, Slovak University of Technology in Bratislava, Bratislava, Slovakia
e-mail: cunderli@svf.stuba.sk



S. Jin, R. Barzaghi (eds.), *IGFS 2014*, International Association of Geodesy Symposia 144,
DOI 10.1007/1345_2015_211

technique uses the fundamental solution of the Laplace equation as its basis functions (Mathon and Johnston 1977). In geodetic community this method is known as "point masses modelling" and many researchers have been applying this approach for gravity field modelling (e.g. Barthelmes and Kautzleben 1983; Barthelmes and Dietrich 1991; Lehmann 1993; Vermeer 1995; Claessens et al. 2001). The crucial point of such modelling is to find an optimal distribution and location (depth) of the source points in order to capture irregularities in the gravity field.

MFS is an efficient numerical technique to process straightforwardly the second derivatives of the unknown geopotential. Since the GOCE observations are sufficiently far from the Earth, the source points can be located directly on the Earth's surface and no singularities appear in the process of deriving the unknown source coefficients. However, to obtain "cm-level" accuracy of numerical solutions, the source points need to be distributed with high-resolution. This yields enormous memory requirements that can be reduced partly by parallel computing and partly by an iterative elimination of far zones' contributions.

The paper presents results of processing 20 datasets of the GOCE measurements observed during its mission, each for different 2 months (≈61 days) period. Their combination results in the final static gravity field model. Our aim is to show whether processing the GOCE measurements by MFS can lead to gravity field models of comparable accuracy that ones obtained by the spherical harmonics (SH) approach.

## 2 MFS for Processing the GOCE Measurements

The gravity field modelling is usually formulated in terms of the Laplace equation for the disturbing potential $T$

$$\nabla^2 T(\boldsymbol{x}) = 0, \quad \boldsymbol{x} \in ext.\ \Omega, \tag{1}$$

where $\Omega$ represents the Earth. An approximate solution by MFS is expressed as a linear combination of the fundamental solutions with respect to different source points

$$T(\boldsymbol{x}^i) = \sum_{j=1}^{N} \alpha_j G(\boldsymbol{x}^i, \boldsymbol{s}^j), \tag{2}$$

where $\boldsymbol{x}^i$ is the $i$-th observation point, $\boldsymbol{s}^j$ is the $j$-th source point, $\alpha_j$ denotes the $j$-th unknown coefficient of the distributed source at $\boldsymbol{s}^j$, $N$ is the number of source points and

$$G(\boldsymbol{x}^i, \boldsymbol{s}^j) = \frac{1}{4\pi |\boldsymbol{x}^i - \boldsymbol{s}^j|}, \tag{3}$$

represents the fundamental solution of the Laplace equation in 3D, which plays a role of the basis functions of the method. Gravity disturbances as the first derivatives of the disturbing potential in the radial direction can be expressed as a linear combination of the first derivatives of the fundamental solution in the radial direction

$$\delta g(\boldsymbol{x}^i) \cong \frac{\partial T(\boldsymbol{x}^i)}{\partial \boldsymbol{r}_{\boldsymbol{x}^i}} = \sum_{j=1}^{N} \alpha_j \frac{\partial G(\boldsymbol{x}^i, \boldsymbol{s}^j)}{\partial \boldsymbol{r}_{\boldsymbol{x}^i}}. \tag{4}$$

Analogously, MFS allows us to express the second derivatives of the disturbing potential provided by GOCE as a linear combination of the second derivatives of the fundamental solution in the corresponding directions.

In this study we focus on processing the radial components $T_{rr}$ of the disturbing tensor in order to derive the unknown coefficients $\{\alpha_j\}$ at the source points $\boldsymbol{s}^j$ using the expression

$$T_{rr}(\boldsymbol{x}^i) = \frac{\partial^2 T(\boldsymbol{x}^i)}{\partial \boldsymbol{r}^2_{\boldsymbol{x}^i}} = \sum_{j=1}^{N} \alpha_j \frac{\partial^2 G(\boldsymbol{x}^i, \boldsymbol{s}^j)}{\partial \boldsymbol{r}^2_{\boldsymbol{x}^i}}, \tag{5}$$

where

$$\frac{\partial^2 G(\boldsymbol{x}^i, \boldsymbol{s}^j)}{\partial \boldsymbol{r}^2_{\boldsymbol{x}^i}} = \frac{\partial^2 G_{i,j}}{\partial \boldsymbol{r}^2_{\boldsymbol{x}^i}} = \frac{1}{4\pi}\left[\frac{1}{d_{ij}^3} - 3\frac{\langle \boldsymbol{d}_{ij}, \boldsymbol{e}_{r_i}\rangle^2}{d_{ij}^5}\right], \tag{6}$$

and $\boldsymbol{e}_{r_i}$ denotes the unit vector in the radial direction at $\boldsymbol{x}^i$, $\boldsymbol{d}_{ij} = \boldsymbol{x}^i - \boldsymbol{s}^j$ and $d_{ij} = |\boldsymbol{d}_{ij}|$ represents the distance between the $i$-th collocation point and the $j$-th source point. By collocating $N$ observation points with respect to $N$ source points, we get the linear system of equations

$$\begin{bmatrix} \frac{\partial G^2_{1,1}}{\partial r^2_1} & \cdots\cdots & \frac{\partial G^2_{1,N}}{\partial r^2_1} \\ \vdots & \ddots & \vdots \\ \vdots & \ddots & \vdots \\ \frac{\partial G^2_{N,1}}{\partial r^2_N} & \cdots\cdots & \frac{\partial G^2_{N,N}}{\partial r^2_N} \end{bmatrix} \cdot \begin{bmatrix} \alpha_1 \\ \vdots \\ \vdots \\ \alpha_N \end{bmatrix} = \begin{bmatrix} T_{rr_1} \\ \vdots \\ \vdots \\ T_{rr_N} \end{bmatrix}. \tag{7}$$

Here the system matrix is created purely by the second radial derivatives of the fundamental solution, which depend solely on geometrical parameters, i.e. on 3D positions of the GOCE observations and source points and their direct distances (see Eq. 6). The linear system of equations (7) represents the standard collocation scheme where the number of observations equals to the number of source points. Obviously, there is a possibility to use a larger number of observations leading to an overdetermined system that can be solved e.g. by a least-square adjustment.

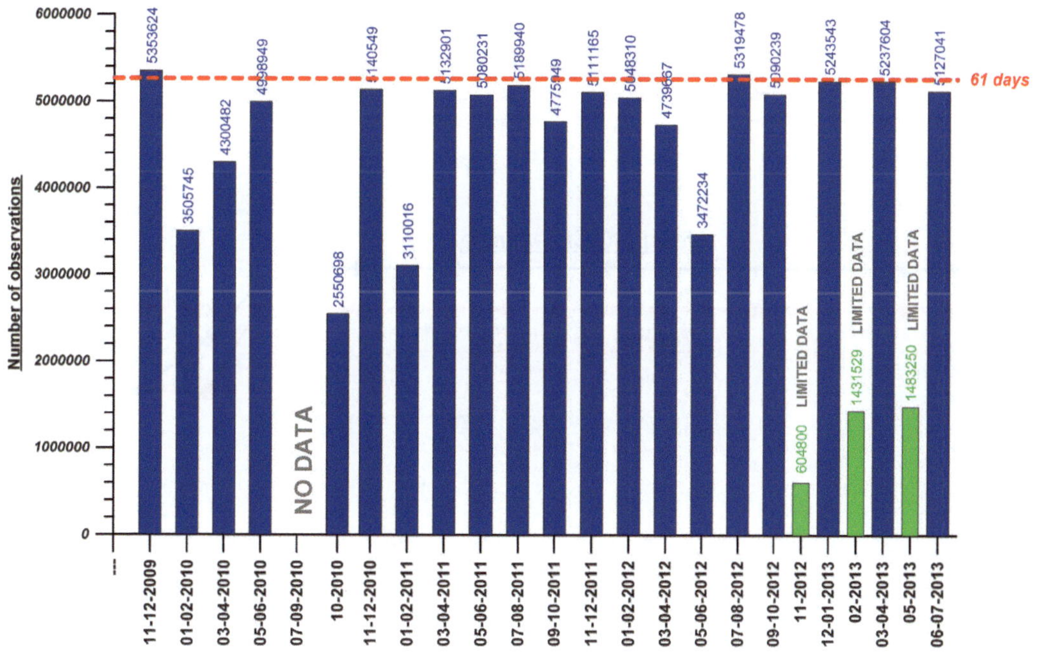


**Fig. 1** Number of the GOCE measurements for the different datasets/periods (only those with Flags_Vzz $\leq 2$)

Since the GOCE observations are given sufficiently far from the Earth (approximately 250 km above the Earth's surface), the source points can be located directly on the Earth's surface considering its complicated topography. Such a configuration does not generate any singularities. The unknown coefficients $\{\alpha_j\}$ can be determined by solving the linear system of equations (7). Afterwards, the disturbing potential or its first derivatives can be easily evaluated anywhere above the Earth's surface using Eqs. (2) or (4). The problem with the singularities appears when computing such quantities directly on the Earth's surface. In this case it is possible to use ideas of the singular boundary method (SBM) (Chen and Wang 2010) that can isolate singularities of the fundamental solution at source points on the Earth's surface. Due to the limited size of this paper we do not provide more details about this approach.

## 3 Input Data from GOCE

The GOCE data available from the EGG_TRF_2 product have been divided into several datasets, each for approximately 2 months observing period. Due to several breaks in the satellite observations, the number of measurements in each dataset varies (Fig. 1). Finally, 20 datasets have been chosen for further processing (Fig. 1, blue bars). They have included the radial components $T_{rr}$ whose flags $\leq 2$. Table 1 summarizes the number of observing days for every dataset, processed and excluding observations (Flags $> 2$), the mean GOCE altitude and the proposed reference altitude (see below).

Before processing the available radial components $T_{rr}$ from the EGG_TRF_2 product it is practically inevitable to filter out the noise. For this purpose we have used the nonlinear diffusion filtering on a closed surface (Čunderlík et al. 2013). Due to the fact that the altitudes of observations vary considerably reaching differences more than 10 km between "neighboring" points of observations (e.g. for distant observing time), we have reduced observed values from the observing altitudes into the reference altitude. A value of the reference altitude has been proposed for every dataset separately, i.e. about 5 km below the mean GOCE altitude for the corresponding period (Fig. 2, Table 1). Corrections from the observed into reference altitude have been evaluated from the GOCE-TIM4 model up to degree 250 (Pail et al. 2011).

The observations reduced into the reference altitude (Fig. 3a) have been filtered on such a closed and smooth surface using the nonlinear diffusion filtering. This approach is based on the regularized surface Perona-Malik model where the diffusion coefficient depends on surface gradients of the filtered function (Čunderlík et al. 2013). To avoid a loss of signal in important structures of the gravity field due to possible smoothing of sharp local extremes by the Perona-Malik model, we have first removed the trend generated from GOCE-TIM4 (Fig. 3b), then filtered the residuals between the reduced observations and this trend (Fig. 3c) and finally restored the input data by adding the trend to the filtered

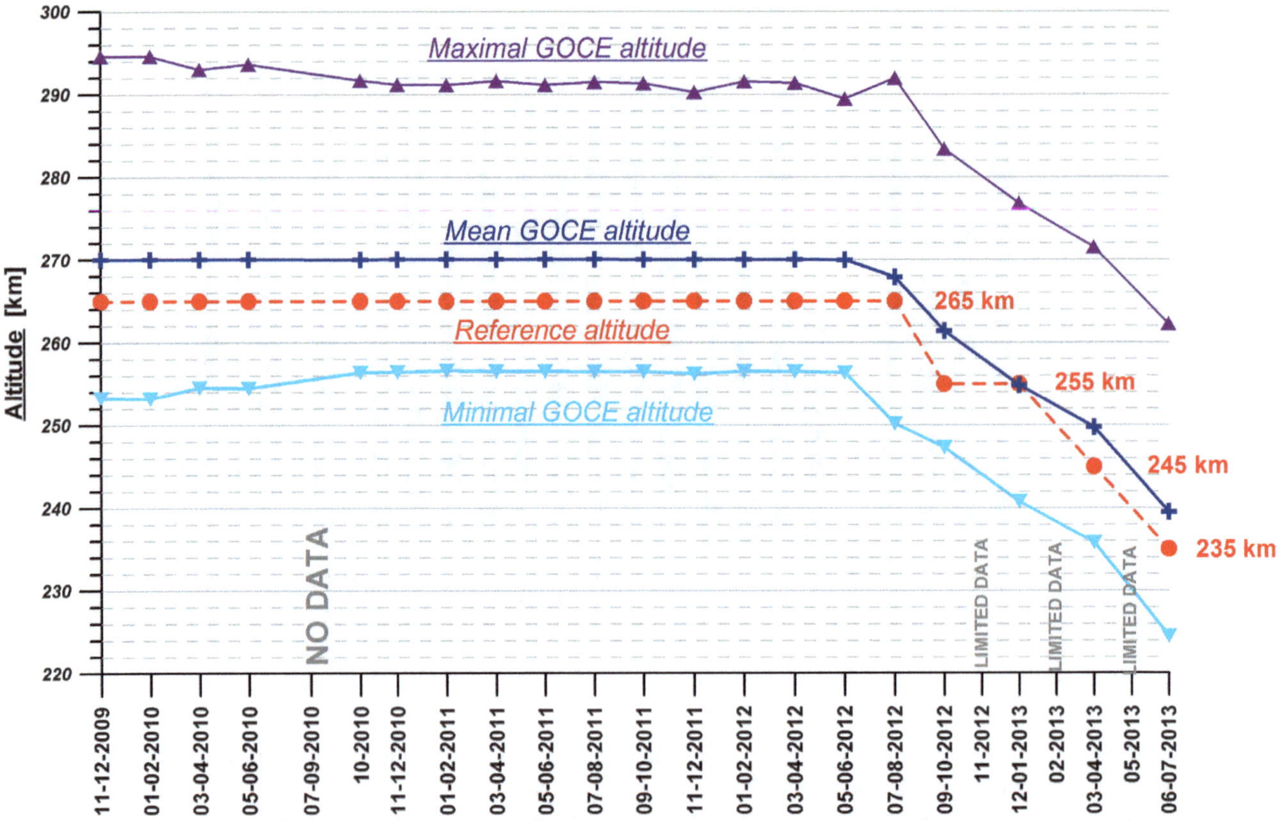


**Fig. 2** The minimal, maximal and mean altitude of the GOCE orbits for the different datasets/periods and the proposed reference altitude

**Table 1** Characteristics for 20 datasets: the period, number of observing days, processed observations (Flags $\leq 2$), excluding observations (Flags $> 2$), the mean GOCE altitude and proposed reference altitude

| Period | Observing Days | Observations (Flags $\leq 2$) | Observations (Flags $> 2$) | Mean altitude (km) | Reference altitude (km) |
|---|---|---|---|---|---|
| 11-12-2009 | 62 | 5,353,624 | 3, 172 | 270.081 | 265 |
| 01-02-2010 | 42 | 3,505,745 | 2, 271 | 270.085 | 265 |
| 03-04-2010 | 50 | 4,300,482 | 2, 318 | 270.065 | 265 |
| 05-06-2010 | 58 | 4,998,949 | 5, 642 | 270.077 | 265 |
| 10-2010 | 26 | 2,550,698 | 14, 648 | 270.001 | 265 |
| 11-12-2010 | 60 | 5,140,549 | 22, 177 | 270.060 | 265 |
| 01-02-2011 | 37 | 3,110,016 | 13, 903 | 270.074 | 265 |
| 03-04-2011 | 59 | 5,132,901 | 24, 442 | 270.058 | 265 |
| 05-06-2011 | 60 | 5,080,231 | 32, 650 | 270.050 | 265 |
| 07-08-2011 | 60 | 5,189,940 | 26, 885 | 270.061 | 265 |
| 09-10-2011 | 56 | 4,775,949 | 22, 575 | 270.041 | 265 |
| 11-12-2011 | 61 | 5,111,165 | 159, 235 | 270.012 | 265 |
| 01-02-2012 | 59 | 5,048,310 | 22, 633 | 270.022 | 265 |
| 03-04-2012 | 57 | 4,739,667 | 24, 731 | 270.003 | 265 |
| 05-06-2012 | 60 | 3,472,234 | 16, 592 | 269.986 | 265 |
| 07-08-2012 | 62 | 5,319,478 | 27, 465 | 267.862 | 265 |
| 09-10-2012 | 60 | 5,090,239 | 32, 257 | 261.378 | 255 |
| 12-01-2013 | 62 | 5,243,543 | 113, 257 | 254.862 | 255 |
| 03-04-2013 | 61 | 5,237,604 | 32, 796 | 249.727 | 245 |
| 06-07-2013 | 61 | 5,127,041 | 98, 965 | 239.516 | 235 |

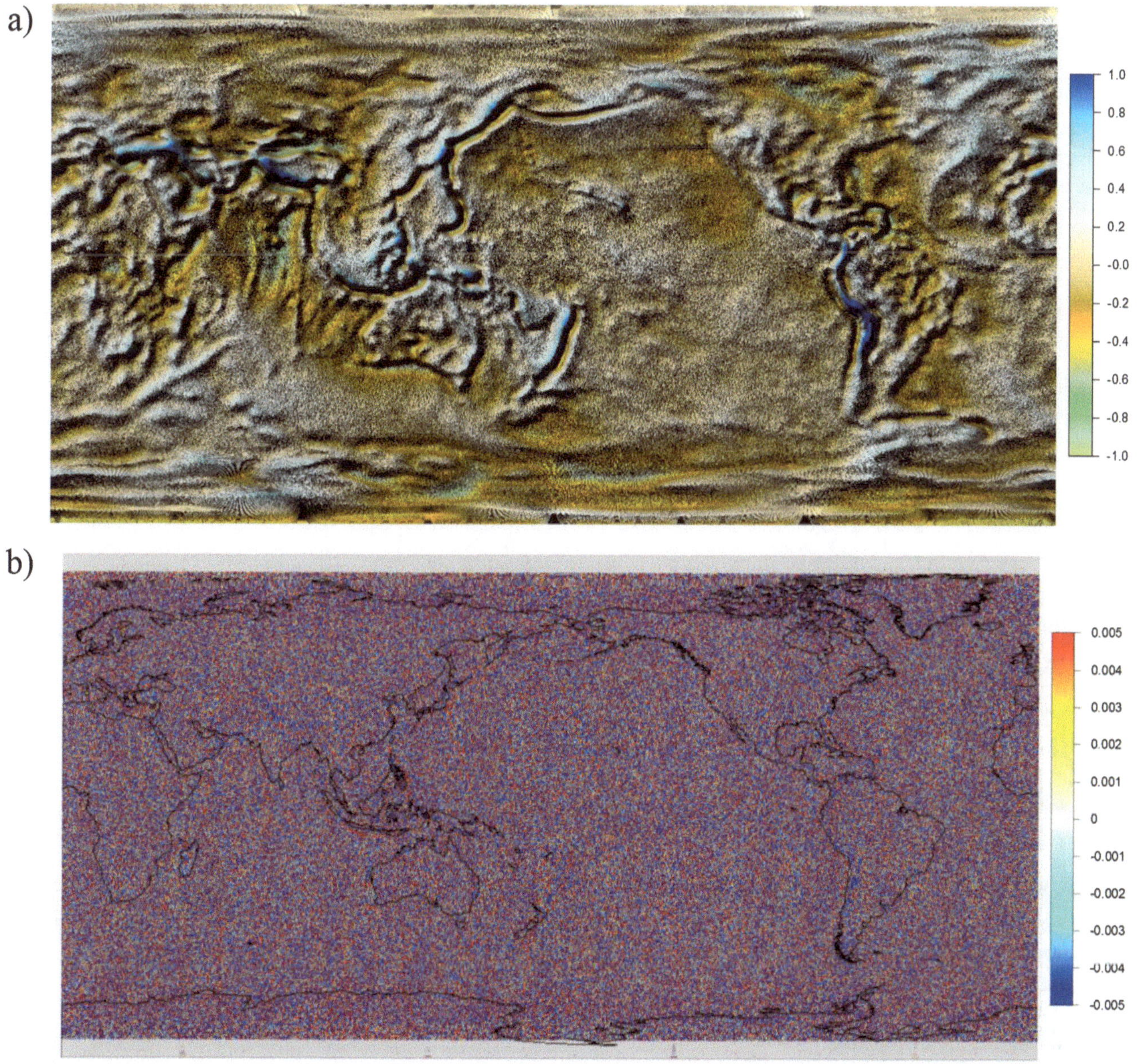


**Fig. 3** (continued)

residuals (Fig. 3d). Standard deviation (STD) of the residuals before and after filtering indicates how the noise included in observed data has been reduced (Fig. 4). STD after filtering (Fig. 4, red bars) also indicates a quality of the input data for every dataset. For example, the dataset "10-2010" includes only 1 month of observations. Such a lack of data yields a strong pattern in filtered data and, consequently, a high value of STD after filtering. On the contrary, its small values (e.g. below 0.001 E) indicate reliable input data for further processing by MFS.

## 4 Global Gravity Field Modelling by MFS

All 20 datasets have been processed separately. In all cases the source points have had the same positions. Namely, they have been located directly on the Earth's surface with a resolution of 0.075°. It has corresponded to 5,760,002 points ($N$) uniformly distributed over the Earth's surface. Their horizontal positions have been adopted from vertices

c)

d)

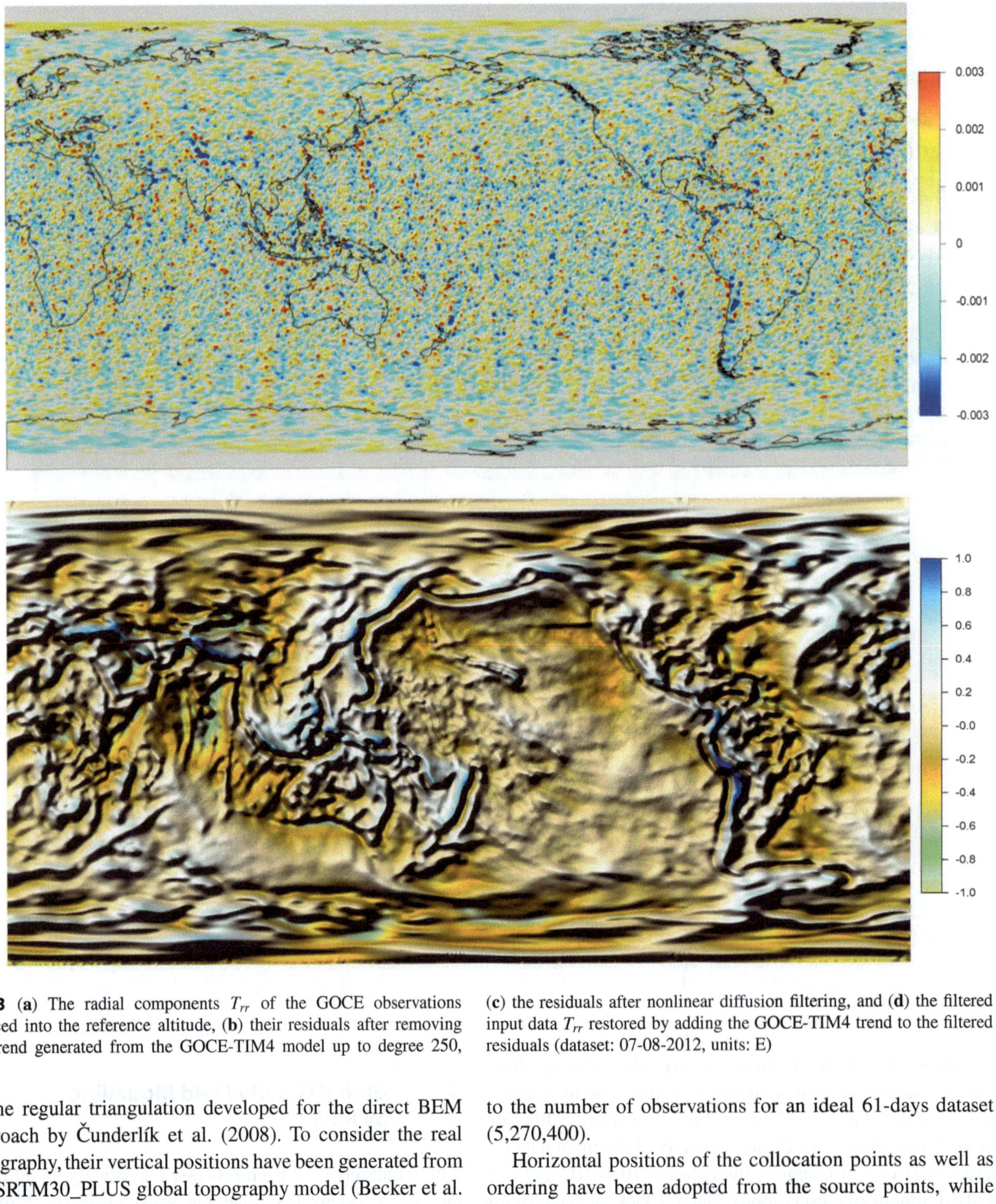


**Fig. 3** (**a**) The radial components $T_{rr}$ of the GOCE observations reduced into the reference altitude, (**b**) their residuals after removing the trend generated from the GOCE-TIM4 model up to degree 250, (**c**) the residuals after nonlinear diffusion filtering, and (**d**) the filtered input data $T_{rr}$ restored by adding the GOCE-TIM4 trend to the filtered residuals (dataset: 07-08-2012, units: E)

of the regular triangulation developed for the direct BEM approach by Čunderlík et al. (2008). To consider the real topography, their vertical positions have been generated from the SRTM30_PLUS global topography model (Becker et al. 2009).

To get the linear system of equations (7), we have chosen the same number of observations (collocation points). So far we have not used an overdetermined system where the number of observations is larger and a least square adjustment can be performed. Instead we have rather preferred to solve the standard collocation scheme and to set the number of source points as large as possible, i.e. to be proportional to the number of observations for an ideal 61-days dataset (5,270,400).

Horizontal positions of the collocation points as well as ordering have been adopted from the source points, while their vertical positions have been prescribed by the chosen reference altitudes (Fig. 2, Table 1). These points on the reference altitude surface have coincided with the computational nodes for the nonlinear diffusion filtering (see the previous section). Hence, the reduced and filtered input values of $T_{rr}$ have been given directly in the collocation points. The missing values in polar gaps (31,330 points $\approx 0.54\%$ of all collocations points) have been generated from the GOCO03S

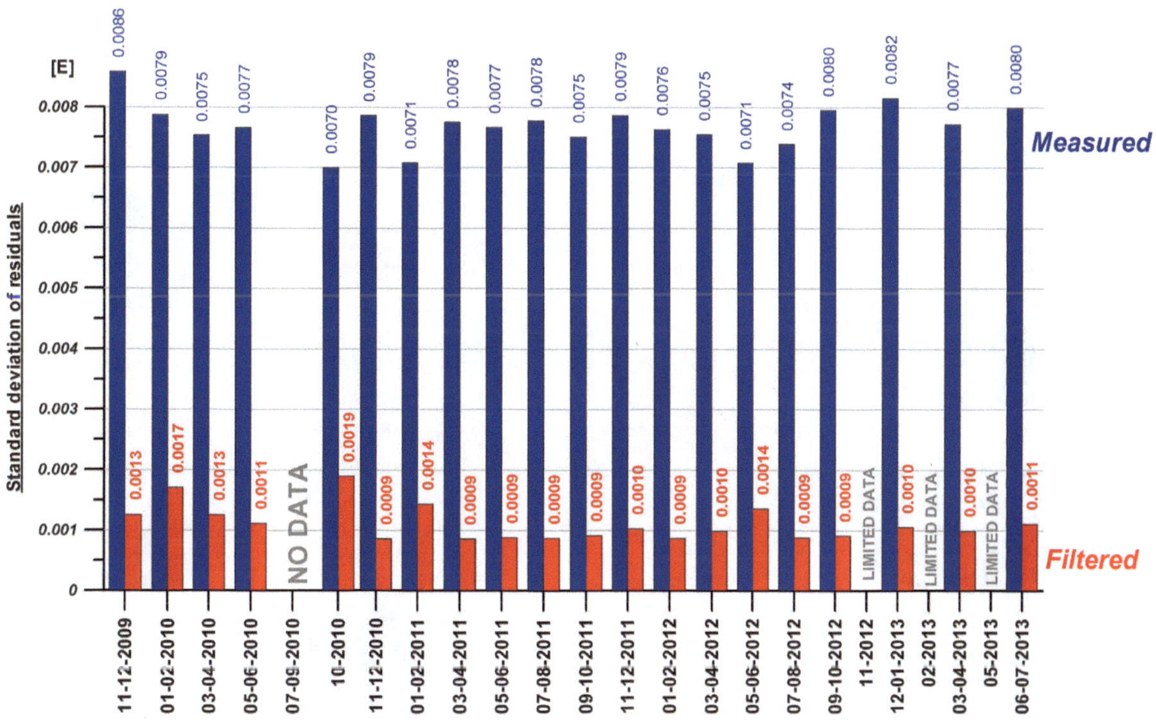


**Fig. 4** STD of the residuals between the radial components $T_{rr}$ of the GOCE observations reduced into the reference altitude and the values generated from GOCE-TIM4 (*blue bars* – before filtering, *red bars* – after nonlinear diffusion filtering)

satellite-only model up to degree 250, which also exploits information from the GRACE, CHAMP and SLR (Mayer-Gürr et al. 2012).

To reduce enormous memory requirements for the full matrix in Eq. (7), an iterative approach has been applied for an elimination of the far zones' contributions (Čunderlík and Mikula 2010). This approach, together with a parallel implementation using the MPI (Message Passing Interface) procedures, has enables us to reach such a high level of resolution. The large-scale parallel computations were performed on the cluster with 1.2 TB of the distributed memory. At first, the unknown coefficients $\{\alpha_j\}$ at the source points have been determined for every dataset by solving the linear system of equations (7). From these coefficients, the disturbing potential has been evaluated for every dataset at the altitude of 235 km above the reference ellipsoid. The particular solutions have been compared with GOCE-TIM4 (Fig. 5). It is evident that the residuals have very similar behavior in many datasets however in some cases they show some large systematic tendencies. Consequently, to get the final static gravity field model as a linear combination of the particular solutions we have used only those datasets/periods whose STD of residuals has been smaller than 0.08 $m^2s^{-2}$ ($\sim$8 mm) (Fig. 5b) and whose STD of input data (after filtering) was smaller than 1 mE (Fig. 4, red bars).

The coefficients $\{\alpha_j\}$ for the final static gravity field model have been computed by simple averaging of the coefficients from the selected 8 datasets. From these coefficients we have evaluated the disturbing potential and gravity disturbances at different altitudes above the reference ellipsoid. Figure 6 depicts the residuals between the obtained static gravity field and the GOCO03S, GOCE-DIR5 (Bruinsma et al. 2014) or GOCE-TIM5 (Brockmann et al. 2014) geopotential models at the altitude of 235 km. Table 2 presents the statistical characteristics of the corresponding residuals. Graphs in Fig. 7 depict how the statistical characteristics of the residuals are changing towards to the Earth's surface.

The comparison of the obtained static gravity field model with the SH-based geopotential models shows a very good agreement at the altitude of the GOCE orbits. Mean values of the residuals are about 0.01 $m^2s^{-2}$ ($\sim$1 mm) and STDs are 0.05 $m^2s^{-2}$ ($\sim$5 mm), while the maximal and minimal residuals do not exceed 0.25 $m^2s^{-2}$ ($\sim$2.5 cm) (Table 2). There is a significant pattern in the equatorial zone corresponding to the ionospheric refraction (Fig. 6). In case of the disturbing potential (Fig. 6a), there are evident large areas of the positive residuals in the northern hemisphere (mainly in Arctic zones, Siberia and Canada) while the negative ones dominate in Antarctica. They can

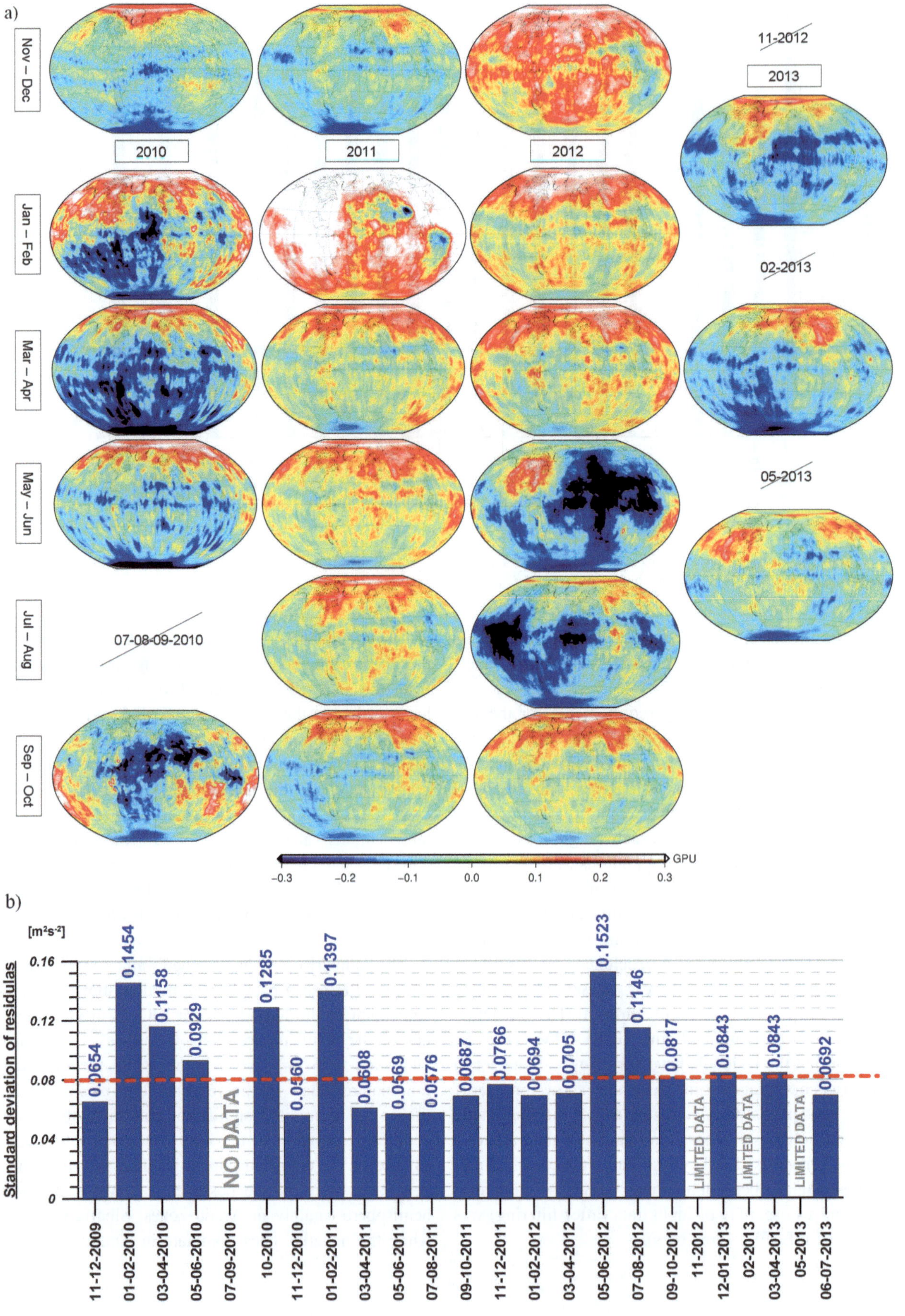


**Fig. 5** (**a**) The residuals between the disturbing potential evaluated by MFS for the different datasets/periods at the altitude of 235 km above the reference ellipsoid and one generated from the GOCE-TIM4 model up to degree 250; (**b**) standard deviations of the residuals

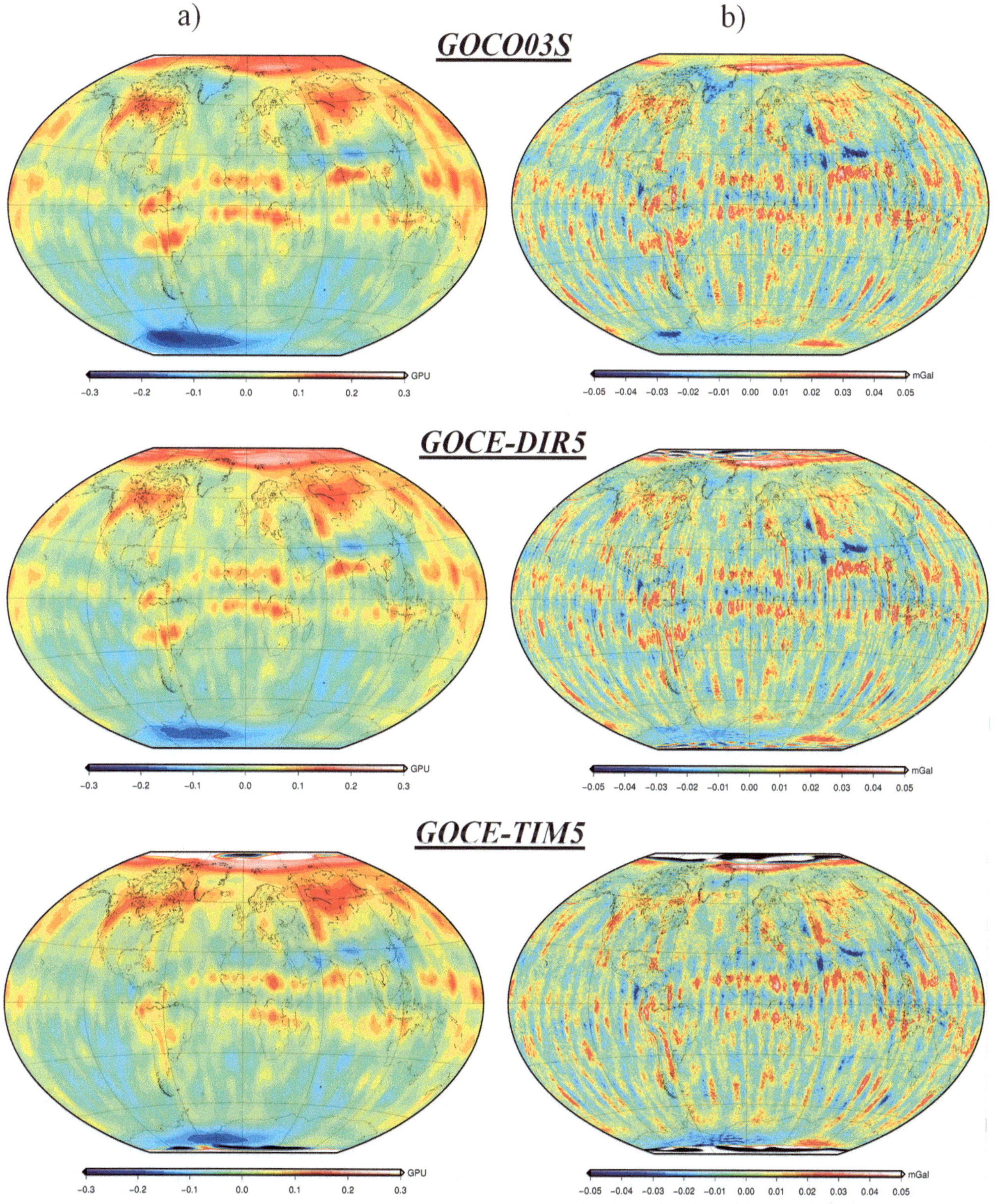


**Fig. 6** Residuals between the static gravity field obtained by MFS at the altitude of 235 km and the GOCO03S, GOCE-DIR5 and GOCE-TIM5 geopotential models; (**a**) the disturbing potential, (**b**) the gravity disturbances

**Table 2** Statistical characteristics of the residuals between the static gravity field obtained by MFS at the altitude of 235 km and the GOCO03S, GOCE-DIR5 or GOCE-TIM5 geopotential models

| GGM | GOCO03S | | GOCE-DIR5 | | GOCE-TIM5[a] | |
|---|---|---|---|---|---|---|
| Quantity | $T$ ($m^2s^{-2}$) | $\delta g$ (mGal) | $T$ | $\delta g$ | $T$ | $\delta g$ |
| Max | 0.198 | 0.046 | 0.232 | 0.141 | 0.246 | 0.049 |
| Min | −0.217 | −0.054 | −0.173 | −0.149 | −0.177 | −0.050 |
| Mean | 0.013 | 0.000 | 0.011 | 0.000 | 0.010 | 0.000 |
| STD | 0.054 | 0.011 | 0.051 | 0.011 | 0.050 | 0.010 |

[a]Excluding polar gaps

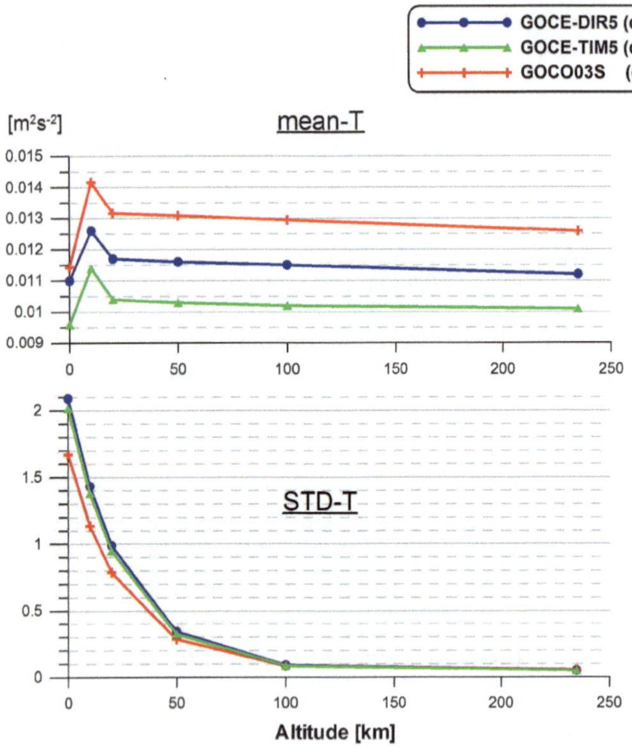


**Fig. 7** Mean values and standard deviations (STDs) of the residuals between the disturbing potential or gravity disturbances evaluated by MFS and the SH-based models (GOCO03S, GOCE-DIR5 or GOCE-TIM5) at different altitudes

indicate some systematic tendencies due to the fact that we have processed purely the GOCE gradients. The apparent strong negative residuals correlating with Himalayas, better visible in case of the gravity disturbances (Fig. 6b), could indicate some loss of masses. On the other hand, they can be also caused by the fact that our source points are located directly on the topography, therefore giving different results than the SH-based approaches in such high mountainous regions.

Comparisons at different altitudes confirm that an agreement between solutions is getting worse when going towards to the Earth's surface (Fig. 7). In addition, the closer to the Earth's surface, the stronger the impact of the singularities becomes and the STD is asymptotically increasing. A special treatment of the singularities by the SBM approach has allowed us to evaluate quantities of the gravity field directly at the source points on the Earth's surface. Here STD of the residuals with respect to GOCO03S is 1.67 $m^2s^{-2}$ (∼16 cm) and 7.2 mGal, with respect to GOCE-DIR5 it is 2.09 $m^2s^{-2}$ (∼21 cm) and 8.8 mGal, and with respect to GOCE-TIM5 it is 2.02 $m^2s^{-2}$ (∼20 cm), and 8.6 mGal.

Finally we would like to remind that the presented static gravity field model is not a GOCE-only solution due to exploiting information from other missions like GRACE or CHAMP to fill the polar gaps (here the input data are generated from GOCO03S) and due to using EGG_TRF_2 data instead of the original gradiometer observations. In fact, an external SH-based model was used to rotate the EGG_TRF_2 gravity gradients from the instrumental reference frame of the GOCE gradiometer to the terrestrial reference frame (Bouman et al. 2011). Moreover, the GOCE-TIM4 model has been used in the process to reduce and filter the input data.

All these issues cause latent dependencies on the used SH-based models.

## 5 Discussion and Conclusions

The presented processing of the GOCE measurements using MFS demonstrates its efficiency. Although this approach requires large-scale parallel computing and the elimination of far zones' contributions, such a numerical technique is very straightforward allowing further refinement of the resolution. The static gravity field obtained by processing purely the radial components $T_{rr}$ available from the EGG_TRF_2 product shows an apparent agreement with the SH-based satellite-only geopotential models like GOCO03S, GOCE-DIR5 or GOCE-TIM5 indicating "cm-level" accuracy at the altitude of 235 km (close to the GOCE orbits). Here we would like to remind that all three SH-based models have been developed by processing also other components of the gravity tensor as well as by considering the GOCE orbits. Moreover, the GOCO03S and GOCE-DIR5 models also include information from GRACE, CHAMP or SLR.

Unfortunately, comparisons between the particular solutions have not shown any significant bimestrial time variations of the gravity field. Strong systematic tendencies, that badly affected almost half of the particular solutions (Fig. 5), are probably due to a lack of the input GOCE observations (Fig. 1). This implies that our choice of dividing the processed observations into the datasets, each for 2-months period, is not optimal. Consequently, in future processing we should take longer period of observations for one dataset avoiding gaps in input data. On the other hand, the comparison of the final static gravity field model with GOCO03S (Fig. 6) clearly demonstrates long period changes in the gravity field, namely a loss of masses in south part of Greenland, in Western Antarctica or Himalayas. Such sensitive results obtained by the presented approach are promising for its further investigations.

**Acknowledgements** The work has been supported by the grant VEGA 1/0714/15 and the project APVV-0072-11

## References

Barthelmes F, Dietrich R (1991). Use of point masses on optimized positions for the approximations of the gravity field. In: Rapp, RH and Sansò, F (eds) Determination of the Geoid-Present and Future, IAG Symp 106:484–493

Barthelmes F, Kautzleben H (1983) A new method of modelling the gravity field of the Earth by point masses. In: Mitt. Zentralinst. Phys. Erde, No. 1214, 18 pp 01/1983

Becker JJ, Sandwell DT, Smith WHF, Braud J, Binder B, Dep-ner J, Fabre D, Factor J, Ingalls S, Kim S-H, Ladner R, Marks K, Nelson S, Pharaoh A, Sharman G, Trimmer R, vonRosenburg J, Wallace G, Weatherall P (2009) Global bathymetry and elevation data at 30 Arc seconds resolution: SRTM30_PLUS. Mar Geod 32(4):355–371

Bouman J, Fiorot S, Fuchs M, Gruber T, Schrama E, Tscherning C, Veicherts M, Visser P (2011) GOCE gravitational gradients along the orbit. J Geod 85:791–805

Brockmann JM, Zehentner N, Höck E, Pail R, Loth I, Mayer-Gürr T, Schuh W-D (2014) EGM_TIM_RL05: an independent geoid with centimeter accuracy purely based on the GOCE mission. Geophys Res Lett 41(22):8089–8099

Bruinsma SL, Förste C, Abrikosov O, Lemoine J-M, Marty J-C, Mulet S, Rio M-H, Bonvalot S (2014) ESA's satellite-only gravity field model via the direct approach based on all GOCE data. Geophys Res Lett 41(21):7508–7514

Chen W, Wang FZ (2010) A method of fundamental solutions without fictitious boundary. Eng Anal Boundary Elem 34(5):530–532

Claessens SJ, Featherstone WE, Barthelmes F (2001) Experiences with point-mass modelling in the Perth region, Western Australia. Geomat Res Aust 75:53–86

Čunderlík R, Mikula K (2010) Direct BEM for high-resolution gravity field modelling. Stud Geophys Geod 54(2):219–238

Čunderlík R, Mikula K, Mojzeš M (2008) Numerical solution of the linearized fixed gravimetric boundary-value problem. J Geod 82:15–29

Čunderlík R, Mikula K, Tunega M (2013) Nonlinear diffusion filtering of data on the Earth's surface. J Geod 87:143–160

Lehmann R (1993) Nonlinear gravity field inversion using point masses – diagnosing nonlinearity. In: H. Montag et al. (eds), Geodesy and Physics of the Earth, IAG Symp 112:256–259

Mathon R, Johnston RL (1977) The approximate solution of elliptic boundary-value problems by fundamental solutions. SIAM J Numer Anal 638–650

Mayer-Gürr T, Rieser D, Hoeck E, Brockmann M, Schuh WD, Krasbutter I, Kusche J, Maier A, Krauss S, Hausleitner W, Baur O, Jaeggi A, Meyer U, Prange L, Pail R, Fecher T, Gruber T (2012) The new combined satellite only model GOCO03s. In: Presented at the GGHS-2012 in Venice, Italy, October 9–12, 2012

Pail R, Bruinsma SL, Migliaccio F, Foerste C, Goiginger H, Schuh WD, Hoeck E, Reguzzoni M, Brockmann JM, Abrikosov O, Veicherts M, Fecher T, Mayrhofer R, Krasbutter I, Sanso F, Tscherning CC (2011) First GOCE gravity field models derived by three different approaches. J Geod 85:819–843

Vermeer M (1995) Mass point geopotential modelling using fast spectral techniques; historical overview, toolbox description, numerical experiment. Manuscr Geod 20:362–378

# Towards a Vertical Reference Frame for South America in View of the GGOS Specifications

Andrea Galudht Santacruz Jaramillo and Sílvio Rogério Correia de Freitas

**Abstract**

One of the objectives of the Geocentric Reference System for the Americas (SIRGAS) is to establish a unified Earth's gravity field-related vertical reference system that meets the Global Geodetic Observing System (GGOS)'s requirements. This statement implies the homogenization and unification of the National Vertical Reference Systems (NVRS) existing in the SIRGAS and their integration into a International Height Reference System (IHRS). At present, different strategies based on the combination of local (e.g. levelling, terrestrial gravity data) and global observations (e.g. global models, GNSS positioning) are under consideration. However, most of them are not immediately applicable in South America due to the differences in the definition and realization of the local height systems. In this work, it is proposed an inventory of the available data and the implementation of a common metadata base in order to provide sufficient information for a consistent characterization of each NVRS. This structure must allow the identification of the different standards and specifications applied for the establishment of those systems and the needs of complementary actions for connecting them, including the standardization of the existing height data. The main goal of this approach is however the minimization of those inconsistencies produced by data unavailability, unknown discrepancies, different data collecting and processing strategies, accuracy, and other non-evaluated errors. Based on results for a case study (or test country) in South America, it is proposed a road map for the inventory of vertical systems established with metadata.

**Keywords**

Metadata in height systems • SIRGAS Vertical Reference System (SVRS) • Vertical datum unification

## 1 Introduction

SIRGAS looks for the definition and realization of a geocentric reference system and a unified Earth's gravity field-related vertical reference system (Sánchez and Brunini 2009). From the organizational point of view, SIRGAS is the Sub-Commission 1.3b (Regional Sub-Commission for South and Central America) of the International Association of Geodesy (IAG) and a Working Group of the Pan-American Institute for Geography and History (PAIGH). Fulfilling the IAG objectives, SIRGAS establishes and maintains a regional reference frame for Earth System research. Satisfying the PAIGH requirements, SIRGAS delivers to the Latin American and Caribbean community the fundamental layer for the Americas' Geospatial Database Infrastructure and other geodetic applications. The SIRGAS activities are performed in three Working Groups: SIRGAS-WGI: Reference System, SIRGAS-WGII: SIRGAS at National

A. Santacruz (✉) • S.R.C. de Freitas
Federal University of Paraná – UFPR, Geodetic Sciences Graduation Course – CPGCG, Curitiba, Brazil
e-mail: andreasantacruzj@gmail.com; sfreitas@ufpr.br


S. Jin, R. Barzaghi (eds.), *IGFS 2014*, International Association of Geodesy Symposia 144,
DOI 10.1007/1345_2015_208

Level, and SIRGAS-WGIII: Vertical Datum. The objective of the SIRGAS-WGIII is the definition and realization of a modern unified SIRGAS Vertical Reference System/Frame (SVRS/SVRF) for the precise unification of existing National Vertical Reference Systems (NVRS) (Drewes et al. 2002). It is well-known that the existing NVRSs do not support the accuracy requirements of modern Geodesy. They refer to local (isolated) levels, are stationary (without considering variations in time), and their realizations contain large uncertainties caused, among others, by systematic errors in levelling, the omission of gravity reductions and non-modeled effects in the height determination. To overcome these deficiencies, the Geodetic Global Observing System (GGOS) of the IAG promotes the standardization of height systems worldwide. The main objectives are to provide a reliable frame for consistent analysis and modelling of global phenomena and processes affecting the Earth's gravity field and the Earth's surface geometry; and to support the precise combination of physical and geometric heights. This allows exploiting the advantages of Satellite Geodesy, for example, by combining satellite positioning and gravity field models for worldwide-unified precise height determination (Sánchez 2012). To advance in this purpose, it is necessary however to know the characteristics of each NVRS, since each of them follows specific features related to its definition and realization as well as to its maintenance. The inventory of these characteristics can be summarized on metadata bases providing reliable input information to identify the best-possible strategies for the unification of the height systems. According to this in the present paper the description of the required data for the detailed inventory of the existing NVRS is studied. As a case study (example of the application of the methodology developed) the status of the NVRS of Ecuador is discussed in detail.

## 2 The SVRS Under the GGOS Specifications

To advance in the definition and realization of a global vertical reference system, GGOS established during its Planning Meeting 2010 (February 1–3, Miami/Florida, USA) the GGOS Theme 1: *Unified Height System*. The main objective is to provide a global gravity field-related vertical reference system that (1) supports a highly-precise combination of physical and geometric heights worldwide, (2) allows the unification of all existing NVRS, and (3) guarantees vertical coordinates with global consistency (the same accuracy everywhere) and long-term stability (the same order of accuracy at any time) (Kutterer et al. 2012). The expected accuracy is one part per billion (1 ppb), i.e. $\sim$1 cm for the vertical coordinates (Plag et al. 2009, p. 18). Activities to be undertaken under the umbrella of the GGOS Theme 1 are understood as the continuation of the work started by the 2007–2011 IAG Inter-Commission Project 1.2 *Vertical Reference Frames* (IAG ICP1.2, Ihde 2007). The main result of the IAG ICP1.2 is the document *Conventions for the Definition and Realization of a Conventional Vertical Reference System* – CVRS (Ihde et al. 2007). The present actions related to the vertical datum homogenization are being coordinated by the working group *Vertical Datum Standardization*, which directly depends on the GGOS Theme 1 and is supported by the IAG Commissions 1 (Reference Frames) and 2 (Gravity Field), as well as by the International Gravity Field Service (IGFS) (Sánchez 2012). The conventions outlined by the IAG ICP1.2 describe the fundamentals to be taken into consideration for the establishment of a vertical reference system fulfilling the requirements outlined by GGOS (Ihde and Sanchez 2005) and are still in development under the cooperation between the working group *Vertical Datum Standardization* (Sánchez 2012) and the *GGOS Bureau for Standards and Conventions* (Hugentobler et al. 2012).

According to Ihde et al. (2007) and Kutterer et al. (2012), the vertical datum for the *GGOS Unified Height System* shall correspond to a level surface of the Earth's gravity field with a given potential value $W_0 = const.$ It is well-known that any $W_0$ value can be adopted for the determination of vertical coordinates. In this context, the challenge is not the selection of the value $W_0$, but its realization, i.e. the estimation of the position and geometry of the correspondent equipotential surface. To provide the correspondence between $W_0$ and the global geoid, both should be estimated from the same geodetic observations and be consistent with other defining parameters of geometric and physical models of the Earth. Usually, $W_0$ was defined to be equivalent to the normal potential $U_0$ generated by a mean Earth level ellipsoid (e.g. GRS80, Moritz 2000). Nowadays, the empirical estimation of $W_0$ is based on the combination of global gravity field models (GGMs) and the geometric representation of the Earth's surface (Sánchez 2012). Since the estimation of $W_0$ depends on the input data and the processing strategies, $W_0$ should be based, like any reference system, on standards and conventions, which guarantee its uniqueness, reliability and reproducibility. However, the present best estimated value of $W_0$ in the IERS Conventions is not reproducible with the necessary accuracy. Otherwise there would be as many $W_0$ reference values (i.e., global zero-height surfaces) as computations.

The SVRS, defined in accordance with the GGOS, includes a geometrical and a physical component to support the precise combination of geometrical and physical heights (see, e.g., Sánchez 2007). The geometrical component is given by a level ellipsoid as a reference surface and ellipsoidal heights ($h$) and their time variation ($dh/dt$) as coordinates. The geometrical vertical coordinates shall be consistent with the realization of the geometrical reference system, i.e. they must refer to the ITRS/ITRF. This is already achieved in South America by the determination, maintenance, and wide-usability of the SIRGAS Reference Frame (Brunini et al. 2012).

The physical component of the SVRS is given in terms of geopotential quantities, i.e. the reference level must be a $W_0$ value, and the vertical coordinates shall be geopotential numbers $C$ (and their variations with time) referred, of course, to this $W_0$ (Sánchez 2009). The basic idea is to introduce the IAG/GGOS $W_0$ as the reference level of the SVRS. Nevertheless, the realization of the equipotential surface defined by $W_0$ (geoid modelling) and the determination of the level differences between the global $W_0$ and the local vertical datums $W_0^i$ remain as the main problems. One possibility to determine the *so-called* vertical datum discrepancies $\delta W_0^i = W_0 - W_0^i$ is the comparison of geopotencial numbers referring to the local vertical datums and those derived after solving the geodetic boundary value problem (GBVP) (Ferreira and de Freitas 2011).

## 3 Basic Strategy for the Vertical Datum Unification

The geopotential number $C_P^i$ at any point $P$ referring to a local vertical datum $W_0^i$ is given by:

$$C_P^i = W_0^i - W_P = \sum_0^P g dn, \tag{1}$$

where $dn$ represents the level difference between two consecutive leveling points and $g$ is the average of the gravity values along the leveling line connecting those points. The potential value $W_P$ can thus be written as

$$W_P = W_0^i - C_P^i. \tag{2}$$

In the frame of the GBVP, the potential value at the same point $P$ is estimated by means of

$$W_P = U_P + T_P, \tag{3}$$

where $U_P$ corresponds to the normal potential generated by a level ellipsoid (e.g. Hofmann-Wellenhof and Moritz 2005, p. 65) and $T_P$ is the anomalous potential. The corresponding geopotential number $C_P$, referring to a global vertical level $W_0$, is then given as:

$$C_P = W_0 - W_P = W_0 - (U_P + T_p). \tag{4}$$

Equation (4) minus Eq. (1) provides the vertical datum discrepancy between the local level $W_0^i$ and the global $W_0$:

$$C_P - C_P^i = W_0 - W_P^i = \delta W_i. \tag{5}$$

In practice, $T_p$ can be determined by solving the GBVP. The usual approach assumes that the vertical position of the boundary surface is unknown (scalar free GBVP; see, e.g., Sacerdote and Sansò 1986; Heck 1989). In this case, the boundary conditions are mainly given by gravity anomalies and geopotentital numbers, both depending on the local vertical datums. Consequently, the estimated unknowns (boundary surface and gravity potential) may also reflect the discrepancies existing between those vertical datums (Rummel and Teunissen 1998; Heck and Rummel 1990; Xu and Rummel 1991).

At present, the availability of satellite altimetry on ocean areas and GNSS positioning on land areas allows a better representation of the Earth's surface geometry, making possible the solution of the fixed GBVP (Heck and Seitz 1993; Sansò 1995; Heck 2011). In this case, the boundary conditions are given by gravity disturbances, which do not depend on the local height datums. The only unknown is the gravity potential and its estimation is free of vertical datum effects.

Recently, a Brovar-type solution (Brovar 1972) to the fixed GBVP has been revisited by Heck (2011) and suggested as the most natural approach in the era of GNSS positioning. In a first approximation, this solution omits non-linear and ellipsoidal terms and interprets ellipsoidal heights as heights above a sphere. The solution given by Heck (2011, Eq. 44) has the same structure as the solution of the scalar free GBVP (aka Molodenskii BVP) and it has the advantage that the well-known computational procedures for the quasi-geoid determination can be applied (see, e.g., Hofmann-Wellenhof and Moritz 2005, pp. 294–303). However, this kind of solution strongly depends on an adequate data distribution.

It must be considered that the Eq. (5) involves several aspects, some of them related to the solution of the GBVP and others related to the practical approaches and existing data. The following aspects must be emphasized as a base for a road map when considering data structures like in South America:

(1) Level differences measured by spirit leveling techniques and referring to the local vertical datums $W_0^i$.
(2) Gravity values along the leveling lines for the determination of $C_P^i$ as Eq. (1). Interpolated gravity values can be used, since a high-precision is not required, i.e. an uncertainty of 1 mGal in the gravity value may cause an error less than 1 mm in the height difference (de Freitas and Blitzkow 1999).

(3) Gravity values with positions referred to the ITRS/ITRF (i.e. SIRGAS in South America) in order to compute the gravity disturbances. Since the solution of the GVBP requires a homogeneous distribution of the boundary conditions on the whole Earth. At least, it is necessary a high-density of gravity/GNSS stations around the evaluation points. In addition, these points have to be connected to the vertical datum $W_0^i$ by means of spirit leveling.

(4) To minimize the radius of the integration area where the gravity and GNSS stations have to be available a spectral decomposition of the gravity disturbances can be used. In this case, the long wavelengths are obtained from a GGM and the short wavelengths from the terrain effects derived from a Digital Elevation Model (DEM). This is the so-called Residual Terrain Model (RTM) technique proposed by Forsberg (1984) and recently extended by Hirt et al. (2010) to make it suitable for regions with poor gravity data distribution. This approach is based on the Remove-Restore technique in accordance with Forsberg and Tscherning (1981), i.e. the spectral components of the gravity disturbances removed before computing $T_P$ are restored as spectral components of $T_P$ after the computation.

(5) It is well-know that gravity values or gravity disturbances are usually not available, but gravity anomalies. In this case, one possibility is the transformation of the existing gravity anomaly datasets, but these anomalies depend on the vertical datum discrepancies that are unknown. Hence, the procedure must be an iterative process and the convergence of the iteration depends on the quality of the data, referred GPS/leveling with gravity measurements and gravity anomalies calculated in points as is presented in Amos and Featherstone (2009).

(6) The treatment of the permanent tide effects play an especial role: usually the GGM are given in the tide-free system (all tide direct and indirect effects are removed) and the terrestrial gravity data are given in zero-tide system (tide direct effects removed, indirect effect retained). Thus, consistency must be guaranteed before computing $T_P$.

(7) The tide system used for the computation of $T_P$ must be the same applied for the computation of the geopotential numbers $C_P^i$, and this system must be consistent with the tide system in which the ellipsoidal height is given. The main reasons are ellipsoidal heights are used in the computation of gravity disturbances and in reducing the normal potential to the Earth's surface.

(8) The basic formulation of the GBVP assumes that the boundary conditions, the boundary surface and the estimated potential are invariant with time, i.e. they are quasi-stationary. Consequently, those observables varying with time shall be given in the same epoch. It must be considered that in a modern view, it is necessary to take in account aspects related with $\partial H/\partial t$ and $\partial h/\partial t$ associated to physical and geometrical heights, respectively. The future mission GRACE-FO is claimed as a powerful tool for this purpose. Some previous results from GRACE mission support this idea.

(9) To guarantee that Eq. (5) provides reliable vertical datum discrepancies $\delta W_i$, it is necessary to remove as far as possible all sources of uncertainties (e.g., all the gravity values shall refer to the same gravity reference system, e.g. IGSN1971). Additionally, all ellipsoidal heights shall refer to the same ITRF and epoch as well as all functional (ellipsoidal heights, geopotential numbers, and anomalous potential) shall be given in the same tide system.

## 4 Inventory of Vertical Reference Frames in South-America

According to the description of the required data for the vertical datum unification presented, one can say that the main limitation in South America is the uncertainty about the characteristics of the existing vertical and gravity data. Consequently, since 1998 SIRGAS promotes the implementation of a detailed inventory of the existing NVRF in South America. This inventory includes vertical datum and networks, gravity networks and co-located gravity/GNSS/leveling observations. The main purpose is to allow each country to supply relevant data to implement a consistent metadata base for identifying the main problems in unifying the NVRS under the GGOS specifications. The NVRS and the current available leveling and gravity data must be analyzed considering their characteristics and possibilities for realizing the SVRF. Preliminary results pointed out the lack of traditional leveling and gravity data on large areas, and the existing heterogeneities in the definition and realization of the existing NVRS. It must be considered that in a next step, it is intended to install a metadata acquisition system in an open access platform for SIRGAS member countries. It will be based on the International Organization for Standardization (ISO) roadmaps for developing and implementing metadata bases (ISO 19115 and ISO 19139). The goal is to implement a metadata profile to facilitate the understanding of the collected data. The proposed methodology for the implementation of the inventory is summarized in Fig. 1 and it is tested in Ecuador as a case study.

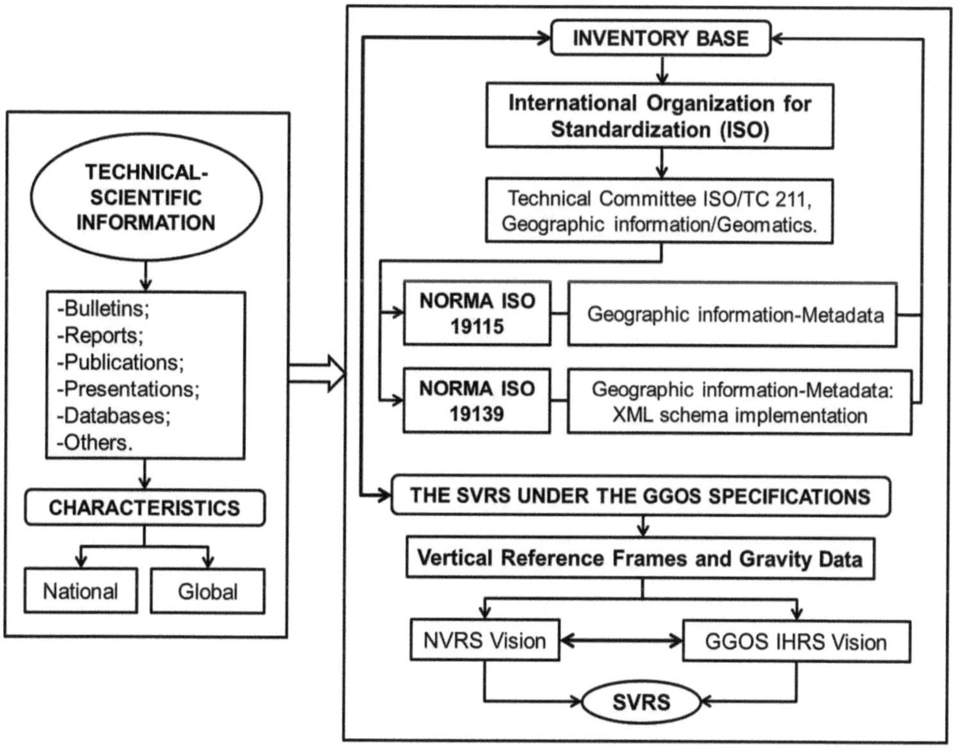


**Fig. 1** Road map for the implementation of vertical datum metadata in South America following GGOS specifications and ISO standards

## 5 Case Study in Ecuador

In the following, the present status of the NVRS of Ecuador is presented in detail as an example for the application of the proposed methodology. The data were classified according to their origin (gravity, spirit leveling, GNSS positioning, and others) with the purpose of analyzing their specific characteristics. This is the first step in identifying and homogenizing the available data according to the GGOS proposal.

### 5.1 Ecuador Geocentric Reference Frame (SIRGAS-ECUADOR)

The Geocentric Reference Frame in Ecuador (Fig. 2) is a densification of the SIRGAS Reference Frame. It is composed by passive and continuously operating reference stations with positions referred to a certain epoch and known velocities. The metadata of this frame is well-documented and transformations can be carried out with high accuracy (at the cm level).

### 5.2 Ecuador Vertical Datum (EVD)

The Ecuador Vertical Datum is realized by the Mean Sea Level (MSL) derived from tide gauge registrations collected during 9 years from 1950 at the tide gauge "*La Libertad*" located in 2°13′8″ S and 80°54′18″ W (Fig. 3).

Strategies for modernizing the EVD:

- Referring the reference tide gauge to the ITRS/ITRF;
- Modelling the time variation of MSL;
- Improving the resolution of available recent GGMs by reducing their omission error (e.g., by RTM and/or gravimetric densification);
- Estimation of the sea surface topography (SSTop) related to the GGOS's IHRS.
- Determination of the potential value at the reference tide gauge $W_0^i$.

### 5.3 Ecuador Vertical Reference Frame (EVRF)

The start point of the EVRF is the benchmark BM3, located near the tide gauge "*La Libertad*" (see Fig. 3), which also serves as a reference point for the tide gauge. The EVRF contains approximately 6,000 first-order (for Ecuador $\pm 4\ \mathrm{mm}\sqrt{\mathrm{km}}$) leveling points. The leveling network is constituted by 62 lines, 20 loops, and 38 nodes. Ellipsoidal heights and gravity values are available at many leveling points. They are spaced by about 1.5 km in 11,000 km along the main roads of the country. This network is being extended to international border lines and to the Amazonas

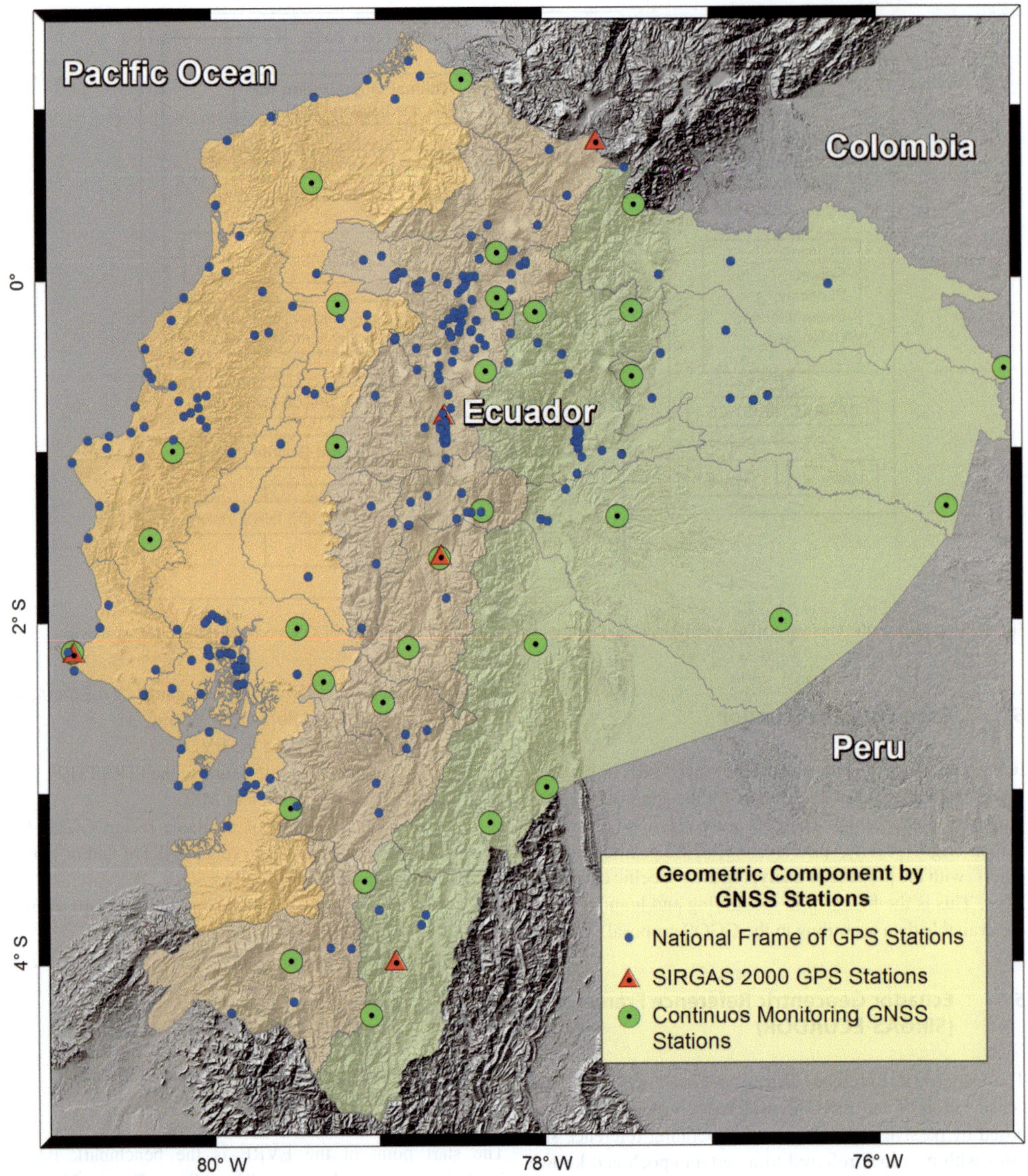


**Fig. 2** Geometric component by GNSS Stations

region. The new lines shall be adjusted in one block with the existing ones.

Strategies for modernizing the EVRF:

- Analysis of leveling errors in leveled heights and gravity data before adjustment;
- Extension of the leveling network to international borders and the Amazonas region, if possible. This is fundamental for determining offsets at border connections (Fig. 3);
- Adjustment on the vertical networks in terms of geopotential numbers;

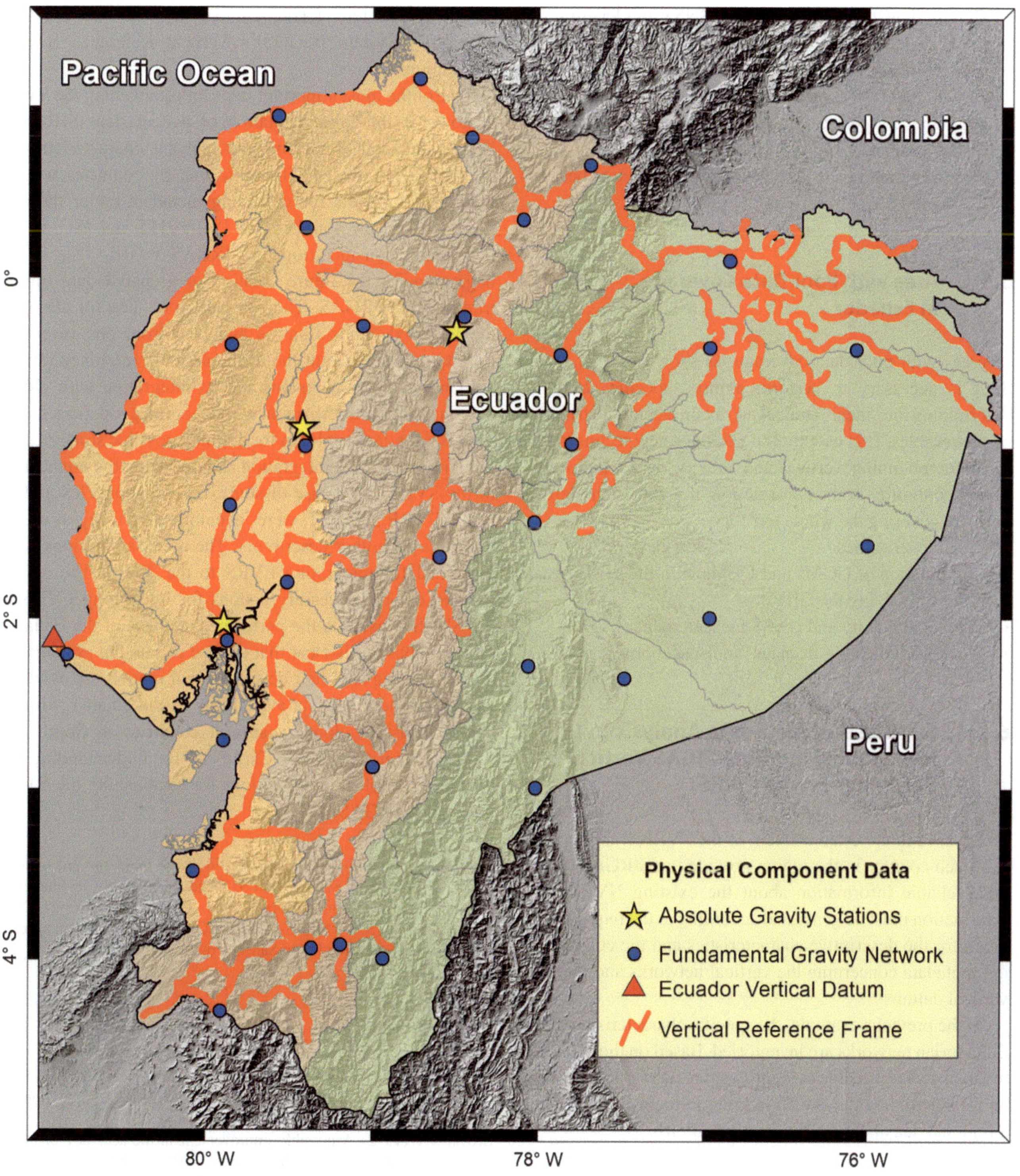


**Fig. 3** Physical component data

## 5.4 Ecuador Gravity Frame (EGF)

The network in use was established using an IGSN71 station located in the city of Quito as reference point. Recently, three absolute gravity stations were established, and they serve as reference for further 39 relative gravity stations. Other gravimetric surveys realized by different institutions and private companies also exist by the whole country.

Strategies for modernizing the EGF:

- Adjustment of EGF with basis on the absolute gravity stations (Fig. 3);

- Gravity densification on existing GNSS and leveling stations for generating geopotential numbers;
- Gravity interpolation along the leveling lines;
- Gravity and GNSS measurements in regions without leveling lines for generating gravity disturbances and geopotential numbers;
- Determination of the transformation parameters between EVRF and SVRF.

## 5.5 Area with Poor Data Coverage in Ecuador

In areas with poor data coverage, as shown in Figs. 2 and 3, it is necessary to perform a more specific study for determining the "ideal" strategy to obtain a better distribution of observables. It is essential to consider new technological tools for obtaining vertical and gravity data which allow giving continuity in the construction of a consistent vertical network covering the whole territory.

Possible strategies:

- Combination of GGMs and DEMs with the existing data bases for solving the GBVP;
- New local gravity and GNSS measurements;
- New data collecting from air- and spaceborne sensors.

# 6 Synthesis of Metadata Analysis for Establishing the SIRGAS Vertical Reference Frame (SVRF)

To advance in the establishment of the SVRF, it is necessary that each country in the region provide the SIRGAS-WGIII with reliable information about the existing NVRS. This information must be generated according to the requirements given for the descriptive characteristics and the corresponding metadata concerning the vertical networks and the local vertical datums.

In the previous example, the specific characteristics of the Ecuadorian networks are inventoried. Based on those characteristics, it was possible to identify strategies for modernizing and standardizing the NVRF in order to realize the SVRF as a regional densification of the global GVRF.

# 7 Summary and Outlook

The aim of the present work was to propose some steps those are considered fundamental for integrating the National Vertical Reference Systems (NVRSs) into the SIRGAS Vertical Reference System (SVRS). A basic conclusion is that SVRS and its realization can be obtained from the integration of each NVRS into a International Height Reference System (IHRS) under the Global Geodetic Observing System (GGOS)'s Vertical Reference System conventions.

One of the most important steps for establishing the SVRS is however the implementation of a consistent SIRGAS-WGIII metadata describing the relevant characteristics of each NVRS provided. The allocation by countries will be facilitated by tools for collecting metadata under the dispositions and the standards of ISO 19115 and ISO 19139 now under consideration by SIRGAS-WGIII. This is the beginning for the future creation of a spatial data infrastructure for SIRGAS. The completion of data for allowing the intended connections is analyzed from one inventory by taking into account the available local reference frames after necessary conversions and its integration with global data including those from new spaceborne geodetic sensors. This is a fundamental step for defining possible strategies for realizing a regional integration and obtaining the connection to the SVRS. As a case of study, the Ecuador VRS and its EVD, EVRF and EGF were inventoried by using available metadata and other descriptors. The necessary actions were discussed for integrating EVRS to the SVRS. The results pointed out the necessary actions for integrating EVRS to the SVRS. Based on the performed analysis of the Ecuador metadata base, it is clear that the use of satellite systems for data collecting and their derived global models are essential for facing the poor data availability in the country. Similar studies must be done in several countries in South and Central America and the Caribbean for understanding the essential complementary actions for establishing the SVRS under the SIRGAS.

**Acknowledgements** The authors would like to thank the 3rd International Gravity Field Service (IGFS) General Assembly in Shanghai, China for the invitation and opportunity. Also many thanks to CNPq trough Grant N. 301797/2008-0, SENESCYT and SIRGAS. In special, thank you very much to Laura Sánchez by the suggestions.

# References

Amos MJ, Featherstone WE (2009) Unification of New Zealand's local vertical datums: iterative gravimetric quasigeoid computations. J Geod 83(1):57–68. doi:10.1007/s00190-008-0232-y

Brovar VV (1972) A possible improvement in the accuracy of gravimetric results in geodesy. Sov Astron 15:1055–1058

Brunini C, Sánchez L, Drewes H, Costa SMA, Mackern V, Martinez W, Seemüller W, Da Silva AL (2012) Improved analysis strategy and accessibility of the SIRGAS reference frame. In: Kenyon S, Pacino MC, Marti U (eds) Geodesy for planet earth, vol 136, IAG Symposia. Springer, Berlin, pp 3–10. doi:10.1007/978-3-642-20338-1_1

De Freitas SRC, Blitzkow D (1999) Altitudes e Geopotencial (In Portuguese). IGeS Bulletin N.9 – International Geoid Service, Junho, Milão, pp 47–62

Drewes H, Sánchez L, Blitzkow D, de Freitas S (2002) Scientific foundations of the SIRGAS vertical reference system. In: Drewes H et al (eds) IAG Symposia, vol 124. Springer, Berlin, pp 297–301

Ferreira VG, de Freitas SRC (2011) Geopotential numbers from GPS satellite surveying and disturbing potential model: a case study of Parana, Brazil. J Appl Geod 5:155–162. doi:10.1515/JAG.2011.016

Forsberg R (1984) A study of terrain reductions, density anomalies and geophysical inversion methods in gravity field modelling, Ohio State University, Report No. 355

Forsberg R, Tscherning CC (1981) The use of height data in gravity field approximation by collocation. J Geophys Res 86:7843–7854

Heck B (1989) A contribution to the scalar free boundary value problem of physical geodesy. Manuscr Geod 14:87–99

Heck B (2011) A Brovar-type solution of the fixed geodetic boundary-value problem. Stud Geophys Geod 55:441–454

Heck B, Rummel R (1990) Strategies for solving the vertical datum problem using terrestrial and satellite geodetic data. IAG Symp Ser 104:116–128. doi:10.1007/978-1-4684-7098-7_14

Heck B, Seitz K (1993) Effects of non-linearity in the geodetic boundary problems, DGK, Reihe A, Heft Nr. 109, München, 74p

Hirt C, Featherstone WE, Marti U (2010) Combining EGM2008 and SRTM/DTM2006.0 residual terrain model data to improve quasigeoid computations in mountainous areas devoid of gravity data. J Geod 84(9):557–567

Hofmann-Wellenhof B, Moritz H (2005) Physical geodesy. Springer, Wien

Hugentobler U, Gruber T, Steigenberger P, Angermann D, Bouman J, Gerstl M, Richter B (2012) GGOS bureau for standards and conventions: integrated standards and conventions for geodesy. In: Kenyon SC, Pacino MC, Marti UJ (eds) Geodesy for planet earth, vol 136, IAG symposia. Springer, Berlin, pp 995–998. doi:10.1007/978-3-642-20338-1_124

Ihde J (2007) Inter-commission project 1.2: vertical reference frames. Final report for the period 2003–2007. In: IAG commission 1 – reference frames, Report 2003–2007. DGFI, Munich. Bulletin No. 20, pp 57–59

Ihde J, Sanchez L (2005) A unified global height reference system as a basis for IGGOS. J Geodyn 40(4–5):400–413. doi:10.1016/j.jog.2005.06.015

Ihde J, Amos M, Heck B, Kersley B, Schöne T, Sánchez L, Drewes H (2007) Conventions for the definitions and realization of a conventional vertical reference system (CVRS). http://www.iho.int/mtg_docs/com_wg/IHOTC/IHOTC8/IHOTC8-3-6-1.pdf. Accessed June 2014

Kutterer H, Neilan R, Bianco G (2012) Global geodetic observing system (GGOS). In: Drewes H, Hornik H, Ádám J, Rózsa S (eds) The geodesist's handbook 2012. Springer, Berlin; J Geod 86(10): 915–926. doi:10.1007/s00190-012-0584-1

Moritz H (2000) Geodetic reference system 1980. J Geod 74:128–133. doi:10.1007/s001900050278

Plag H-P, Altamimi Z, Bettadpur S, Beutler G, Beyerle G, Cazenave A, Crossley D, Donnellan A, Forsberg R, Gross R, Hinderer J, Komjathy A, Mannucci C, Ma AJ, Noll C, Nothnagel A, Pavlis EC, Pearlman M, Poli P, Schreiber U, Senior K, Woodworth PL, Zerbini S, Zuffada C (2009) The goals, achievements, and tools of modern geodesy. In: Plag H-P, Pearlman M (eds) Global geodetic observing system. Meeting the requirements of a global society on a changing planet in 2020. Springer, Berlin, p 18

Rummel R, Teunissen P (1998) Height datum definition, height datum connection and the role of the geodetic boundary value problem. Bull Géod 62:477–498. doi:10.1007/BF02520239

Sacerdote F, Sansò F (1986) The scalar boundary value problem of physical geodesy. Manuscr Geod 11:15–28

Sánchez L (2007) Definition and realization of the SIRGAS vertical reference system within a globally unified height system. In: Tregoning P, Rizos C (eds) Dynamic planet, vol 130, IAG Symposia. Springer, Berlin, pp 638–645

Sánchez L (2009) Strategy to establish a global vertical reference system. In: Drewes H (ed) Geodetic reference frames, vol 134, IAG Symposia. Springer, Berlin, pp 273–278. doi:10.1007/978-3-642-00860-3_42

Sánchez L (2012) Towards a vertical datum standardisation under the umbrella of global geodetic observing system. J Geod Sci. doi:10.2478/v10156-012-0002-x

Sánchez L, Brunini C (2009) Achievements and challenges of SIRGAS. In: Drewes H (ed) Geodetic reference frames, vol 134, IAG Symposia. Springer, Berlin, pp 161–166. doi:10.1007/978-3-642-00860-3_25

Sansò F (1995) The long road from measurements to boundary value problems in physical geodesy. Manuscr Geod 20(5):326–344

Xu P, Rummel R (1991) A quality investigation of global vertical datum connection. Netherlands Geodetic Commission. Publ Geod 34

# An Ellipsoidal Analogue to Hotine's Kernel: Accuracy and Applicability

Otakar Nesvadba and Petr Holota

**Abstract**

In this paper a mathematical apparatus is discussed that involves effects of the flattening of the Earth in the determination of the gravity potential. It rests on the use of Green's function of the second kind (Neumann's function) constructed for Neumann's boundary value problem in the exterior of an oblate ellipsoid of revolution. The apparatus has a natural tie to the reproducing kernel of Hilbert's space of functions harmonic in the solution domain considered. For at least one of the points (arguments of the kernel) inside the solution domain an expression of the reproducing kernel is developed. However, for both the points (arguments) on the ellipsoidal boundary a practical use of the kernel represented by means of series of ellipsoidal harmonic is not possible. Therefore the application of an approximate closed formula as the integral kernel is discussed and tested. A quality enhancement, if compared with the use of the spherical apparatus, is demonstrated by means of closed loop simulations. The paper contributes to methods related to the geoid (quasigeoid) computations.

**Keywords**

Gravimetric boundary value problem • Neumann's function • Oblate spheroid • Reproducing kernel

## 1 Introduction

Methods applied in solving geodetic boundary value problems often rest on a mathematical apparatus that actually was developed for a spherical boundary. This has some traditions and seemingly also an advantage in a relatively easy construction of the respective integral kernels and in the possibility to express them in closed form. Nevertheless, considering the real shape of the Earth, the use of kernels constructed for a sphere is limited under higher accuracy requirements. The aim of this paper is to discuss further the results obtained in Holota (2011) and Holota and Nesvadba (2014b), which mainly concern the construction of the reproducing kernel for the solution domain given by the exterior of an oblate ellipsoid of revolution and the tie of the kernel to Neumann's function. In the first stage the kernel was expressed by means of a series of ellipsoidal harmonics. In Holota and Nesvadba (2014b) the series summation was treated and closed expressions were found that under some approximations represent the reproducing kernel and Neumann's function. The quality of the approximate kernel was subjected to numerical tests. The exact kernel represented by series of ellipsoidal harmonics was compared with the approximate kernel expressed in closed form, but only in the case that one of the points, i.e. arguments of the kernel, is inside the solution domain. When both the points (arguments) are on the boundary of the domain, the convergence

O. Nesvadba (✉)
Land Survey Office, Pod Sídlištěm 9, 182 00 Prague, Czech Republic
e-mail: nesvadba@sky.cz

P. Holota
Research Institute of Geodesy, Topography and Cartography, Zdiby 98, 250 66, Prague-East, Czech Republic
e-mail: petr.holota@pecny.cz



S. Jin, R. Barzaghi (eds.), *IGFS 2014*, International Association of Geodesy Symposia 144,
DOI 10.1007/1345_2015_133

of the series that represents the kernel is very slow and the numerical comparison with the approximate kernel is hardly possible. The quality check of the approximate kernel in this particular case is discussed in this paper.

Our considerations are related to the linear gravimetric boundary value problem (LGBVP), see Koch and Pope (1972), Bjerhammar and Svensson (1983) or Holota (1997),

$$\nabla^2 T(\mathbf{x}) = 0 \quad \text{for} \quad \mathbf{x} \in \Omega_E \,, \tag{1}$$

$$\langle \nabla T(\mathbf{x}), \nabla U(\mathbf{x}) \rangle = -\,|\nabla U(\mathbf{x})| \delta g(\mathbf{x}) \,, \text{for } \mathbf{x} \in \partial\Omega_E \,, \tag{2}$$

where $\mathbf{x}$ is a point with Cartesian coordinates $(x_1, x_2, x_3)$ in three-dimensional Euclidean space $\mathbb{R}^3$, $\langle \mathbf{x}, \mathbf{y} \rangle = \sum_{i=1}^{3} x_i y_i$ denotes the scalar product, $|\mathbf{x}| = \sqrt{\langle \mathbf{x}, \mathbf{x} \rangle}$, $\nabla$ is the nabla operator and $T$ represents the *disturbing potential* defined as a difference $T(\mathbf{x}) = W(\mathbf{x}) - U(\mathbf{x})$ between the true gravity potential $W$ and a normal gravity potential $U$. Finally, $\delta g$ is the *gravity disturbance*, $\delta g(\mathbf{x}) = |\nabla W(\mathbf{x})| - |\nabla U(\mathbf{x})|$.

We focus on a special case, when the solution domain $\Omega_E$ represents the exterior of an oblate ellipsoid of revolution. For its parameters $a, E$, $e = \frac{E}{a}$, $b = \sqrt{a^2 - E^2}$ we take the values given by GRS80, Moritz (1980b). Moreover, denoting by $\partial\Omega_E$ the surface of the ellipsoid, we suppose that $\nabla U(\mathbf{x}) = -\mathbf{n}_x |\nabla U(\mathbf{x})|$ for any $\mathbf{x} \in \partial\Omega_E$, where $\mathbf{n}_x$ represents the unit outer normal of $\partial\Omega_E$ at the point $\mathbf{x}$, which holds true in the case that $U$ is the Somigliana-Pizzetti normal potential related to the level ellipsoid $\partial\Omega_E$.

*Remark 1* An apparatus applicable for the analytical continuation of $\delta g$ up to the boundary $\partial\Omega_E$ can be constructed in a similar way as in the spherical case, see Moritz (1980a). However, an approach based on a straightforward use of an ellipsoidal analogue to Poisson's formula can not be applied. Details were discussed in Holota and Nesvadba (2014a).

## 2 Reproducing Kernel and Neumann's Function

Since we deal with Neumann's problem for Laplace's partial differential equation, we can express the solution $T$ with the aid of *Green's function of the second kind* $G(\mathbf{y}, \mathbf{x})$, i.e. Neumann's function as

$$T(\mathbf{y}) = \int_{\partial\Omega_E} G(\mathbf{y}, \mathbf{x}) \delta g(\mathbf{x}) \, \mathrm{d}_x S \quad \mathbf{y} \in \Omega_E \,, \tag{3}$$

see e.g. Kellogg (1953).

Now, in parallel, $\mathscr{H}^{1,2}(\Omega_E)$ be Hilbert's space of functions harmonic in $\Omega_E$ equipped with inner product

$$\langle u, v \rangle_{\Omega_E} = \int_{\Omega_E} \langle \nabla u, \nabla v \rangle \, \mathrm{d}\Omega \,, \quad u, v \in \mathscr{H}^{1,2}(\Omega_E) \,. \tag{4}$$

One can show, there exists a kernel $K_E = K_E(\mathbf{y}, \mathbf{x})$ which is an element of $\mathscr{H}^{1,2}(\Omega_E)$ for every $\mathbf{y} \in \Omega_E$ (or symmetrically for every $\mathbf{x} \in \Omega_E$) and has the *reproducing property*, i.e.

$$v(\mathbf{y}) = \langle K_E(\mathbf{y}, \cdot), v \rangle_{\Omega_E} \,, \quad \forall v \in \mathscr{H}^{1,2}(\Omega_E) \,, \tag{5}$$

see Meschkowski (1962) or Holota (2004). Assuming that $v \in \mathscr{H}^{1,2}(\Omega_E)$ is sufficiently smooth and rewriting the right hand side of Eq. (5) with the aid of Green's first identity, we get

$$v(\mathbf{y}) = \langle K_E(\mathbf{y}, \cdot), v \rangle_{\Omega_E} = \int_{\partial\Omega_E} K_E(\mathbf{y}, \mathbf{x}) \, \langle \nabla v(\mathbf{x}), \mathbf{n}_x \rangle \, \mathrm{d}_x S \,. \tag{6}$$

Obviously, Eq. (6) holds also for $v = T$. Therefore, using our assumption $\nabla U = -\mathbf{n}|\nabla U|$ on $\partial\Omega_E$ and recalling Eqs. (2) and (3) we get

$$G(\mathbf{y}, \mathbf{x}) = K_E(\mathbf{y}, \mathbf{x}) \,, \quad \forall \mathbf{x} \in \partial\Omega_E \,. \tag{7}$$

Hence

$$T(\mathbf{y}) = \int_{\partial\Omega_E} K_E(\mathbf{y}, \mathbf{x}) \delta g(\mathbf{x}) \, \mathrm{d}_x S \,. \tag{8}$$

## 3 Reproducing Kernel for Ellipsoidal Domain

The reproducing kernel $K_E$ has been constructed in Holota (2004, 2011) and further developed in Holota and Nesvadba (2014b). Using these references we can write

$$K_E(\mathbf{y}, \mathbf{x}) = \frac{1}{4\pi b} \sum_{n=0}^{\infty} \sum_{m=0}^{n} (2 - \delta_{0,m})(2n+1) \frac{(n-m)!}{(n+m)!} K_{nmxy} \times \\ \times P_{nm}(\sin\beta_y) P_{nm}(\sin\beta_x) \cos m(\lambda_x - \lambda_y) \,, \tag{9}$$

where $\delta_{0,0} = 1$, $\delta_{0,m} = 0$ for $m \neq 0$,

$$K_{nmxy} = \frac{iEb}{a^2} \frac{Q_{nm}(i\frac{u_x}{E})}{Q_{nm}(i\frac{b}{E})} \frac{Q_{nm}(i\frac{u_y}{E})}{Q_{nm}(i\frac{b}{E})} \frac{Q_{nm}(i\frac{b}{E})}{Q'_{nm}(i\frac{b}{E})} \,, \tag{10}$$

$i = \sqrt{-1}$ and $P_{nm}$, $Q_{nm}$ are Legendre functions of the first and the second kind, respectively. The position of the points

$\mathbf{x}, \mathbf{y}$ is expressed in oblate spheroidal coordinates $(u, \beta, \lambda)$ which are related to $(x_1, x_2, x_3)$ by the equations

$$x_1 = \sqrt{E^2 + u_x^2} \cos\beta_x \cos\lambda_x ,$$
$$x_2 = \sqrt{E^2 + u_x^2} \cos\beta_x \sin\lambda_x ,$$
$$x_3 = u_x \sin\beta_x .$$

### 3.1 Numerical Evaluation of $K_E$

Here we refer to the notation used in Nesvadba and Holota (2015). This also means we apply the substitution

$$\rho = \frac{a}{\sqrt{E^2 + u^2}}, \qquad \tau = \sin\beta \tag{11}$$

and in the sequel we work with the modified spheroidal coordinates $(\rho, \tau, \lambda)$.

Employing the fully normalised Legendre functions $\overline{P}_{nm}$, $\overline{P}_{nm}(\tau) = \sqrt{(2 - \delta_{0,m})(2n+1)\frac{(n-m)!}{(n+m)!}}\, P_{nm}(\tau)$ and putting $\lambda_{xy} = \lambda_x - \lambda_y$ we can rewrite Eq. (9) as

$$K_E(\mathbf{y}, \mathbf{x}) = \frac{1}{4\pi b} \sum_{n=0}^{\infty} \sum_{m=0}^{n} K_{nmxy} \overline{P}_{nm}(\tau_y) \times \\ \times \overline{P}_{nm}(\tau_x) \cos m\lambda_{xy} . \tag{12}$$

Moreover, we replace $\overline{P}_{nm}(\tau)$ by

$$p_{n,m}(\tau, \rho) = \rho^{n+1} \overline{P}_{nm}(\tau) \tag{13}$$

and instead of $Q_{nm}(z)$ with $z = i\frac{u}{E} = i\sqrt{\rho^{-2}e^{-2} - 1}$ we use

$$q_{n,m}(\rho) = \rho^{-n-1} \frac{Q_{nm}(z)}{Q_{nm}(i\frac{b}{E})} . \tag{14}$$

In consequence it results from Eq. (14) that

$$\left.\frac{Q'_{nm}(z)}{Q_{nm}(i\frac{b}{E})}\right|_{z=i\frac{b}{E}} = \left.\frac{\mathrm{d}q_{n,m}(\rho)\rho^{n+1}}{\mathrm{d}\rho} \frac{\mathrm{d}\rho}{\mathrm{d}z}\right|_{\rho=1, z=i\frac{b}{E}} = \\ = \frac{iEb}{a^2}\left(q'_{n,m}(1) + n + 1\right) , \tag{15}$$

where $q'_{n,m}(1) = h_{n+1,m}$ with

$$h_{n,m} = i\frac{E}{b}(n - m)\frac{Q_{nm}(i\frac{b}{E})}{Q_{n-1,m}(i\frac{b}{E})} \tag{16}$$

according to Nesvadba and Holota (2015).

Note that for $\rho_x\rho_y < 1$ (i.e. at least one of the points $\mathbf{x}, \mathbf{y}$ lies inside $\Omega_E$) and any given numerical accuracy we can find an upper bound $N$ for the summation indices $n, m$ such that the infinite series in Eq. (12) may be substituted by a finite sum

$$K_E(\mathbf{y}, \mathbf{x}) = \frac{1}{4\pi b} \sum_{m=0}^{N} \cos m\lambda_{xy} \sum_{n=m}^{N} q_{n,m}(\rho_x) p_{n,m}(\tau_x, \rho_x) \times \\ \times q_{n,m}(\rho_y) p_{n,m}(\tau_y, \rho_y) \frac{1}{n + 1 + h_{n+1,m}} , \tag{17}$$

where in addition we changed the order of the summation indices and inserted from Eqs. (13)–(16). This enables an efficient numerical treatment of $K_E$ by means of the Clenshaw summation technique, as discussed in Nesvadba and Holota (2015).

## 4 Approximate Kernel and Its Closed Formula

As shown in Holota (2011) or Holota and Nesvadba (2014b), in a close neighbourhood of $\partial\Omega_E$ we can put approximately

$$K_{nmxy} \approx \rho_{xy}^{n+1} \frac{1}{n+1} \left(1 - \frac{E^2}{ab} \frac{(n+1)^2 - m^2}{(2n+3)(n+1)}\right) , \tag{18}$$

where $\rho_{xy} = \rho_x\rho_y = a^2(E^2 + u_x^2)^{-\frac{1}{2}}(E^2 + u_y^2)^{-\frac{1}{2}}$.

In consequence, after some algebra as in Holota and Nesvadba (2014b), one can find that $K_E$ approximately equals to $\tilde{K}_E$,

$$\tilde{K}_E(\mathbf{y}, \mathbf{x}) = \frac{1}{4\pi b}\left(K^{sph} - \frac{E^2}{ab} K^{ell}\right) , \tag{19}$$

where

$$K^{sph} = \frac{2\rho_{xy}}{L} - \ln\frac{L + \rho_{xy} - \cos\psi}{1 - \cos\psi} , \tag{20}$$

$$\cos\psi = \tau_x\tau_y + \sqrt{(1 - \tau_x^2)(1 - \tau_y^2)} \cos\lambda_{xy} , \tag{21}$$

$$L = \sqrt{1 - 2\rho_{xy}\cos\psi + \rho_{xy}^2} \tag{22}$$

and

$$K^{ell} = \left[a_{xy} + b_{xy}\cos\psi\right] K^{sph} - \\ - c_{xy}\left(\frac{\rho_{xy}}{L} - 2\sqrt{1 - \frac{4\rho_{xy}}{(1+\rho_{xy})^2}\cos^2\frac{\psi}{2}}\right) - \\ - c_{xy}\frac{2}{\sqrt{\rho_{xy}}} \mathrm{E}\left(2\arctan\sqrt{\rho_{xy}} \middle| \cos^2\frac{\psi}{2}\right) + \\ + \left[c_{xy} - 2b_{xy}\right]\frac{1}{\sqrt{\rho_{xy}}} \mathrm{F}\left(2\arctan\sqrt{\rho_{xy}} \middle| \cos^2\frac{\psi}{2}\right) + \\ + b_{xy}\left[4 + \ln 2 - \ln(L + 1 - \rho_{xy}\cos\psi)\right] \tag{23}$$

with

$$a_{xy}\sin^2\psi = (1-\tau_x^2)(1-\tau_y^2)\sin^2\lambda_{xy}\,, \tag{24}$$

$$b_{xy}\sin^2\psi = \tau_x\tau_y + (2a_{xy}-1)\cos\psi\,, \tag{25}$$

$$c_{xy} = 3a_{xy} + 2b_{xy}\cos\psi - 1\,, \tag{26}$$

while $\mathrm{F}(\cdot|\cdot)$ and $\mathrm{E}(\cdot|\cdot)$ are Legendre's incomplete elliptic integrals of the first and second kind, respectively, see Abramowitz and Stegun (1964). Nevertheless, in our implementation of $\tilde{K}_E$ we used the Carlson form of elliptic integrals, Carlson (1979), since it allows very efficient evaluation of $K^{ell}$ based on the duplication theorem.

Formally the term $K^{sph}$ has the same structure as the reproducing kernel for a spherical domain, while $K^{ell}$ is induced by effects of spheroidal flattening and makes the kernel anisotropic.

Let us mention that a direct 'naïve' floating-point implementation of $\tilde{K}_E$ could be unstable for $\sin\psi \to 0$, cf. Eq. (20) or Eqs. (24) and (25). Our solution of this technical problem rests on the application of the special elementary functions $\exp(x)-1$ and $\ln(x+1)$, see Nesvadba (2010).

## 5 Restriction to $\partial\Omega_E$

The formula given by Eq. (8) can be used, for instance in combination with the method of analytical continuation, as a natural starting point for computing $T$ on $\partial\Omega_E$ in the geoid (quasigeoid) determination. However, the absolute convergence of the series in Eq. (9) is not guaranteed for $\mathbf{x},\mathbf{y} \in \partial\Omega_E$, thus the evaluation of $K_E$ based on Eq. (17) is not applicable. A possible treatment of this problem offers the use of $\tilde{K}_E$ restricted to $\mathbf{x}$ and $\mathbf{y}$ on the boundary $\partial\Omega_E$.

For $\mathbf{x},\mathbf{y} \in \partial\Omega_E$ we have $\rho_{xy} = 1$, $L = \sqrt{2-2\cos\psi} = 2\sin\frac{\psi}{2}$ and can verify that for the $K^{sph}$ and $K^{ell}$ terms we have

$$K^{sph} = \frac{1}{\sin\frac{\psi}{2}} - \ln\left(\frac{1}{\sin\frac{\psi}{2}}+1\right) \tag{27}$$

and

$$\begin{aligned} K^{ell} &= \left[a_{xy} + b_{xy}\cos\psi\right]K^{sph} - \\ &- c_{xy}\left(\frac{1}{2\sin\frac{\psi}{2}} - 2\sin\frac{\psi}{2} + 2\mathrm{E}\left(\cos^2\frac{\psi}{2}\right)\right) + \\ &+ \left[c_{xy} - 2b_{xy}\right]\mathrm{K}\left(\cos^2\frac{\psi}{2}\right) + \\ &+ b_{xy}\left[4 - \ln\left(\sin\frac{\psi}{2} + \sin^2\frac{\psi}{2}\right)\right], \end{aligned} \tag{28}$$

where $\mathrm{K}(\cdot) = \mathrm{F}\left(\frac{\pi}{2}|\cdot\right)$ and $\mathrm{E}(\cdot) = \mathrm{E}\left(\frac{\pi}{2}|\cdot\right)$ now represent complete elliptic integrals of the first and second kind, respectively. For details see the explanation in Holota and Nesvadba (2014b). Note also that $K^{sph}$ term given by Eq. (27) is often called the *Hotine-Koch function* in physical geodesy, cf. Hofmann-Wellenhof and Moritz (2005).

## 6 Experiments

A direct numerical comparison of $K_E$ and $\tilde{K}_E$ was given in Holota and Nesvadba (2014b), but only in case when $u_y > b$, i.e. for $\mathbf{y} \in \Omega_E$. Since we focus on a more sensitive situation when $\mathbf{x},\mathbf{y} \in \partial\Omega_E$, we applied another idea to check the applicability of the approximate kernel $\tilde{K}_E$ in computations based on Eq. (8). In what follows, we performed a closed loop test of the kernel $\tilde{K}_E$ restricted to $\partial\Omega_E$.

### 6.1 Simulated Input Data $\delta g$

In our experiments $W$ means the potential function from EGM2008, see Pavlis et al. (2012, 2013), represented by spherical harmonics expansion up to degree $N = 2190$,[1] i.e.

$$\begin{aligned} W(\tilde{\rho},\tilde{\tau},\lambda) &= \sum_{n=0}^{N}\sum_{m=0}^{n}\left(\overline{C}_{n,m}\cos m\lambda + \overline{S}_{n,m}\sin m\lambda\right) \\ &\times p_{n,m}(\tilde{\tau},\tilde{\rho}) + \frac{1}{2}\omega^2\frac{a^2}{\tilde{\rho}^2}(1-\tilde{\tau}^2)\,. \end{aligned} \tag{29}$$

Note that in this formula $\tilde{\rho} = \frac{a}{r}$ and $\tilde{\tau} = \sin\phi$ substitute the spherical radius $r$ and spherical latitude $\phi$; angular velocity $\omega$ is the same as in GRS80.

For the normal gravity potential $U$ the GRS80 parameters are used, cf. Moritz (1980b) and $|\nabla U|$ on $\partial\Omega_E$ is given by the Somigliana-Pizzetti formula

$$|\nabla U(\tau)| = \frac{a|\nabla U(1)|\tau^2 + b|\nabla U(0)|(1-\tau^2)}{a\sqrt{1-e^2(1-\tau^2)}}\,. \tag{30}$$

Thus the gravity disturbance $\delta g$ can be computed from

$$\begin{aligned} \delta g(\tau,\lambda) &= \left|\nabla W\left(\frac{1}{\sqrt{1-e^2\tau^2}}, \frac{b}{a\sqrt{1-e^2\tau^2}}\tau, \lambda\right)\right| - \\ &- |\nabla U(\tau)| \end{aligned} \tag{31}$$

at any point $(1,\tau,\lambda)$ of $\partial\Omega_E$. The global map of the input $\delta g$ on $\partial\Omega_E$ in mGal units[2] is shown in Fig. 1.

[1] Note that the original EGM2008 coefficients $\overline{C}_{n,m}$, $\overline{S}_{n,m}$ need to be rescaled by $\frac{GM}{\tilde{a}}\left(\frac{\tilde{a}}{a}\right)^n$, $GM = 3.986004415 \cdot 10^{14}\,\mathrm{m^3\,s^{-2}}$, $\tilde{a} = 6378136.3$ m, to comply with the notation used in Eq. (29).

[2] $1\,\mathrm{mGal} = 10^{-5}\,\mathrm{m\,s^{-2}}$.

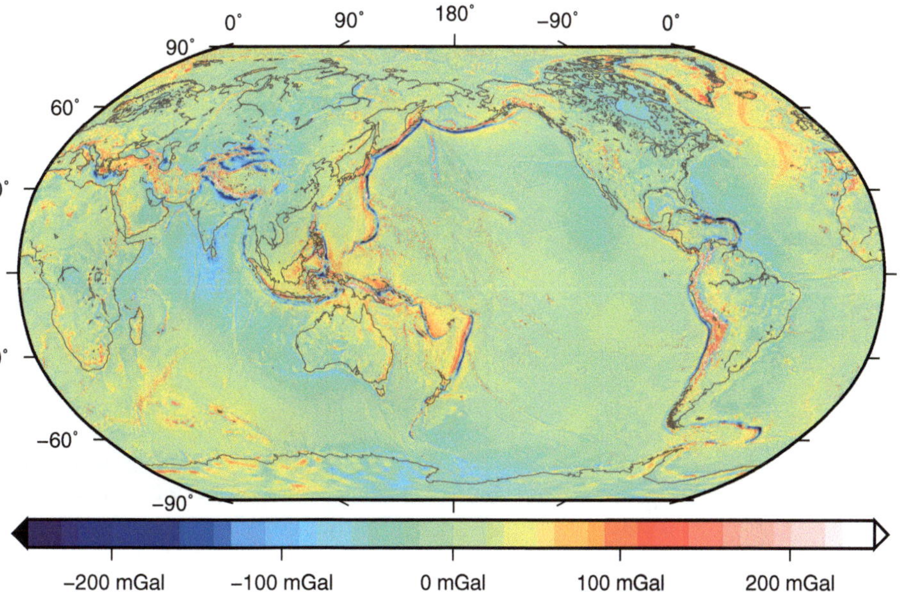


**Fig. 1** Input gravity disturbance $\delta g$ on $\partial\Omega_E$ (GRS80): $\min \delta g = -405$ mGal, $\max \delta g = 977$ mGal, $\overline{\delta g} = -1$ mGal, rms $\delta g = 44$ mGal

## 6.2 Numerical Integration

Here we return to Eq. (8) and replace the integral kernel $K_E$ by its approximate form $\tilde{K}_E$,

$$T(\mathbf{y}) = \int_{\partial\Omega_E} \tilde{K}_E(\mathbf{y},\mathbf{x})\,\delta g(\mathbf{x})\,\mathrm{d}_x S = \\ = \int_{-1}^{1}\int_{-\pi}^{\pi} \tilde{K}_E(\mathbf{y},(1,\tau,\lambda))\delta g(\tau,\lambda)\,\mathrm{d}\lambda\; J(\tau)\,\mathrm{d}\tau\,, \tag{32}$$

where $J(\tau) = a^2\sqrt{1-e^2(1-\tau^2)}$.

Rotational symmetry of $\tilde{K}_E$ in $\lambda_{xy}$ allows to interpret the inner integral in Eq. (32) as a *circular convolution* of $\tilde{K}_E$ with the data $\delta g$ along the parallel $\tau_x = const.$ The convolution method of rotationally symmetric integrals in a similar application has been treated e.g. in Haagmans et al. (1993). For the variable $\lambda_y \in (-\pi,\pi)$ we can successively write

$$\int_{-\pi}^{\pi} \tilde{K}_E\left((\rho_y,\tau_y,0),(1,\tau_x,\lambda_{xy})\right)\delta g(\tau_x,\lambda_x)\,\mathrm{d}\lambda_x = \\ = \tilde{K}_E\left((\rho_y,\tau_y,0),(1,\tau_x,\cdot)\right) * \delta g(\tau_x,\cdot) = \\ = \mathscr{F}^{-1}\left\{\mathscr{F}\left\{\tilde{K}_E\left((\rho_y,\tau_y,0),(1,\tau_x,\cdot)\right)\right\}\mathscr{F}\left\{\delta g(\tau_x,\cdot)\right\}\right\}, \tag{33}$$

where $*$ means the convolution and $\mathscr{F}\{f(\cdot)\}$ denotes the periodic Fourier transformation of a function $f$.

To realise the integration in Eq. (32) by numerical quadrature, $\tilde{K}_E$ and $\delta g$ were discretised in $\lambda,\beta$ (or $\tau$). If we suppose a regular sampling in $\lambda_{xy}$, the discrete Fourier transformation based on FFT algorithm, makes the numerical evaluation of Eq. (33) very efficient, speeding up the quadrature of Eq. (32) significantly.

*Richardson Extrapolation* In order to increase the accuracy of the numerical quadrature in Eq. (32) one can densify the discretisation or increase the order of the quadrature method. Nonetheless, as the kernel $\tilde{K}_E$ is weakly singular for $\mathbf{x}\to\mathbf{y}$, we followed a low-order quadrature method (e.g. Simpson's rule) but applied another approach to increase the accuracy. We constructed a sequence $h_i = \frac{h_0}{d^i}$, $i \in \{0,\ldots,k\}$, $d > 1$ and subsequently also a low-order numerical estimates $T(\mathbf{y};h_i)$ of Eq. (32), where $h_i$ denotes a particular discretisation step, i.e. $\mathrm{d}\beta_i \times \mathrm{d}\lambda_i$. Finally, the sequence $T(\mathbf{y};h_i)$ has been accelerated to an estimate of its limit $\lim_{h_i\to 0} T(\mathbf{y};h_i)$ by means of Richardson's extrapolation, see Richardson and Gaunt (1927).

The extrapolation method saves us from extensive evaluation of $\tilde{K}_E$ and $\delta g$ in dense grids and allows to overcome difficulties connected with the kernel singularity, supposing $|\lambda_{xy}| \geq \frac{1}{2}\mathrm{d}\lambda$. In the following calculations we put $d = 4$, $h_0 = \mathrm{d}\beta_0 \times \mathrm{d}\lambda_0 = 8' \times 10.55'$ and $k = 2$.

## 6.3 First Experiment

In order to verify the applicability of the approximate kernel $\tilde{K}_E$ for $\mathbf{x},\mathbf{y} \in \partial\Omega_E$ we conducted an experiment, where the simulated data $\delta g$ as in Sect. 6.1 are used for the computation of $T$ according to formula (32). The resulting $T$ on $\partial\Omega_E$ is shown in Fig. 2.

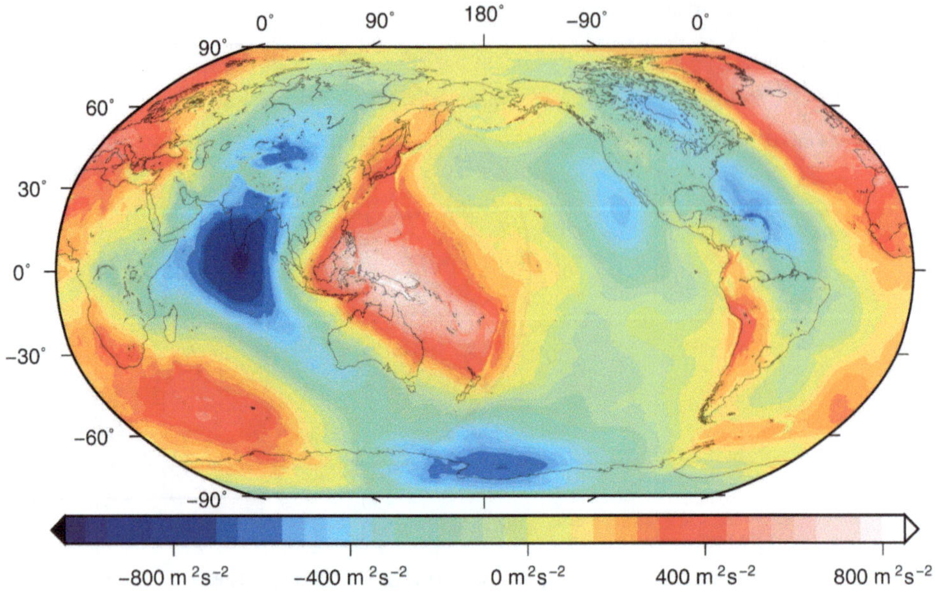


**Fig. 2** $T$ resulting from Eq. (32) on $\partial\Omega_E$: $\min T = -1{,}051\,\mathrm{m^2\,s^{-2}}$, $\max T = 844\,\mathrm{m^2\,s^{-2}}$, $\overline{T} = -17\,\mathrm{m^2\,s^{-2}}$, $\mathrm{rms}\,T = 287\,\mathrm{m^2\,s^{-2}}$

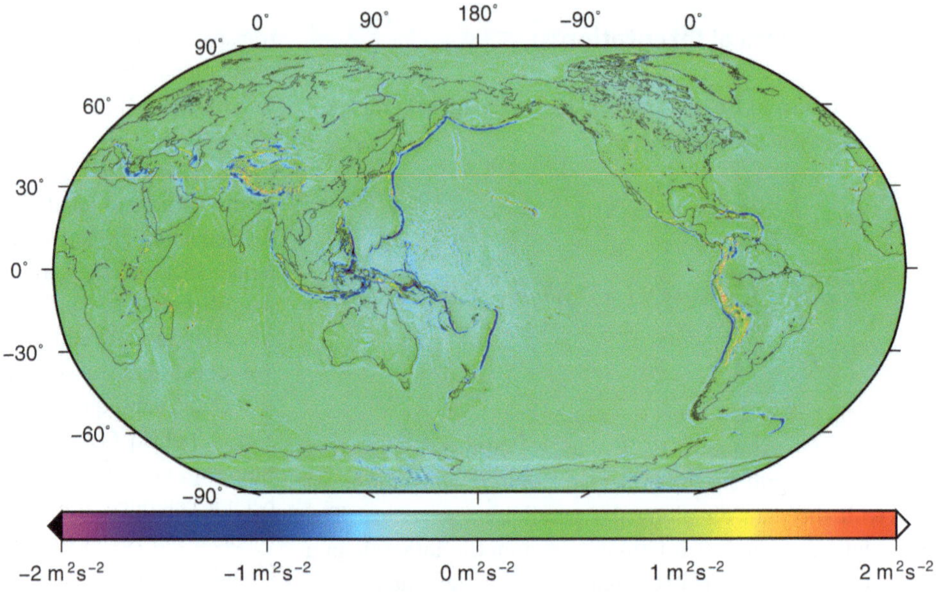


**Fig. 3** Residuals $\delta T$ resulting from Eq. (34): $\min \delta T = -3.1\,\mathrm{m^2\,s^{-2}}$, $\max \delta T = 5.9\,\mathrm{m^2\,s^{-2}}$, $\overline{\delta T} = 0.00\,\mathrm{m^2\,s^{-2}}$, $\mathrm{rms}\,\delta T = 0.26\,\mathrm{m^2\,s^{-2}}$

Following the principle of the closed loop test, we compared $T$ computed above [from Eq. (32)] with $T = W - U$ deduced directly from our source models, i.e. we formed the difference

$$\delta T(\mathbf{y}) = W(\mathbf{y}) - U_0 - \int_{\partial\Omega_E} \tilde{K}_E(\mathbf{y},\mathbf{x})\,\delta g(\mathbf{x})\,\mathrm{d}_x S\,, \quad (34)$$

where $U(\mathbf{y}) = U_0 = const.$ for all $\mathbf{y} \in \partial\Omega_E$. The residuals $\delta T$ on $\partial\Omega_E$ are shown in Fig. 3. They are fairly small and their mean equals zero. The cause of nonzero $\delta T$ can be seen in:

(i) the use of the approximate kernel $\tilde{K}_E$ in Eq. (32) instead of $K_E$,

(ii) errors in numerical quadrature of Eq. (32) (amplified especially in locations of strong or rapidly changing $\delta g$, e.g. rifts and tectonic plate boundaries),

(iii) the linearisation error associated with the LGBVP formulation [Eq. (2)].

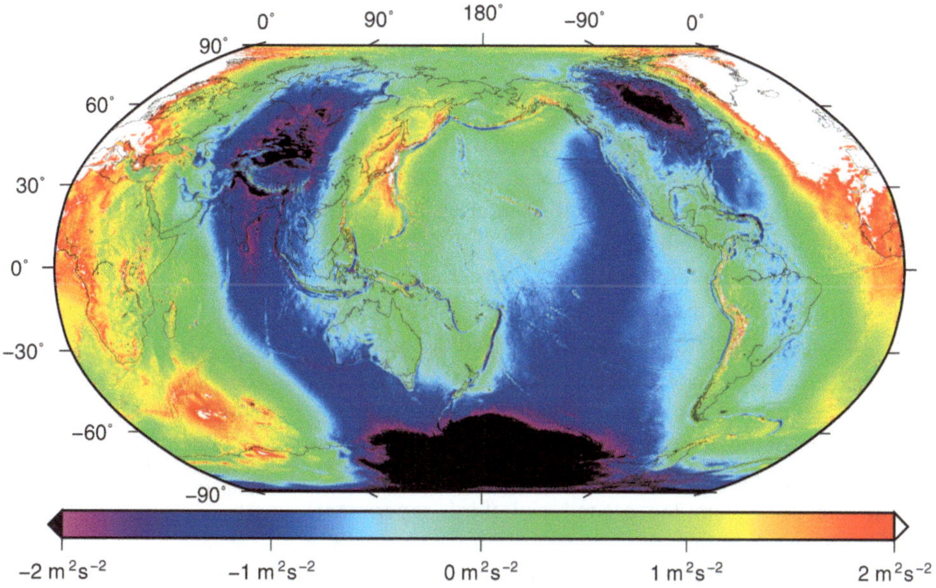


**Fig. 4** Residuals $\delta T$ resulting from Eq. (35): $\min \delta T = -5.0\,\mathrm{m}^2\,\mathrm{s}^{-2}$, $\max \delta T = 8.2\,\mathrm{m}^2\,\mathrm{s}^{-2}$, $\overline{\delta T} = -0.09\,\mathrm{m}^2\,\mathrm{s}^{-2}$, $\mathrm{rms}\,\delta T = 1.39\,\mathrm{m}^2\,\mathrm{s}^{-2}$

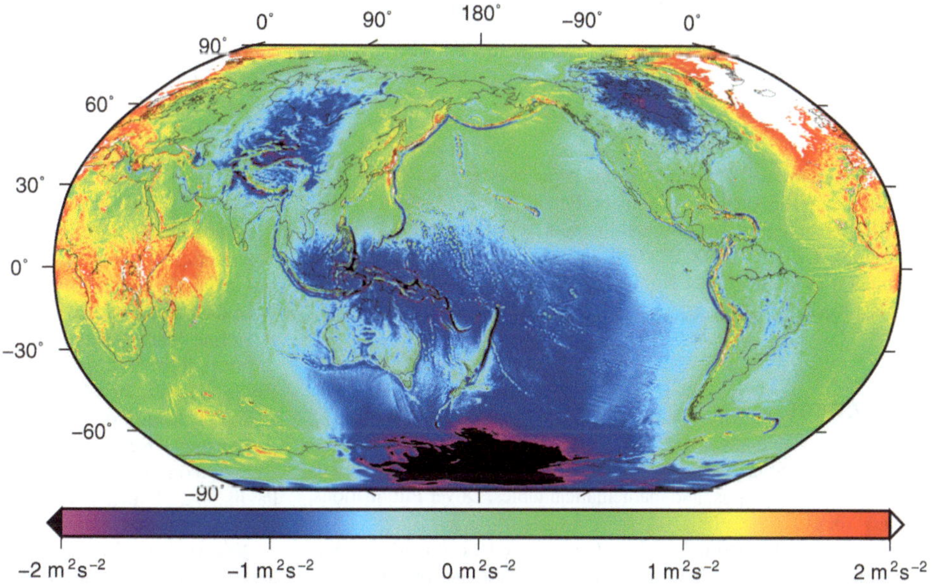


**Fig. 5** Residuals $\delta T$ resulting from Eq. (36): $\min \delta T = -3.9\,\mathrm{m}^2\,\mathrm{s}^{-2}$, $\max \delta T = 7.1\,\mathrm{m}^2\,\mathrm{s}^{-2}$, $\overline{\delta T} = -0.05\,\mathrm{m}^2\,\mathrm{s}^{-2}$, $\mathrm{rms}\,\delta T = 0.98\,\mathrm{m}^2\,\mathrm{s}^{-2}$

## 6.4 Second Experiment: $T$ Computed by Means of $K^{sph}$

To demonstrate better the importance of the ellipsoidal part of the kernel $\tilde{K}_E$, we arranged the second experiment, where just the spherical part $K^{sph}$, i.e. Hotine's function, is used in the integral formula (32). The resulting residuals

$$\delta T(\mathbf{y}) = W(\mathbf{y}) - U_0 - \frac{1}{4\pi b} \int_{\partial\Omega_E} K^{sph}(\mathbf{y},\mathbf{x})\, \delta g(\mathbf{x})\, \mathrm{d}_x S \quad (35)$$

on $\partial\Omega_E$ are shown in Fig. 4. In comparison with $\delta T$ obtained in the first experiment (Fig. 3) we clearly see a significant degradation of the result. As the other sources of the error remain the same, it leads us to the conclusion, that $K^{ell}$ component plays an important role in the approximation of $K_E$. The magnitude of the residuals in this experiment makes it evident that the classical spherical approximation is not sufficient for a precise quasigeoid determination.

## 6.5 Third Experiment: $T$ Computed by $K^{sph}$ Rescaled

This experiment is almost identical to the second experiment. Instead of $b$ we took $R = \frac{2a+b}{3} = 6{,}371{,}008.7714\,\mathrm{m}$. The resulting

$$\delta T(\mathbf{y}) = W(\mathbf{y}) - U_0 - \frac{1}{4\pi R} \int_{\partial\Omega_E} K^{sph}(\mathbf{y},\mathbf{x})\, \delta g(\mathbf{x})\, \mathrm{d}_x S \quad (36)$$

for $\mathbf{y} \in \partial\Omega_E$ is shown in Fig. 5. One can see that despite the scaling, $K^{sph}$ did not provide the accuracy of $T$ comparable with the first experiment in Sect. 6.3.

## 7 Conclusions

For the determination of $T$ the Green function method was applied in our ellipsoidal domain. We used the trace identity between Green's function of the second kind (Neumann's function) and the reproducing kernel $K_E$ constructed for the ellipsoidal solution domain.

An approximate representation of the kernel $K_E$ has been realised by means of the closed formula $\tilde{K}_E$. In the closed loop test with EGM2008 potential we verified the quality of $\tilde{K}_E$ for $\mathbf{x}, \mathbf{y} \in \partial\Omega_E$. The residuals from the first experiment, see Fig. 3, confirmed that $\tilde{K}_E$ is applicable to high accuracy computations of the disturbing potential or the related height anomaly in the quasigeoid determination.

The outcome from the second and third experiment, see Figs. 4 and 5, shows a significant improvement of $\tilde{K}_E$ over $K^{sph}$ and thus demonstrate the enhancement when $K^{ell}$ extends the basic spherical structure of the kernel.

The achieved results give us a driving impulse for further investigations concerning the summation of the series representing the integral kernels discussed.

**Acknowledgements** The work on this paper was partly supported by the European Regional Development Fund (ERDF), project "NTIS – New Technologies for Information Society", European Centre of Excellence, CZ.1.05/1.1.00/02.0090 and also by the Czech Science Foundation through Project No. 14-34595S. All this support is gratefully acknowledged.

## References

Abramowitz M, Stegun IA (1964) Handbook of mathematical functions with formulas, graphs, and mathematical tables. Dover Publications, New York

Bjerhammar A, Svensson L (1983) On the geodetic boundary value problem for a fixed boundary surface – a sattelite approach. Bull Géod 57(1–4):382–393

Carlson BC (1979) Computing elliptic integrals by duplication. Numer Math 33(1):1–16

Haagmans R, de Min E, van Gelderen M (1993) Fast evaluation of convolution integrals on the sphere using 1D FFT and a comparison with existing methods for Stoke's integral. Manuscr Geodaet 18(5):227–241

Hofmann-Wellenhof B, Moritz H (2005) Physical geodesy. Springer, New York

Holota P (1997) Coerciveness of the linear gravimetric boundary-value problem and a geometrical interpretation. J Geodesy 71(10):640–651

Holota P (2004) Some topics related to the solution of boundary-value problems in geodesy. In: Sansò F (ed) Proceedings from V-th Hotine-Marussi symposium on mathematical geodesy, Matera, Italy, 17–21 June 2002, vol 127. Springer, Berlin/Heidelberg, pp 189–200

Holota P (2011) Reproducing kernel and Galerkin's matrix for the exterior of an ellipsoid: application in gravity field studies. Stud Geophys Geod 55(3):397–413

Holota P, Nesvadba O (2014a) Analytical continuation in physical geodesy constructed by means of tools and formulas related to an ellipsoid of revolution. In: EGU general assembly, April 2014

Holota P, Nesvadba O (2014b) Reproducing kernel and Neumann's function for the exterior of an oblate ellipsoid of revolution: application in gravity field studies. Stud Geophys Geod 58(4):505–535

Kellogg OD (1953) Foundations of potential theory. Dover Publications, New York

Koch KR, Pope AJ (1972) Uniqueness and existence for the geodetic boundary-value problem using the known surface of the Earth. Bull Géod 106:467–476

Meschkowski H (1962) Hilbertsche Räume mit Kernfunktion. Springer, Berlin

Moritz H (1980a) Advanced physical geodesy. Wichmann, Karlsruhe

Moritz H (1980b) Geodetic reference system 1980. Bull Géod 54(3):395–405

Nesvadba O (2010) Reproducing kernels in harmonic spaces and their numerical implementation. In: EGU general assembly, April 2010

Nesvadba O, Holota P (2015) An OpenCL implementation of ellipsoidal harmonics. In: Rizos C, Willis P (eds) Proceedings of the VIII. Hotine-Marussi symposium, vol 142. Springer, Berlin/Heidelberg (in print). http://link.springer.com/chapter/10.1007/1345_2015_59

Pavlis NK, Holmes SA, Kenyon SC, Factor JK (2012) The development and evaluation of the Earth Gravitational Model 2008 (EGM2008). J Geophys Res Solid Earth 117(B04406): 1–38. http://onlinelibrary.wiley.com/doi/10.1029/2011JB008916/abstract

Pavlis NK, Holmes SA, Kenyon SC, Factor JK (2013) Correction to "The development and evaluation of the Earth Gravitational Model 2008 (EGM2008)". J Geophys Res Solid Earth 118(2633)

Richardson LF, Gaunt JA (1927) The deferred approach to the limit. Part I. Single lattice. Part II. Interpenetrating lattices. Philos Trans R Soc Lond Ser A, Containing Papers of a Mathematical or Physical Character 226(636–646):299–361

# Evaluation of GOCE/GRACE GGMs Over Attica and Thessaloniki, Greece, and Wo Determination for Height System Unification

G.S. Vergos, V.D. Andritsanos, V.N. Grigoriadis, V. Pagounis, and I.N. Tziavos

**Abstract**

Within the frame of the *Elevation* project, recently acquired collocated GPS/Leveling observations over trigonometric benchmarks (BMs) have been used for the evaluation of the recent GOCE/GRACE Global Geopotential Models (GGMs) and the unification of the Greek Local Vertical Datum (LVD). To this extent all available satellite-only and combined GOCE/GRACE GGMs were evaluated to conclude on the possible improvement brought by GOCE in the determination of the geoid over Greece. At a second stage, the present work focuses on the determination of the zero-level geopotential value $W_0^{LVD}$ for the Greek LVD. The estimation of $W_0^{LVD}$ was carried out using a least squares adjustment of Helmert orthometric heights, surface gravity disturbances and geopotential values computed from EGM2008 and GOCE/GRACE GGMs over the available GPS/Levelling BMs. Moreover, given that the BMs used belong to two distinct areas, i.e., one over Attica and another in Thessaloniki, the $W_0^{LVD}$ determination was carried out for each region separately, to conclude on the possible biases of the Hellenic LVD itself. From the evaluation of the GOCE/GRACE models it was concluded that the latest releases provide a significant, compared to EGM2008, improvement in the comparisons with the GPS/Levelling data, by as much as 3 cm, in terms of the standard deviation. Furthermore, the $W_0^{LVD}$ determined for the Greek LVD indicates a bias of about −4.95 $m^2/s^2$ compared to the conventional value of 62,636,856.0 $m^2/s^2$.

**Keywords**

Global geopotential models • GOCE • GPS/levelling BMs • LVD • Validation • Zero-level geopotential

## 1 Introduction

With the GOCE mission having reached its end in late 2013, the unprecedented contribution of the first mission to carry-on gradiometric observations in space was and is still being evaluated. GOCE contributed significantly not only in the field of geodesy, where its impact on gravity field and geoid modelling was long expected, but to oceanography, geophysics and even time-variable gravity field modelling. Its contribution to geodesy has been predominant, since

G.S. Vergos (✉) • V.N. Grigoriadis • I.N. Tziavos
Department of Geodesy and Surveying, Aristotle University of Thessaloniki, University Box 440, 54124 Thessaloniki, Greece
e-mail: vergos@topo.auth.gr

V.D. Andritsanos • V. Pagounis
Department of Civil Engineering and Surveying and Geoinformatics Engineering, Technological and Educational Institute of Athens, Athens, Greece



S. Jin, R. Barzaghi (eds.), *IGFS 2014*, International Association of Geodesy Symposia 144,
DOI 10.1007/1345_2015_53

GOCE provided improved representations of the Earth's gravity field especially in the long-to-medium and medium wavelengths of the spectrum. The improvements in the entire spectral band, and especially between degree and order (d/o) 210–240, contributed to improved geodetically-derived dynamic ocean topography (DOT) models (Albertella et al. 2012; Knudsen et al. 2011; Tziavos et al. 2013), new insights in the Earth's interior by modelling the Moho discontinuity (Reguzzoni et al. 2013) and even the identification of time-variable gravity changes due to seismic events (Fuchs et al. 2013). In the pure geodetic context, the contribution of GOCE is viewed in the improved representation of the Earth's gravity field functionals and especially gravity anomalies and geoid heights. These improvements are commonly viewed in terms of the differences with external validation datasets, such as terrestrial gravity anomalies and GPS/Levelling geoid heights (Gruber et al. 2011; Hirt et al. 2011; Šprlák et al. 2012; Tocho et al. 2014; Tziavos et al. 2015; Vergos et al. 2014) on trigonometric benchmarks (BMs), compared to the best available combined Global Geopotential Model (GGM) from the pre-GOCE era, i.e., EGM2008 (Pavlis et al. 2012). The latter dataset is very useful for such GOCE GGM validation experiments since it offers an independent source of information that is not included in the development of GGMs. Moreover, GOCE data are now commonly used for the determination of the zero-level geopotential value towards the unification of Local Vertical Datums (LVD) to a global one (Grigoriadis et al. 2014; Gruber et al. 2012; Hayden et al. 2012; Sánchez et al. 2014).

The focus of this work is twofold. The first part is devoted to the evaluation of almost all currently available GOCE, GOCE/GRACE and combined GGMs from the first release to the fifth release of the models. The second part is devoted to the determination of the zero-level geopotential value of the Greek LVD $W_0^{LVD}$. Both are carried out through a dedicated set of collocated GPS/Levelling geoid heights, which has been collected in the frame of the "Elevation" project (Anastasiou et al. 2013) for the regions of Thessaloniki and Attica (Athens region).

## 2 Methodology, GGMs and Local Data

### 2.1 GOCE GGM Validation Methodology

In order to evaluate the geoid undulations derived from the GOCE GGMs ($N^{GOCE}$) an external dataset of "geometric" geoid heights from collocated GPS and spirit levelling observations on trigonometric BMs ($N^{GPS/Levelling}$) has been used. The residual geoid heights have been evaluated following a spectral enhancement approach as:

$$\Delta N = N^{GPS/Levelling} - N^{GOCE}\Big|_2^{n_1} - N^{EGM2008}\Big|_{n_1+1}^{2160} - N^{RTM} - N_0, \quad (1)$$

where, $\Delta N$ denotes the geoid heights differences between the GPS-derived and GGM-derived geoid heights. The latter are denoted as $N_{n_i}^{GOCE}$ and are determined after evaluating first heights anomalies from the GOCE and GOCE/GRACE GGMs and then converting them to geoid heights through the use of Bouguer anomalies and orthometric heights (Heiskanen and Moritz 1967, Eqs. 8.100–8.102). Based on the spectral enhancement approach (Gruber et al. 2011; Tocho et al. 2014; Vergos et al. 2014) the GOCE/GRACE GGMs contribute up to some maximum degree and order (d/o) of expansion $n_I$ ($N^{GOCE}\big|_2^{n_1}$), and then EGM2008 is used as a fill-in information for the rest of the geoid signal from degree $n_I + 1$ to degree 2,160 ($N^{EGM2008}\big|_{n_1+1}^{2160}$) along with Residual Terrain Model (RTM) effects on geoid heights ($N^{RTM}$) to represent the topographic signal above degree 2,160. The estimation of the RTM effects on geoid heights comes from an SRTM-based 3 arcsec digital terrain model (Tziavos et al. 2010), so that the geoid spectrum represented is equivalent to d/o 216,000. Therefore the geoid omission error is very small (mm-level), so it can be neglected in the formed differences. It should be pointed out that the way the spectral enhancement of the low-degree GOCE-based GGMs is performed, is by computing the contribution from each GGM solely and then adding them together, i.e., no or minimal coefficient patching is performed. A more elaborate GGM combination around each degree $n_I$ (say for instance for ±5–10 d/o), would be to combine the coefficients of each model using as a weight their errors or error degree variances. Then, a spectrally consistent merging of the GGMs might yield better results and no discontinuities in the GGM merging degrees. In the present case, this would yield too much computational burden, given the amount of GGMs studied, with no statistically significant improvement in the final combined GGM-derived geoid heights. For the evaluation of the zero-degree geoid term ($N_o$) GRS80 was used as a reference ellipsoid and the computation was performed as in Heiskanen and Moritz (1967, Eq. 2.182). Finally, all computations have been performed in the Tide Free (TF) system with any conversions from the Zero Tide and the Mean Tide system to the TF being performed according to Ekman (1989). The evaluation scheme has been carried out for all d/o of each GGM up to their $n_{max}$ with an increment step of 10°. This increment step is sufficient in order to conclude on the spectral range that the GOCE GGMs perform better than EGM2008. For a more elaborate discussion of the followed methodology and conventions Gruber et al. (2011), Tocho et al. (2014), Tziavos et al. (2015) and Vergos et al. (2014) should be consulted.

## 2.2 $W_0^{LVD}$ Estimation Methodology Over Greece

For the estimation of the zero-level geopotential value over Greece, we followed the methodology outlined in Grigoriadis et al. (2014) and Tocho and Vergos (in press). The methodology is based on a combination, through Least-Squares (LS), of available Helmert orthometric heights, surface gravity data and geopotential values on the trigonometric BMs, the latter two estimated from the available GGMs. This is one of the possible options for the unification of a LVD and its connection to a world height system, while the second one employs the formulation of a geodetic boundary value problem (GBVP) as the fixed gravimetric GBVP at sea and the scalar-free GBVP on land (Sánchez 2009). For the marine areas knowledge of the mean dynamic ocean topography is needed, which can be obtained through a mean sea surface model and a GOCE-based GGM. For the continental part, the observations include the usual gravity anomalies, potential differences, deflections of the vertical (Sánchez 2009; Tenzer et al. 2013).

In the present study, the observation equation is based on the definition of orthometric heights in Heiskanen and Moritz (1967) as:

$$H_i^{Helmert} = \frac{W_0^{LVD} - W_i}{\overline{g}_i^{Helmert}}, \tag{2}$$

where $H_i^{Helmert}$ is the known Helmert orthometric height at the BM w.r.t. the LVD, $W_i$ is the actual gravity potential at the BM, and $\overline{g}_i^{Helmert}$ is the mean value of gravity. The generic estimation of $W_0^{LVD}$ can then be carried out as:

$$\widehat{W}_0^{LVD} = \frac{\sum_{i=1}^{m} p_i \left( \overline{g}_i^{Helmert} H_i^{Helmert} + W_i \right)}{\sum_{i=1}^{m} p_i}, \tag{3}$$

where the available observations are assigned positive weights $p_i$, so that the residuals after the adjustment are minimized. It is acknowledged that the estimation of $W_0^{LVD}$ following the proposed approach is susceptible to the inherent uncertainties in the determination of Helmert orthometric heights and the accumulated errors in leveling data. In Eq. (3) the gravity potential $W_i$ has been synthesized, according to the IERS conventions (2010), from the gravitational potential part $V_i$, obtained by the GGM spherical harmonic coefficients, and a centrifugal part $\Phi_i$ using the benchmark's known spatial position and the Earth's conventional rotational velocity. The mean gravity value $\overline{g}_i$ along the plumb line between the LVD's zero-height equipotential reference surface and the Earth's surface was estimated according to the Poincare-Prey reduction scheme (Heiskanen and Moritz 1967, Eqs. 4–24). Further details can be found in Grigoriadis et al. (2014). For the zero-level geopotential estimation, GRS80 has been used as the reference ellipsoid (Moritz 1992), while the IERS conventions (2010) for the Earth's geocentric gravitational constant $GM$ and the gravity potential at the geoid $W_0$ have been followed, so that $GM = 398{,}600.4418 \times 10^9\ \mathrm{m^3 s^{-2}}$ and $W_o = 62{,}636{,}856.00\ \mathrm{m^2 s^{-2}}$.

## 2.3 GOCE/GRACE GGMs and External Data for Validation and $W_0^{LVD}$ Estimation

Twenty-one GGMs up to their maximum degree and order were used in this study for evaluation. Table 1 summarizes the GGMs evaluated with some of them being satellite-only ones and others being generated with combined information. In Table 1 we present the abbreviation of the GGM names that will be used in the present work, their official names as listed at the International Centre for Global Earth Models (ICGEM) service, their maximum d/o of expansion and a descriptive information on the data used for their development. EGM2008 (Pavlis et al. 2012) complete to d/o 2,160 is used throughout as reference against which all GOCE/GRACE based ones are evaluated. For the GOCE and GOCE-GRACE models, their basic categorization is from the methodology used in their development (time-wise approach for TIM, direct approach for DIR, GOCE and GRACE combined for GOCO, etc.). Moreover, their designation as first, second, third, fourth and fifth release (R1, R2, R3, R4 and R5) refers to the effective volume of GOCE data used in their development, i.e., 2, 8, 12, 26.5 months of data and the entire mission (including the lower orbit data), respectively.

For the evaluation of the GGMs, the local data used refer to 230 collocated GPS/Levelling observations on BMs over the regions of Thessaloniki and Attica (see Fig. 1). This set is based on historical orthometric heights from the HMGS (Hellenic Military Geographic Service) and ellipsoidal heights collected by the research teams during the "Elevation" project (Anastasiou et al. 2013). The same set of BMs will be used for the estimation of the zero-level geopotential value of the Greek LVD. It should be noted that this is a completely independent set of GPS/Levelling observations than the usual one used during the latest GGM evaluation over Greece (see e.g., Tziavos et al. 2015; Vergos

**Table 1** GOCE/GRACE GGMs used for evaluation

| Models | $n_{max}$ | Data | ICGEM name | References |
|---|---|---|---|---|
| EIGEN-51C | 360 | S(GRACE, LAGEOS),G,A | EIGEN-51C | Förste et al. 2008 |
| EIGEN-6C | 1,420 | S(GOCE, GRACE, LAGEOS),G,A | EIGEN-6C | Förste et al. 2011 |
| EIGEN-6C2 | 1,949 | S(GOCE, GRACE, LAGEOS),G,A | EIGEN-6C2 | Förste et al. 2012 |
| EIGEN-6C3stat | 1,949 | S(GOCE, GRACE, LAGEOS),G,A | EIGEN-6C3stat | Förste et al. 2012 |
| DIR-R1 | 240 | S(GOCE) | GO_CONS_GCF_2_DIR_R1 | Bruinsma et al. 2010 |
| DIR-R2 | 240 | S(GOCE) | GO_CONS_GCF_2_DIR_R2 | Bruinsma et al. 2010 |
| DIR-R3 | 240 | S(GOCE, GRACE, LAGEOS) | GO_CONS_GCF_2_DIR_R3 | Bruinsma et al. 2010 |
| DIR-R4 | 260 | S(GOCE, GRACE, LAGEOS) | GO_CONS_GCF_2_DIR_R4 | Bruinsma et al. 2013 |
| DIR-R5 | 300 | S(GOCE, GRACE, LAGEOS) | GO_CONS_GCF_2_DIR_R5 | Bruinsma et al. 2013 |
| TIM-R1 | 224 | S(GOCE) | GO_CONS_GCF_2_TIM_R1 | Pail et al. 2010 |
| TIM-R2 | 250 | S(GOCE) | GO_CONS_GCF_2_TIM_R2 | Pail et al. 2011 |
| TIM-R3 | 250 | S(GOCE) | GO_CONS_GCF_2_TIM_R3 | Pail et al. 2011 |
| TIM-R4 | 250 | S(GOCE) | GO_CONS_GCF_2_TIM_R4 | Pail et al. 2011 |
| TIM-R5 | 280 | S(GOCE) | GO_CONS_GCF_2_TIM_R5 | Pail et al. 2011 |
| GOCO01S | 224 | S(GOCE, GRACE) | GOCO01S | Pail et al. 2011 |
| GOCO02S | 250 | S(GOCE, GRACE) | GOCO02S | Goiginger et al. 2011 |
| GOCO03S | 250 | S (GOCE, GRACE) | GOCO03S | Mayer-Gürr et al. 2012 |
| ITG-GOCE02 | 240 | S (GOCE) | ITG-GOCE02 | Schall et al. 2014 |
| GOGRA02S | 230 | S(GOCE, GRACE) | GOGRA02S | Yi et al. 2013 |
| JYY-GOCE02S | 230 | S (GOCE) | JYY-GOCE02S | Yi et al. 2013 |
| EGM2008 | 2,160 | S(GRACE),G,A | EGM2008 | Pavlis et al. 2012 |

*S* satellite tracking data, *G* gravity data, *A* altimetry-derived gravity data, *GRACE* gravity recovery and climate experiment, *CHAMP* challenging mini-satellite payload, *GOCE* gravity field and steady state ocean circulation explorer, *LAGEOS* laser geodynamics satellite, *SLR* satellite laser ranking

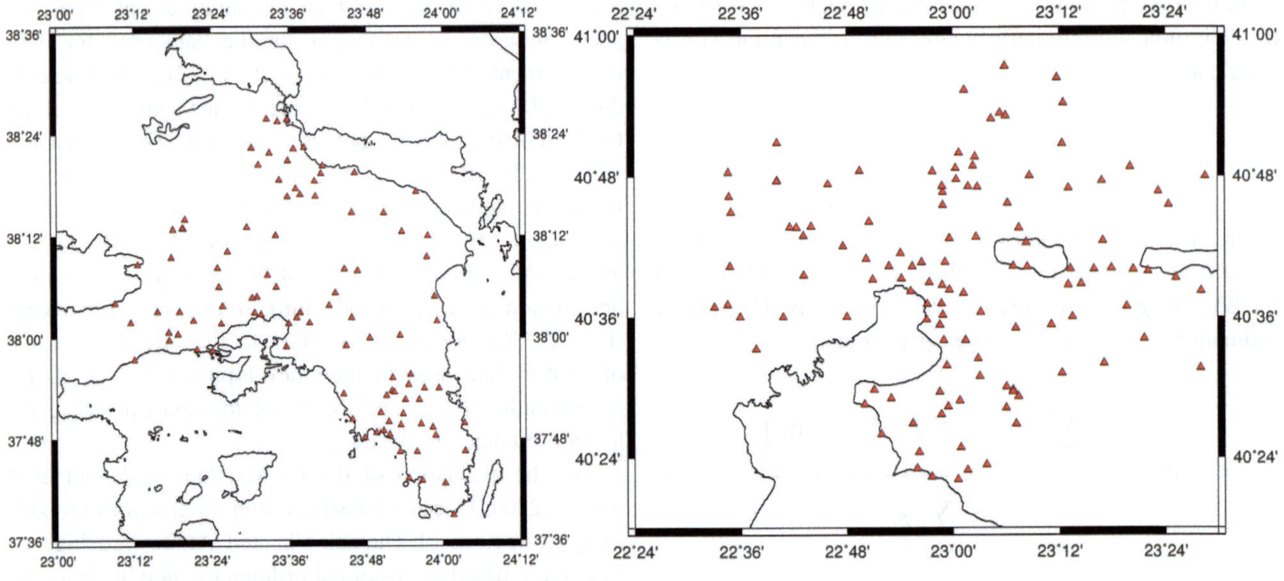


**Fig. 1** Distribution of local GPS/Levelling data in Athens (*left*) and Thessaloniki (*right*) for GOCE GGM validation

et al. 2014) therefore the results acquired will provide a new independent look on the GOCE GGM performance. The main difference of this new dataset is that longer GPS observations (larger than 2 h compared to 1 h) have been carried out, while spirit levelling campaigns between the BMs, wherever possible due to the distance limitations, were performed to validate the available orthometric heights. The two areas under study have rather small extents ($\sim 1° \times 1°$), so they will provide little insight in the validation of the longer wavelengths of the GGMs. For the specifications of

**Table 2** Statistics of the ellipsoidal ($h$), orthometric ($H$), GPS/Levelling ($N^{GPS/Lev}$) geoid heights and RTM effects ($N^{RTM}$) on the BMs (total of 230 BMs)

| | Max | Min | Mean | Std |
|---|---|---|---|---|
| h | 1,231.429 | 38.021 | 248.724 | ±209.418 |
| H | 1,189.398 | 0.681 | 208.782 | ±209.360 |
| $N^{GPS/Lev}$ | 42.763 | 34.752 | 39.942 | ±1.848 |
| $N^{RTM}$ | 0.077 | −0.019 | 0.018 | ±0.019 |

Units: (m)

the reference systems and tide conventions of the local data as well as the definition of the Greek LVD, Grigoriadis et al. (2014) and Vergos et al. (2014) should be consulted.

## 3 GGM External Validation

Following the spectral enhancement approach for the GOCE GGMs, outlined previously, the geoid height differences for all models with the local GPS/Levelling data were evaluated with an incremental step of 10°. Table 2 presents the statistics of the available GPS ellipsoidal heights, Helmert orthometric heights, the formed GPS/Levelling geometric "geoid" heights and the RTM effects on geoid heights for the network of the 230 BMs. Table 3 presents the statistics of the differences between the available local data and the synthesised GGM and RTM contribution. In Table 3, we report the statistics of the differences only for the GGM d/o $n_l$ (see Eq. 1) that provides the smallest standard deviation (std) of the differences with the GPS/Levelling data. The reference model, i.e., EGM2008 provides a std of 15.9 cm which is outperformed by the GOCE GGMs when the latter are used up to d/o 140. It should be pointed out, as outlined in Vergos et al. (2014), that Greece is a peculiar case for the validation of GOCE-based GGMs, since most of the country's gravity data have been used in the development of EGM2008. Therefore, an improvement by GOCE compared to EGM2008, even if marginal, can be regarded as significant.

The combined EIGEN-models improve the std by 3 cm, for the latest 6C3stat version, while the improvement between the three versions. i.e., 6C, 6C2 and 6C3stat is at the mm level. The DIR models, based on both GOCE and GRACE data, show an improvement of 3.1 cm compared to EGM2008, while they reduce the range (difference between the minimum and maximum values of the differences) by 11.3–12.4 cm. The latter is of importance as well, since the largest and smallest difference are found at the BMs with the highest elevation, hence it can be concluded that the GOCE/GRACE GGMs manage to provide more uniform information for the geoid signal. One interesting point is that DIR-R1, which is based on the fewer GOCE data, performs equally well with the latest R5 model, but this is attributed to the fact that during its development, a priori information from EIGEN-51C has been used. The TIM models show a similar performance with a reduction of the std and range between 2.8–3.1 cm and 11.4–12.7 cm. As expected, the GOCO GGMs being based on the first three releases of the GOCE data show inferior performance w.r.t. the R4 and R5 DIR and TIM models. Finally, GOGRA02s and JYY-GOCE02s have an astonishing performance, given that they are based on the second release of GOCE data. The std they offer is similar to that of the latest TIM GGM and they provide the same if larger amount of range reduction. ITG-GOCE02s provides the overall best std at the 12.7 cm and a range reduction of 12.1 cm, indicating promising prospects for the followed short-arc approach, when the R4 and R5 version of GOCE data are included.

**Table 3** Statistics of the differences between GPS/levelling and geoid heights from the GGMs

| | n1 | Range | Mean | Std |
|---|---|---|---|---|
| EGM2008 | 2,160 | 0.849 | −0.505 | ±0.159 |
| EIGEN-51C | 90 | 0.823 | −0.471 | ±0.152 |
| EIGEN-6C | 140 | 0.727 | −0.467 | ±0.131 |
| EIGEN-6C2 | 140 | 0.727 | −0.467 | ±0.130 |
| EIGEN-6C3stat | 140 | 0.731 | −0.464 | ±0.129 |
| DIR-R1 | 120 | 0.725 | −0.460 | ±0.128 |
| DIR-R2 | 140 | 0.732 | −0.469 | ±0.129 |
| DIR-R3 | 140 | 0.729 | −0.475 | ±0.131 |
| DIR-R4 | 140 | 0.733 | −0.467 | ±0.129 |
| DIR-R5 | 140 | 0.736 | −0.469 | ±0.128 |
| TIM-R1 | 120 | 0.722 | −0.491 | ±0.129 |
| TIM-R2 | 140 | 0.731 | −0.477 | ±0.128 |
| TIM-R3 | 140 | 0.728 | −0.478 | ±0.130 |
| TIM-R4 | 140 | 0.729 | −0.471 | ±0.131 |
| TIM-R5 | 140 | 0.735 | −0.465 | ±0.131 |
| GOCO01s | 120 | 0.726 | −0.476 | ±0.133 |
| GOCO02s | 140 | 0.734 | −0.473 | ±0.132 |
| GOCO03s | 140 | 0.734 | −0.473 | ±0.131 |
| ITG-GOCE02s | 140 | 0.728 | −0.469 | ±0.127 |
| GOGRA02s | 140 | 0.733 | −0.473 | ±0.133 |
| JYY-GOCE02s | 140 | 0.723 | −0.464 | ±0.130 |

Units: (m)
$n_l$ denotes the maximum d/o that the GOCE GGMs are used, whilst above that they are complemented with EGM2008 and RTM

The progressive evolution of the GOCE GGMs as more GOCE data are incorporated in their development is evidenced from Table 3 and Fig. 2. A better insight in that respect can be gained from Fig. 2 where the std of the differences per d/o for the DIR-R1, R2, R3, R4 and R5 models is presented. Ignoring DIR-R1 which is a combined model, the DIR GGMs are better than EGM2008 up to d/o

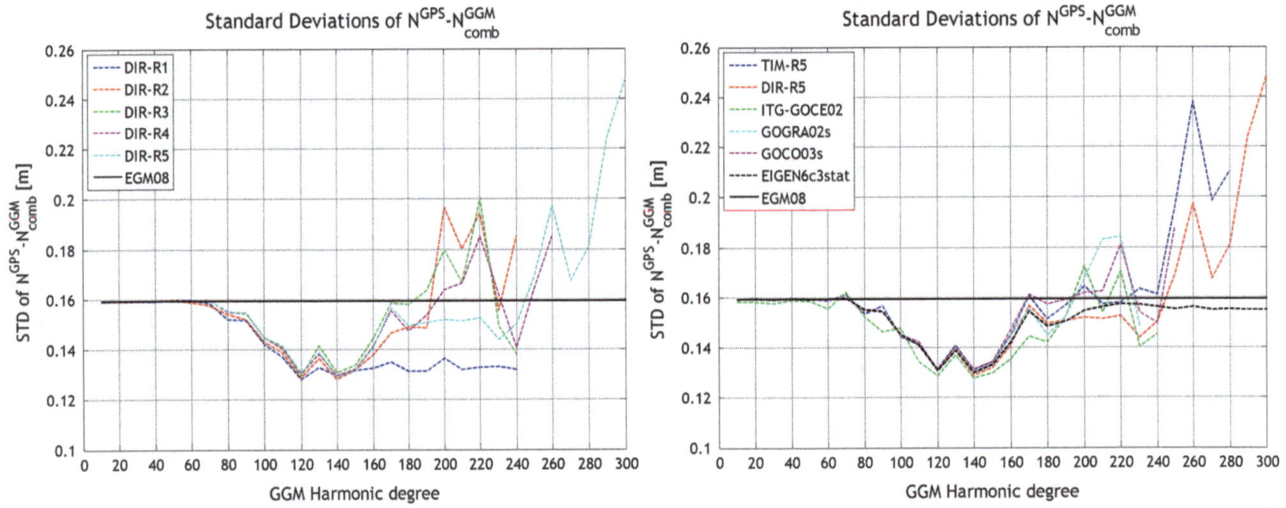


**Fig. 2** Standard deviation of the differences between the five releases of the DIR GGMs (*left*; R1 *normal line*, R2 *dashed line*, R3 *dash-dot line*, R4 *dotted line* and R5 *line with x-marker*) and the latest GGMs (*right*; TIM-R5 *normal line*, DIR-R5 *dashed line*, ITG-GOCE02 *dash-dot line*, GOGRA02s *dotted line*, GOCO03s *line with x-marker*, and EIGEN6c3stat *line with circle*) with the GPS/Levelling geoid heights for various degrees of expansion

190, 180, 200 and 250 for the R2, R3, R4 and R5 models. DIR-R5 is the only model within the DIR family of GGMs that remains below EGM20008 for its entire spectrum up to d/o 250, showing the great value of the additional low-orbit GOCE data during the end of the mission. Comparing the different GOCE models (see Fig. 2, right) TIM-R5 is better than EGM2008 to d/o 195, GOCO03S to d/o 190, and GOGRA02s to d/o 195. ITG-GOCE02s is surprisingly good again, being better than EGM2008 to d/o 195 and better than most of the other GOCE GGMs up to that d/o. At specific degrees, e.g., degree 170, it is better by ~1.5 cm compared to the other GGMs, even though ITG_GOCE02 is based on fewer GOCE observables. The overall best GGM is DIR-R5 since it manages to show a very good std and range with the GPS/Levelling data, but also the largest useful spectral band for the geoid spectrum spanning from d/o 40 to 250.

## 4 $W_0^{LVD}$ Estimation Results for Greece

The estimation of the zero-level geopotential value of the Hellenic LVD is based on the same GPS/Levelling BMs employing Eqs. (2) and (3). In principle, physical heights in the Hellenic LVD were modeled as Helmert orthometric heights, while an unknown $W_o$ value is associated with the LVD. The Helmert orthometric heights refer to the tide gauge (TG) station at the Piraeus port (Athens) where the local MSL was computed from sea level measurements over the period 1933–1978.

The true accuracy of the Hellenic LVD leveling network is largely unknown. First an un-weighted ($p_i = 1$) Least squares (LS) estimation of the $W_0^{LVD}$ has been performed (see Table 4), employing data from EGM2008, DIR-R4, DIR-R5, TIM-R4 and TIM-R5 all evaluated to their $n_{max}$. The EGM2008 estimated $W_0^{LVD}$ is 62,636,860.77 m$^2$/s$^2$ and forms the initial reference for the GOCE GGMs. DIR-R4 and DIR-R5 provide different zero-level geopotential values by 8.7 cm and 7.6 cm, while TIM-R4 and TIM-R5 are close at the 2.7 and −1.5 cm. The latter is peculiar deviating significantly from the other GOCE models and can be probably attributed to the higher harmonics (higher than d/o 240) not being modelled properly in TIM-R5. This can be justified from Fig. 2 as well, since above d/o 240, the differences of TIM-R5 with the local GPS/Leveling data increase by 4–8 cm. In any case, from this first estimate it can be concluded that the GOCE estimates do not seem robust among each other.

One the other hand, when employing the enhanced GOCE GGMs, i.e., GOCE up to some d/o and then EGM2008, the results improve significantly. In this approach, the GOCE GGMs are used up to their degree that provides the best std (see also Table 3) and then EGM2008 is used as fill-in information. The new $W_0^{LVD}$ are reported again in Table 4, where it can be seen that their differences are 3.8, 3.5, 3.4 and 3.8 cm for DIR-R4, DIR-R5, TIM-R4 and TIM-R5, respectively. Between them, the new GOCE estimates differ only by 0.1–0.4 cm, showing very good robustness. The std of the height residuals of the system observation equation (3) are, before the synthesis, 17.2, 36.3, 39.5, 41.4, and 40.4 cm for EGM2008, DIR-R4, DIR-R5, TIM-R4 and TIM-R5, respectively, while after the combination with EGM2008 they reduce to 13.7 cm for DIR-R4, 13.6 cm for DIR-R5, 13.9 cm for TIM-R4 and 13.7 for TIM-R5 models.

From the analysis of these residuals it was found that a height correlation exists, so that the zero-level geopotential

**Table 4** Estimates of the zero-height geopotential value for Greece

| | $\widehat{W}_0^{LVD}$ [m$^2$s$^{-2}$] | |
|---|---|---|
| | Un-weighted | *With λ estimation* |
| EGM2008 (2,160) | 62,636,860.77 ± 0.04 | |
| EGM2008 (260) | 62,636,859.43 ± 0.04 | |
| DIR-R4 (250) | 62,636,859.90 ± 0.04 | |
| DIR-R4 (140) + EGM2008 | 62,636,860.39 ± 0.04 | 62,636,860.92 ± 0.008 |
| DIR-R5 (300) | 62,636,860.01 ± 0.04 | |
| DIR-R5 (140) + EGM2008 | 62,636,860.42 ± 0.04 | 62,636,860.95 ± 0.008 |
| TIM-R4 (250) | 62,636,860.50 ± 0.04 | |
| TIM-R4 (140) + EGM2008 | 62,636,860.43 ± 0.04 | 62,636,860.96 ± 0.008 |
| TIM-R5 (280) | 62,636,860.92 ± 0.04 | |
| TIM-R5 (140) + EGM2008 | 62,636,860.39 ± 0.04 | 62,636,860.93 ± 0.008 |

determined from each BM had a decreasing values with increasing height. This meant that the highest BM were providing a value close to 62,636,856 m$^2$/s$^2$ while the lowest ones a value as high as 62,636,865 m$^2$/s$^2$. To overcome this problem, a revised model of Eqs. (2) and (3) was tested, where a height-dependent parameter λ was estimated along-side the $W_0^{LVD}$ for the combined with EGM2008 GOCE GGMs (see Table 4), as:

$$H_i^{Helmert} = \frac{W_0^{LVD} - W_i}{\overline{g}_i^{Helmert}} + \lambda H_i^{Helmert}. \qquad (4)$$

The estimated height-dependent parameters λ were for all GOCE GGMs of the order of $(2.559 \pm 0.008) \times 10^{-4}$ m$^2$/s$^2$, while for EGM2008 it was two orders of magnitude smaller and was deemed insignificant. The new GOCE-based $W_0^{LVD}$ shows even more robustness, since they differ to each other by $\delta\widehat{W}_0^{LVD} = 0.3 - 0.4$ cm only. This is a very good proof not only of the appropriateness of the proposed methodology (with its inherent uncertainties due to the use of orthometric heights), but of the capability of GOCE GGMs to be used for height system control, and vertical reference system unification, since they manage to "pick-up" the height dependency of the high-elevation BMs, something that could not be done by EGM2008. The latter is due to the fact that EGM2008 is based on local gravity data to model the medium part of the gravity field spectrum, which are very sparse over mountain ranges. The std of the height residuals reduces further to 12.6 cm for DIR-R4, 12.5 cm for DIR-R5, 12.8 cm for TIM-R4 and 12.7 cm for TIM-R5.

Finally, two weighted LS adjustment schemes were investigated, the first one using the inverse of the BM orthometric height as the observation weight and the second one using the inverse of the spherical distance $L_i$ of the BM from the tide-gauge at the Piraeus harbor, which serves as the origin of the Greek LVD. These two new scenarios were carried out for DIR-R5 only, since it was the GGM with the overall smallest std against the GPS/Levelling BMs and the one with the smallest height residuals during the $W_0^{LVD}$ determination. When using a weight of $p_i = 1/H_i^{helm}$ the estimated $W_0^{LVD}$ is 62,636,860.94 ± 0.002 m$^2$/s$^2$ and with a weight of $p_i = 1/L_i$ the estimated $W_0^{LVD}$ is 62,636,860.92 ± 0.002 m$^2$/s$^2$. Therefore, it becomes apparent that the extra weights are redundant and their significance is little, if any.

The finally proposed $\hat{W}_0^{LVD}$ for the Greek territory, based on the available data, is that of the combined DIR-R5 model with the estimation of the height-dependent parameter, i.e., $\widehat{W}_0^{LVD} = 62636860.95 \pm 0.008\ m^2/s^2$. A final estimation is worthy, i.e., that of estimating the $\hat{W}_0^{LVD}$ for the areas of Attica and Thessaloniki separately. If this is done with the selected DIR-R5 + EGM2008 combination, then the resulting values are $\widehat{W}_0^{Attica} = 62636860.31 \pm 0.021\ m^2/s^2$ and $\widehat{W}_0^{Thessaloniki} = 62636860.79 \pm 0.021\ m^2/s^2$. Their difference is 4.8 cm, showing that there is indeed a bias between the various stations of the Greek LVD. The difference between the tide-gauge records situated at the Piraeus harbor (Attica) and the harbor of Thessaloniki is 2.5 cm, indicating that their LVD bias cannot be attributed solely to the properties of the sea, but probably also to the geoid commission error in both DIR-R5 and EGM2008 (mostly the latter) and inconsistencies in the Greek LVD. The latter being a point that needs attention when existing orthometric heights are to be implemented for $\hat{W}_0^{LVD}$ determination.

## 5 Conclusions

In this work a detailed evaluation of the latest complete set of GOCE, GOCE/GRACE and combined GGMs has been presented employing a local set of collocated GPS/Levelling observations. From the results acquired it can be concluded that as the GOCE models progress from the first to the fifth release the useful spectrum is getting larger. Being limited up to d/o 180–200 for the first releases it reaches d/o 245 for DIR-R5, with significant improvement in the spectral range between d/o 185–230. The latest releases of the GOCE/GRACE GGMs are better as much as 3.2 cm in terms of the std and 12.6 cm in terms of the range, compared to EGM2008.

Moreover, an estimation of the zero-level geopotential value of the Greek LVD was carried out, based on a LS adjustment scheme. From that analysis it was concluded that the combined GOCE GGM and EGM2008 $W_0^{LVD}$ is very robust showing differences of the order of 0.1–0.4 cm. When including in the adjustment a parameter to absorb the dependency with height, then height residuals of about 12.5 cm are determined for DIR-R4 compared to 17.2 cm for EGM2008. The use of observation weights, either the inverse of the height of the BM or the inverse of the distance from the tide-gauge that serves as the origin of the Greek LVD, did not alter the results and was deemed insignificant. When evaluating the $W_0^{LVD}$ for the two regions separately, a bias between the local vertical datums of Thessaloniki and Attica of the order of 4.8 cm was found. Part of it, about 2.5 cm, can be attributed to the dynamic ocean topography difference between the two regions, as realized by the difference in the mean sea level records, while the rest is due to the GGM commission error and the inherent inconsistences of the Greek LVD.

**Acknowledgements** The authors wish to acknowledge the funding provided for this work, in the frame of the "Elevation" project, by the E.U. (European Social Fund) and Hellenic national funds under the Operational Program "Education and Lifelong Learning 2007–2013", action "Archimedes III – Funding of research groups in T.E.I.".

## References

Albertella A, Savcenko R, Janjić T, Rummel R, Bosch W, Schröter J (2012) High resolution dynamic ocean topography in the southern ocean from GOCE. Geophys J Int 190:922–930

Anastasiou D, Gaifillia D, Katsdourou A, Kolyvaki E, Papanikolaou X, Gianniou M, Vergos GS, Pagounis V (2013) First validation of the Hellenic vertical datum as a prerequisite for the efficient disaster and resources management: the "Elevation" project. FIG Commission 3 "Spatial Information, Informal Development, Property and Housing", December 11–12, Athens, Greece

Bruinsma SL et al (2010) GOCE gravity field recovery by means of the direct numerical method. Presented at the ESA Living Planet Symposium, Bergen, Norway, 27 June–2 July

Bruinsma S et al (2013) The new ESA satellite-only gravity field model via the direct approach. Geophys Res Lett 40(14):3607–3612. doi:10.1002/grl.50716

IERS Conventions (2010). In: Petit G, Luzum B(eds) IERS technical note 36. Frankfurt am Main: Verlag des Bundesamts für Kartographie und Geodäsie, p 179. ISBN 3-89888-989-6

Ekman M (1989) Impacts of geodynamic phenomena on systems for height and gravity. Bull Geod 63(3):281–296

Förste C et al (2008) EIGEN-GL05C – a new global combined high-resolution GRACE-based gravity field model of the GFZ-GRGS cooperation. Geophys Res Abst 10:EGU2008-A-03426, SRef-ID: 16077962/gra/EGU2008-A-03426

Förste C et al (2011) EIGEN-6 – a new combined global gravity field model including GOCE data from the collaboration of GFZ-Potsdam and GRGS-Toulouse. Geophys Res Abst 13:EGU2011-3242-2 (EGU General Assembly)

Förste C et al (2012) A preliminary update of the direct approach GOCE processing and a new release of EIGEN-6C. Presented at the AGU Fall Meeting 3–7 Dec 2012 San Francisco, Abstract No. G31B-0923

Fuchs MJ, Bouman J, Broerse T, Visser P, Vermeersen B (2013) Observing coseismic gravity change from the Japan Tohoku-Oki 2011 earthquake with GOCE gravity gradiometry. J Geophys Res 118(10):5712–5721

Goiginger H et al (2011) The combined satellite-only global gravity field model GOCO02S; presented at the 2011 General Assembly of the European Geosciences Union, Vienna, Austria, April 4–8

Grigoriadis VN, Kotsakis C, Tziavos IN, Vergos GS (2014) Estimation of the geopotential value Wo for the local vertical datum of continental Greece using EGM08 and GPS/leveling data. IAG Symp 141:249–255

Gruber T, Visser PNAM, Ackermann C, Hosse M (2011) Validation of GOCE gravity field models by means of orbit residuals and geoid comparisons. J Geod 85(11):845–860

Gruber T, Gerlach C, Haagmans R (2012) Intercontinental height datum connection with GOCE and GPS-levelling data. J Geod Sci 2(4):270–280. doi:10.2478/v10156-012-0001-y

Hayden T, Amjadiparvar B, Rangelova E, Sideris MG (2012) Evaluation of W0 in Canada using tide gauges and GOCE gravity field models. J Geod Sci 2(4):257–269. doi:10.2478/v10156-012-0008-4

Heiskanen WA, Moritz H (1967) Physical geodesy. W.H. Freeman, San Francisco, p 895

Hirt C, Gruber T, Featherstone WE (2011) Evaluation of the first GOCE static gravity field models using terrestrial gravity, vertical deflections and EGM2008 quasi geoid heights. J Geod 85(10):723–740

Knudsen P, Bingham R, Andersen OB, Rio M-H (2011) A global mean dynamic topography and ocean circulation estimation using a preliminary GOCE gravity model. J Geod 85(11):861–879

Mayer-Gürr T et al (2012) The new combined satellite only model GOCO03S. Presented at the IAG Commission 2 "gravity, geoid and height systems GGHS2012" conference, October 9th–12th, Venice, Italy

Moritz H (1992) Geodetic reference system 1980. Bull Geod 66:187–192

Pail R et al (2010) Combined satellite gravity field model GOCO01S derived from GOCE and GRACE. Geophys Res Lett 37:L20314. doi:10.1029/2010GL044906

Pail R et al (2011) First GOCE gravity field models derived by three different approaches. J Geod 85(11):819–843

Pavlis NK, Holmes SA, Kenyon SC, Factor JK (2012) The development and evaluation of the earth gravitational model 2008 (EGM2008). J Geophys Res 117:B04406. doi:10.1029/2011JB008916

Reguzzoni M, Sampietro D, Sans F (2013) Global Moho from the combination of the crust 2.0 model and GOCE data. Geophys J Int 195(1):222–237

Sánchez L (2009) Strategy to establish a global vertical reference system. In: Drewes H (ed) Geodetic reference systems. International association of geodesy symposia, vol 134. Springer, Switzerland, pp 273–278. doi:10.1007/978-3-642-00860-3_42

Sánchez L, Dayoub N, Čunderlík R, Minarechová Z, Mikula K, Vatrt V, Vojtíšková M, Šíma M (2014) W0 estimates in the frame of the GGOS working group on vertical datum standardisation. IAG Symp 141:203–210

Schall J, Eicker A, Kusche J (2014) The ITG-Goce02 gravity field model from GOCE orbit and gradiometer data based on the short arc approach. J Geod 88(4):403–409. doi:10.1007/s00190-014-0691-2

Šprlák M, Gerlach C, Pettersen PR (2012) Validation of GOCE global gravity field models using terrestrial gravity data in Norway. J Geod Sci 2(2):134–143

Tenzer R, Dayoub N, Abdalla A (2013) Analysis of a relative offset between vertical datums at the North and South Islands of New Zealand. Appl Geomat 5:133–145

Tocho C, Vergos GS, Pacino MC (2014) Evaluation of the latest GOCE/GRACE derived global geopotential models over Argentina with collocated GPS/levelling observations. IAG Symp 141: 75–83

Tziavos IN, Vergos GS, Grigoriadis VN (2010) Investigation of topographic reductions and aliasing effects to gravity and the geoid over Greece based on various digital terrain models. Surv Geophys 31(3):23–67. doi:10.1007/s10712-009-9085-z

Tziavos IN, Vergos GS, Mertikas SP, Daskalakis A, Grigoriadis VN, Tripolitsiotis A (2013) The contribution of local gravimetric geoid models to the calibration of satellite altimetry data and an outlook of the latest GOCE GGM performance in GAVDOS. Adv Space Res 51(8):1502–1522. doi:10.1016/j.asr.2012.06.013

Tziavos IN, Vergos GS, Grigoriadis VN, Tzanou EA, Natsiopoulos DA (in press) Validation of GOCE/GRACE satellite only and combined global geopotential models over Greece, in the frame of the GOCE-SeaComb Project. Accepted for Publication to the IAG Scientific Assembly 2013, International Association of Geodesy Symposia Vol. 143, Springer International Publishing Switzerland

Vergos GS, Grigoriadis VN, Tziavos IN, Kotsakis C (2014) Evaluation of GOCE/GRACE global geopotential models over Greece with collocated GPS/levelling observations and local gravity data. IAG Symp 141:85–92

Yi W, Rummel R, Gruber T (2013) Gravity field contribution analysis of GOCE gravitational gradient components. Stud Geophys Geod 57(2):174–202. doi:10.1007/s11200-011-1178-8

# The DTU13 MSS (Mean Sea Surface) and MDT (Mean Dynamic Topography) from 20 Years of Satellite Altimetry

Ole Andersen, Per Knudsen, and Lars Stenseng

**Abstract**

The DTU13MSS is the latest release of the global high resolution mean sea surface (MSS) from DTU Space. The new MSS is based on multi-mission satellite altimetry from 10 different satellites. Three major advances have been made in order to release the new MSS. The time series have been extended to 20 years from 17 years used for DTU10MSS creating the first multi-decadal MSS. Secondly, the DTU13MSS ingest Cryosat-2 LRM and SAR data as well as 1 year of Jason-1 geodetic mission as part as it end-of-life mission between May 2012 and June 2013. Finally, the availability of Cryosat-2 SAR altimetry enables the determination of sea level in leads in the ice, which has enabled us to derive an accurate MSS all the way to 88°N.

With the availability to determine the geoid with higher accuracy than ever before due to the launch of the GRACE and GOCE satellites, is hence become possible to derive a satellite only mean dynamic topography (MDT) from the difference between the MSS and the geoid. Here the DTU13MSS and DTU13MDT are presented and we demonstrate how these can be used to derive realistic geostrophic currents in the world's ocean comparable to oceanographic derived MDT.

**Keywords**

Satellite altimetry • Mean sea surface • Mean dynamic topography

## 1 Introduction

Since the release of the DTU10 MSS in 2010, the amount of geodetic mission altimetry has nearly tripled. The Cryosat-2 satellite has provided new altimetric data along its 369 day near repeat since 2010 and since May 2012 the Jason-1 satellite has been operating in geodetic mission as part as its end of life mission and it continued in this until it was decommissioned in June 2013.

With release of satellite gravity data from the GRACE mission and the completion of the ESA Gravity and steady-state Ocean Circulation Explorer (GOCE) during 2009–2013 accurate and higher resolution global picture of the Earth's gravity field and its geoid are becoming available. This is revealing new details of the ocean dynamic topography as predicted by (Johannessen et al. 2003; Hughes and Bingham 2008)

In the study we will initially describe the development of the DTU13MSS which is derived from 9 different satellites (TOPEX/POSEIDON, Jason-1, Jason-2, Geosat, ICESAT, ERS-1, ERS-2, ENVISAT and Cryosat-2). Subsequently the derivation of the DTU13MDT using a GOCE based geoid will be presented. Evaluation of the MDT will be performed using derived surface geostrophic currents through the comparison with the MDT by Maximenko et al. (2009) derived from oceanographic in-situ data.

O. Andersen (✉) • P. Knudsen • L. Stenseng
National Space Institute, Technical University of Denmark, DTU Space, 2800 Lyngby, Denmark
e-mail: oa@space.dtu.dk





S. Jin, R. Barzaghi (eds.), *IGFS 2014*, International Association of Geodesy Symposia 144,
DOI 10.1007/1345_2015_182

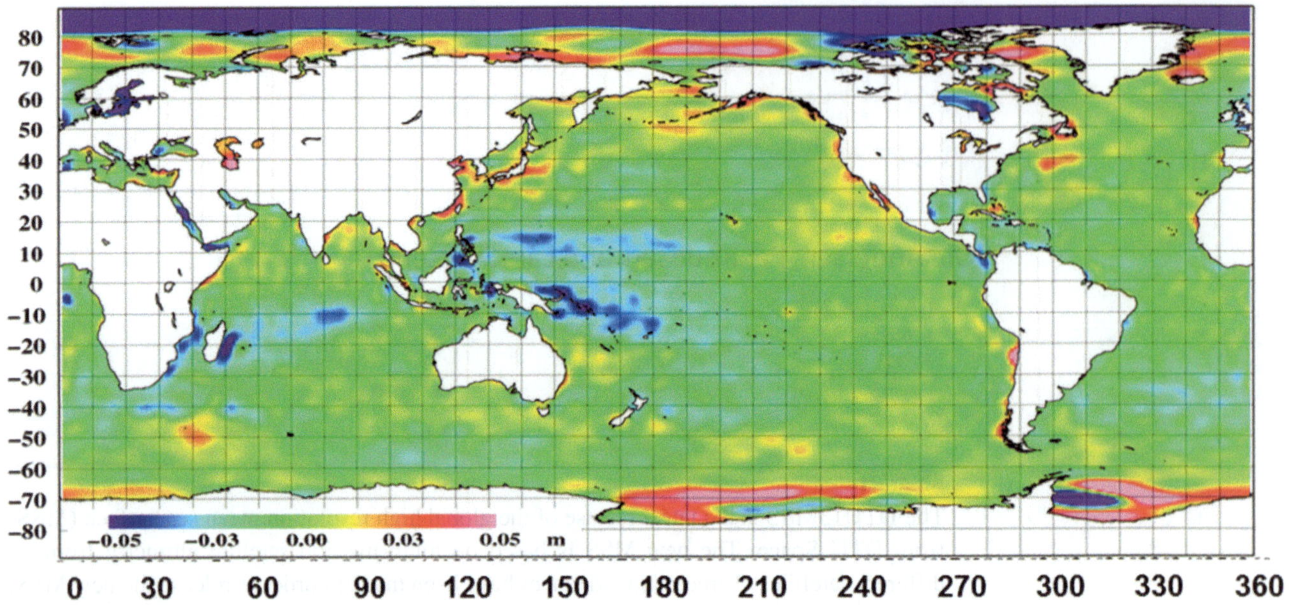


**Fig. 1** Long wavelength difference between DTU10MSS and DTU13MSS. The difference is computed as (spherical harmonic degree and order 80 of 20 years joint TP-Jason1-Jason-2 mean relative to DTU10MSS. Outside the 65°parallel, 3-year extended joint ERS-1&ERS-2/ENVISAT time series have been used. North of 80°N the effect of 1 year of Cryosat-2 altimetry can be seen. Here the former DTU10MSS was based on the considerably less accurate ICESat data

**Table 1** Altimeters and averaging periods for recent DTU and DNSC (former DTU) mean sea surfaces. Also differences in range corrections that might affect the MSS are shown naturally

| | DTU13 | DNSC08MSS | DTU 10 |
|---|---|---|---|
| Reference period | 1993–2012 | 1993–2004 | 1993–2009 |
| Satellites | TP/J1/J2/ERS-1/ERS-2/ENVISAT/ICESat/Geosat/GFO/J-1 EOL/Cryosat-2 | TP/J1/J2/ERS-1/ERS-2/ENVISAT/ICESat/Geosat/GFO | TP/J1/J2/ERS-1/ERS-2/ENVISAT/ICESat/Geosat/GFO |
| Dynamic atmosphere | MOG-2D_IB | IB (1,013 mbar) | MOG-2D_IB |
| Ocean tides | GOT 4.8 | GOT 00.2 | GOT 4.7 |
| Sea State Bias | Non-PARAM | BM4 | Non-PARAM |

## 2 The DTU13 Mean Sea Surface

Previous studies by Andersen and Knudsen (2009) described in detail the development of the global DNSC08MSS. This model was during 2010 updated to the DTU10MSS with an extended time-series, which was documented in Andersen (2010). Here we focus on the new steps taken to derive the DTU13 MSS from the DTU10MSS. Particularly, the fact that both Cryosat-2 and Jason-1 are new generations of satellite altimeters offering increased range precision compared with the older ERS-1 and Geosat generation of satellites (Raney 1998).

Initially a twenty year (1993–2012) long wavelength MSS is derived from the TOPEX/JASON mean profiles within the 65° parallels. Between the 65° and 82° parallels 20 years of ERS-1/ERS-2/ENVISAT mean profiles were used. However, these mean profiles has limited spatial resolution due to the ground track spacing so they only provide wavelength longer than 500 km being twice the ground track spacing. The effect of adding four additional years compared to what was used for the DTU10MSS (1993–2009) is illustrated in Fig. 1. Here the long wavelength difference between 20 years mean profiles of joint TP-Jason1-2 time series and the DTU10MSS based on 17 years as seen in Table 1. North of 80°N and south of 60°S the effect of including 1 year of Cryosat-2

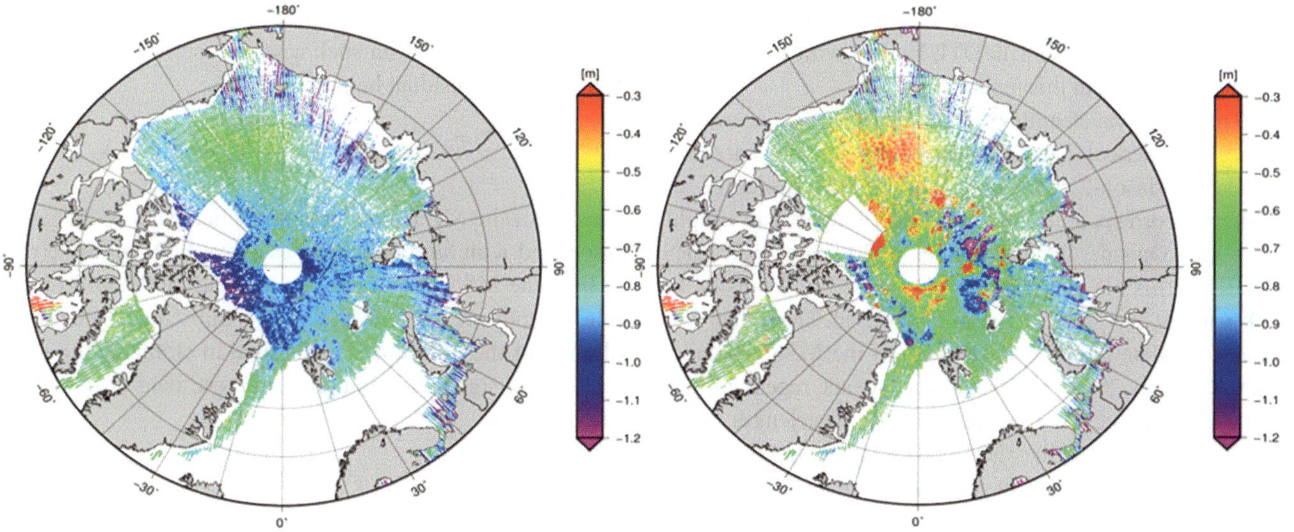


**Fig. 2** 20 Hz Cryosat-2 SAR altimetry characterized as leads in the Arctic Ocean. There is no data in the Barents Sea as this region is generally operated in LRM mode. The *left panel* shows the difference between observed SSH and the DTU10MSS during 2012. The *right panel* shows the difference between the observed SSH and the UCL04 MSS. The 70 cm SAR range bias has not been applied in this figure

altimetry can be seen. Generally the differences are on the order of a few cm however at high latitudes the effect of including Cryosat-2 can be larger. Temporal variation in sea level change will be the major contributor to this difference but particularly changes in El Nino pattern is visible.

The altimeter onboard Cryosat-2 operates in three different modes determined by a mask defined by ESA (see earth.esa.int). The major part of the open ocean is operated in conventional mode (now called LRM or low resolution mode) like all altimeters flown prior to Cryosat-2 (i.e., Jason, ERS/Envisat). However Cryosat-2 also operate in Syntetic Apertur Radar mode (SAR) where the along track resolution is improved from 7 km to 300 m and in SAR-interferometric mode where two receiving antennas operate simultaneous (Wingham et al. 2006). The majority of the Arctic Ocean is covered with SAR altimetry.

In the Arctic Ocean usable SSH observations will generally come from either open ocean or from water between ice-floes also called sea ice leads. For a conventional altimeter (e.g. ERS-1/2 or ENVISAT) the separation of return waveforms, into sea ice leads and sea ice floes, can be performed using the pulse peakiness parameter. The pulse peakiness parameter is defined as the maximal sampled power scaled by a constant and divided by the summed power of all useable samples in the waveform and thus indicates how spiky the main return of the waveform is. A SAR altimeter can discriminate much more accurately between open ocean and leads because SAR data are multi-looked. This means that the same location is observed and stacked from multiple observations (Wingham et al. 2006). Consequently a SAR waveform has additional information about the stack standard deviation associated with it and this parameter is very powerful in discriminating between different surface types like ocean or lead (Stenseng and Andersen 2012). Here we analyzed all Arctic Cryosat-2 20-Hz SAR observations during 2012 and used those characterized as leads to update the MSS. The data are shown in Fig. 2 relative to the DTU10MSS (left) panel and relative to the UCL04 MSS (ESA and UCL, 2013) which is default in the Cryosat-2 products. A 70 cm SAR range bias for the Baseline-B data used here (ESA and Parrinello, 2013) has not been corrected for in the figure. The absence of data in most of the Barents Sea is due to the fact that this region is generally operated in LRM mode.

In Fig. 2 all altimetric data from Cryosat-2 during 2012 in the Arctic Ocean have been plotted to illustrates how the DTU13MSS performs in the Arctic Ocean and how accurate Cryosat-2 SAR determines sea surface height (Jain et al. 2015). Generally the noise of the observations is relatively small seen by the scatter in the sea surface height anomaly maps. The largest differences are seen in the Beaufort Gyre and in the pack-ice north of Greenland. The increase SSH in recent years the Beaufort Gyre is a consequence of increased intensity of the gyre over 2002–2012 period as also pointed out by (Giles et al. 2008; 2012) from ENVISAT data. As the averaging period for the DTU13MSS is 1993–2012 and as this Cryosat-2 derived signal only represent an average over the year of 2012 we decided to exclude data within this region not to compromise the averaging period. North of Greenland Cryosat-2 finds an averaged SSH several decimeters lower than DTU10MSS as also shown in Fig. 2. This can be explained from the fact that the DTU10MSS was derived from 17 individual month of ICESat between 82 N and 86 N. ICESat provided much less accurate SSH than

Cryosat-2 and a close inspection of the left panel in Fig. 2 reveals a lot of problems in DTU10MSS with ICESat track related structures in this region.

The long wavelength part of the MSS was computed using a spherical harmonic expansion of 20 years joint TP-Jason1-Jason-2 residuals to DTU10MSS to degree and order 80 corresponding to wavelength longer than roughly 250 km. Outside the 65°parallel a 20-years joint ERS-1&ERS-2/ENVISAT time series was used and outside the 80° parallel one year of Cryosat-2 was used. These data were used to update the long wavelength of the MSS (Andersen and Knudsen 2009) as coarse track exact repeat mission data (ERM) does not represent short wavelength across track. Subsequently, the short wavelength corrections were updated using data from all geodetic missions (Geosat, ERS-1, Jason-1 and Cryosat-2). All data for the MSS uses state of the art range corrections as available in the Radar Altimetry Database System (RADS, rads.tudelft.nl)(Andersen and Scharroo, 2011). Again the method applied was similar to the method applied in Andersen and Knudsen (2009) with the only difference being longer time series and more data. The globe was divided into processing tiles of 2° by 5° regions and these were finally tiled together to give the short-wavelength contribution.

## 3 The DTU13 Mean Dynamic Topography

The practical task of computing a MDT from a MSS and a geoid is conceptually very simple; however there are several issues that must be considered in order to obtain a good MDT product. Both the MSS and the geoid must be represented relative to the same reference ellipsoid and in the same tidal system. Then the MDT is expressed by

$$\zeta = \overline{h} - N \tag{1}$$

where $\overline{h}$ is the MSS height and $N$ is the geoid height both relative to the same reference ellipsoid. The MSS is associated with a specific time period. When using the MDT together with satellite altimetry to reference the Sea level anomalies (SLA), it is important that the altimetry used for the MSS in the MDT calculation has the same corrections applied as the altimetry that is used for the computation of the sea level anomalies. Also, it is important that the reference time periods match.

The largest difficulty in carrying out the operation of creating a MDT occur because the MSS and geoid height does not resolve the same scales because they are provided with different resolution. Global gravity field models such as the GOCE models are normally represented in terms of spherical harmonic coefficient up to a certain harmonic degree and order $L$ which is generally much lower than the harmonic degree used for the MSS. Hence, when subtracting a geoid model based on such a set of coefficients from the MSS, then the residual heights

$$\Delta h = \overline{h} - N_L = \zeta + N - N_L = \zeta + \Delta N_L \tag{2}$$

consists of the MDT plus the unmodelled parts of the geoid associated with harmonic degrees above L. Naturally errors in both the MSS and in the gravity field model will play a role, but they are ignored at this stage. Subsequently, a proper filtering of the differences is required to eliminate the short scale geoid signals to obtain a useful estimate of the MDT. That is

$$\widehat{\zeta} = F \circ (\zeta + \Delta N_L) \tag{3}$$

where MDT estimate is obtained by applying a filter $F$ on the height residuals in Eq. (2). The filter is not defined nor used over land. The best estimate in a least squares sense

$$\begin{aligned} \|\zeta - F \circ (\zeta + \Delta N_L)\| &= \|\zeta - F \circ \zeta - F \circ \Delta N_L\| \\ &\leq \|\zeta - F \circ \zeta\| + \|F \circ \Delta N_L\| \end{aligned} \tag{4}$$

is obtained when the filtering does little harm to the MDT and minimizes the short scale geoid signals.

This filtering of the difference between the MSS and the geoid may be carried out in either the space domain, where the MSS is usually represented, or in the spectral domain where global geoid models are usually represented. Both methods have their advantages and their disadvantages. In both cases, we found that it is recommended to extend the GOCE spherical harmonic series using other higher degree harmonic expansions of the gravity field to reduce the magnitude of the short scale geoid signal in the MSS which you need to filter out to make a consistent MDT. Research within the ESA GOCE User Toolbox study (GUTS) (Benveniste et al. 2007) looked at several procedures for determining the MDT, applying both space domain and spectral domain methodologies. This was subsequently used to derive MDTs (which initially was presented in Knudsen et al. 2011; Knudsen and Andersen (2013). Hughes and Bingham (2008), Bingham and Haines (2006), Bingham et al. (2008) and Bingham et al. (2011) found that the spectral method is most efficient in removing the short scale geoid signals. Surface geostrophic currents are associated with the slope of the MDT. If accelerations and friction terms are neglected and horizontal pressure gradients in the atmosphere are absent, then the components of the surface geostrophic currents ($u$,$v$) are obtained from the MDT by

$$u = \frac{-\gamma}{f\ R}\frac{\partial \zeta}{\partial \phi},\quad v = \frac{\gamma}{f\ R\, cos\phi}\frac{\partial \zeta}{\partial \lambda} \tag{5}$$

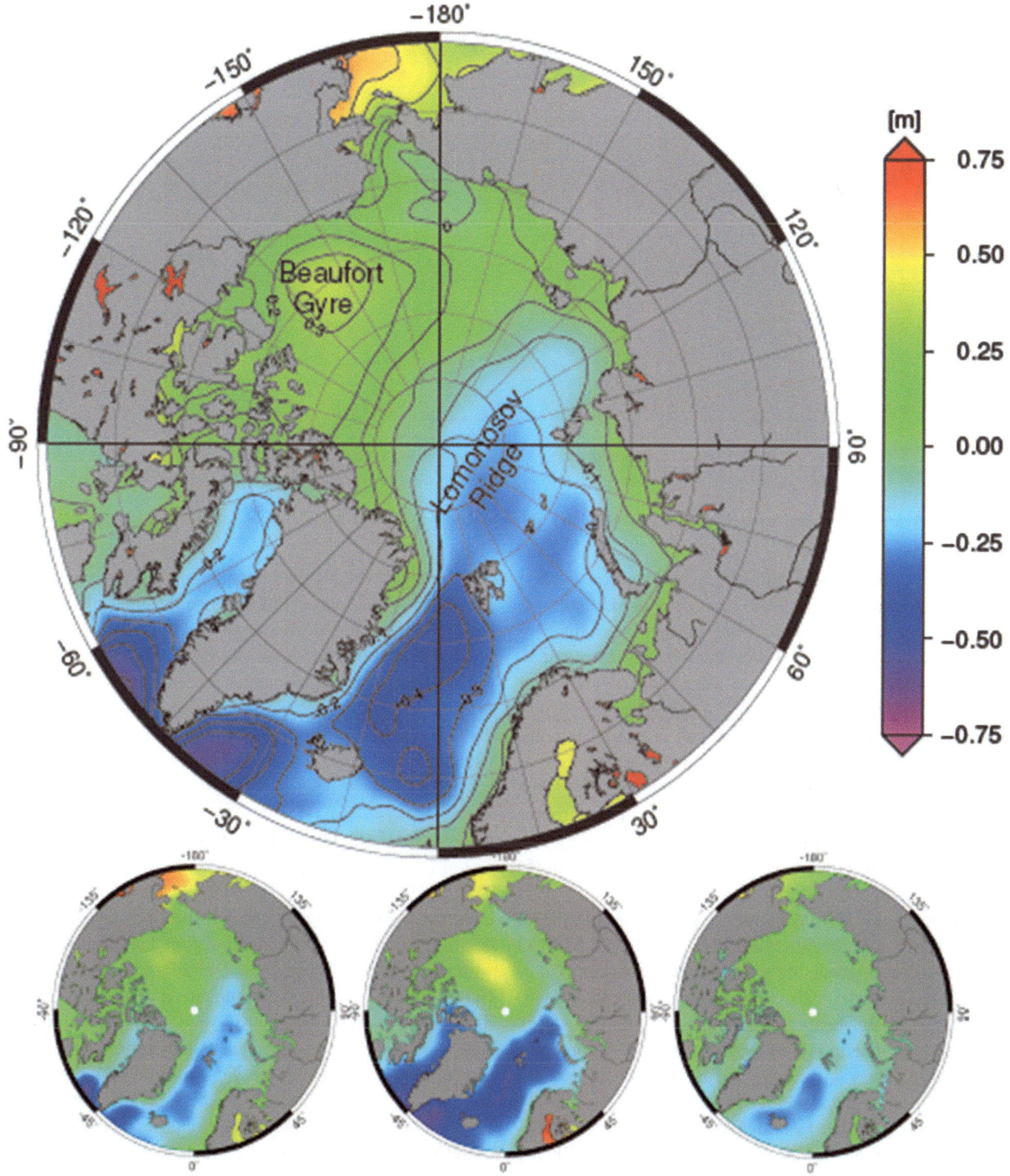


**Fig. 3** The DTU13MDT (height in meters) for the Arctic Ocean (*upper figure*). The lower figure shows the MDT computed from 3 years average of the GECCO, the MICOM, and University of Washington PIO (*bottom left* to *right*) hydrodynamic MDT

where $f = 2\omega_e \sin\varphi$ is the Coriolis force coefficient, $\omega_e$ is the angular velocity of the Earth, $R$ is the mean radius of the Earth, $\varphi$ is the latitude, $\lambda$ is the longitude, and $\gamma$ is the normal gravity.

In this study the combination model EIGEN-6C3 (Förste et al. 2011) is used together with the DTU13MSS. Consequently, the un-modelled parts of the geoid is much smaller because EIGEN-6C3 is a combination model where,

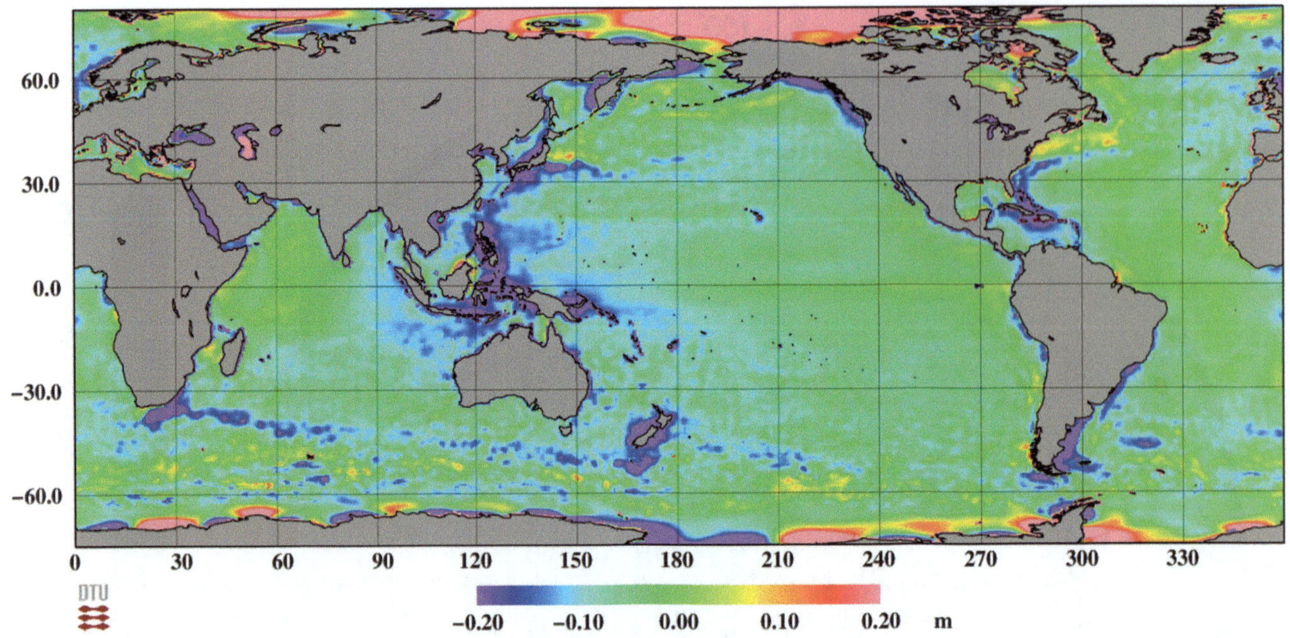


**Fig. 4** Difference between the Maximenko and the DTU13MDT. The color scale is −20 til 20 cm

e.g., GOCE, GRACE and surface gravity based on satellite altimetry (mainly DTU10GRA) have been used. Naturally, the use of altimetric gravity over the oceans will not improve the estimation of the MDT but less filtering is required. In this computation an isotropic truncated Gaussian filter with a half-width at half-maximum of 0.75 spherical degrees was used. This is particularly important in coastal regions. Approaching the Equator an an-isotropic filter was used to overcome problems with small GRACE related north-south geoid stripes. Furthermore, the computation of geostrophic current components, especially the North-south velocity, was regularized at the Equator as to be able to evaluate Eq. (5) which has a singularity.

The updated DTU13MDT in the Arctic Ocean is shown in Fig. 3 for the Arctic Ocean along with the German ECCO (http://www.ecco-group.org/), the Miami Isopycnic Coordinate Ocean Model (MICOM, http://rda.ucar.edu/datasets/ds287.1/), and University of Washington PIO (http://psc.apl.uw.edu/) bottom left to right) hydrodynamic MDT for visual comparison. The same color scale is used for all plots in order to illustrate the similarity between the synthetic MDT from satellite and the hydrodynamic MDT's. The averaging periods for the various models are not identical being (1993–2003) for GECCO; (1995–2005) for MICOM_F04 and 2003–2006 for University of Washington PIO model. Consequently a detailed comparison is hard to perform. A description of the various versions of the models can be found in Forsberg et al. (2007). Also different offsets have been added to the different models. The figure visually illustrates, that the agreement among hydrodynamic models is the same as the agreement between hydrodynamic and satellite derived models. Satellite derived model can then for the first time be used to validate hydrodynamic models in the Arctic and vice versa.

## 4 Surface Geostrophic Currents

The DTU13MDT surface geostrophic current speeds and direction shown in Figs. 4 and 5 respectively, display more details about the mean ocean circulation than what was seen with previous models like DTU10 and former releases of the GOCE models. In Fig. 4 the magnitude of the geostrophic flows of the Gulf Stream, the Kuroshio, the Agulhas, and the ACC systems are clearly depicted with their flows in the right directions (Fig. 5). In addition, Fig. 4 displays the Equatorial currents very well. Especially in the Equatorial Pacific the Westward flow of the Equatorial current and the Eastern flow of the North Equatorial Pacific current are clearly seen. In the next section these findings are addressed in more details.

Figure 4 performs a direct comparison with the MDT model based on GRACE and oceanographic in-situ data

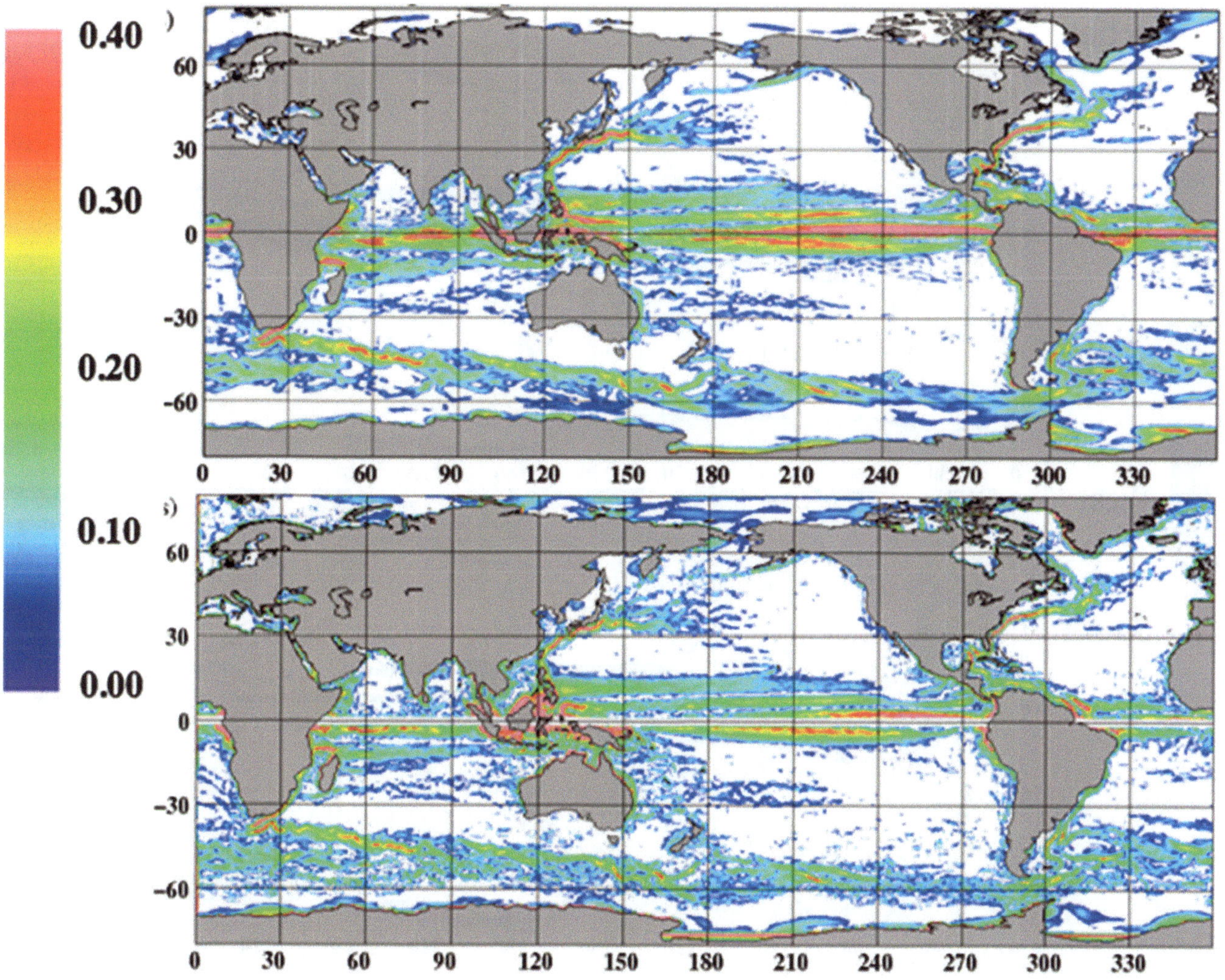


**Fig. 5** Surface geostrophic current speed (in m/s) from DTU13MDT (*upper panel*) and from Maximenko (*lower panel*). Currents weaker than 5 cm/s are not shown

constructed by Maximenko et al. (2009). It is clear that when trying to compare two MDT models which are not derived in the same reference frame we added an offset of 40 cm to the difference to center this around zero.

The largest difference will be seen in the coastal zone where satellite derived MDT will have a hard time due to the filtering. Also the difference in the averaging period which is different for the two models is clearly seen. Also there is relatively large differences in the Arctic Ocean where there is no drifter data to constrain the MDT by Maximenko.

In order to perform a detailed validation of the model it was chosen to investigate the associated surface geostrophic currents. Figures 5 and 6 in the lower panels. The comparison was carried out using the associated geostrophic surface current components as this gave more interesting visible results, since the two MDT models appear nearly identical on a ±1.5 m height scale.

In Figs. 5, 6, and 7 we decided to whiten out all currents weaker than 5 cm/s to highlight the agreement between major currents. Even though the color scale in Figs. 5 and 7 starts from 0 cm/s no purple or dark blue colors are shown in the figure due to removal of weaker currents in the figures.

Going into a more detailed comparison of the recovered sub-current systems and their different flow paths, the DTU13MDT derived surface geostrophic flow in the North Atlantic Ocean is shown in Fig. 7. Besides the high agreement in the Gulf Stream it is also seen that several sub-branches within the Gulf Stream extension in the Atlantic Ocean is now revealed from satellite and that they are in agreement with the Maximenko MDT.

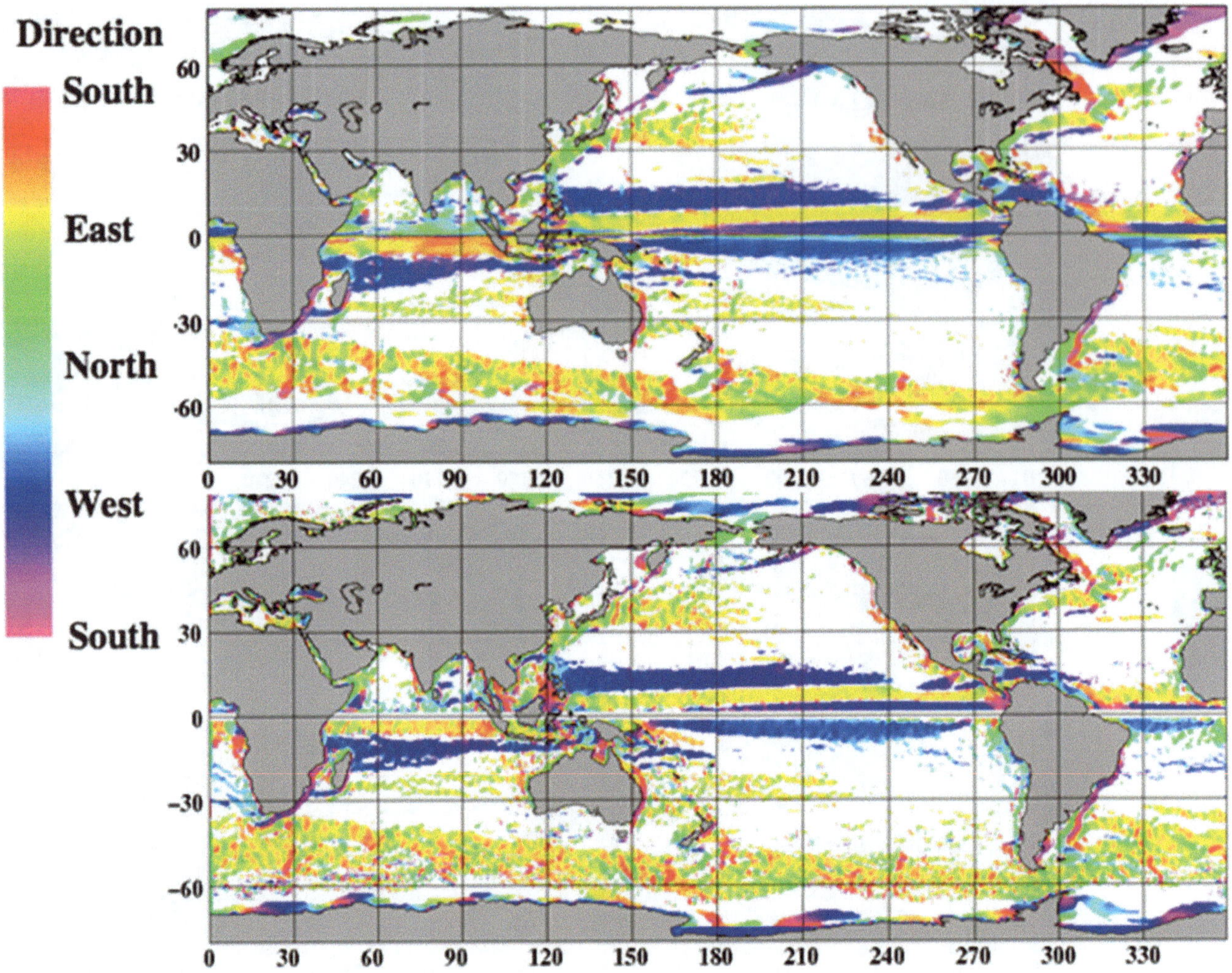


**Fig. 6** The direction of the surface geostrophic currents from DTU13MDT (*upper panel*) and from Maximenko (*lower panel*). Currents weaker than 5 cm/s are not shown

## 5 Summary

The development of the DTU13MSS and associated DTU13MDT has been presented in this paper. Both are the latest release of the global high resolution models MSS from DTU Space. The new MSS is based on two decades of multi-mission satellite altimetry from 9 different satellites. By combining with state of the art geoid derived with the use of data from the ESA satellite GOCE the DTU13MDT was derived. We demonstrated how the DTU13MDT gives very realistic surface geostrophic currents and how it compares with oceanographic derived MDT.

**Acknowledgement** The authors would like to thank the space agencies for the EO missions. Part of the work is a contribution to the ESA supported Sea level CCI and ESA supported sequel of GUT projects.

**Data Availability**

The DTU13MSS and DTU13MDT are both available for free download via ftp.space.dtu.dk/pub/DTU13. MSS and MDT models are available in resolution ranging from 1min to 1/8° along with other models in the suite of DTU high resolution geophysical model. Various formats as well as software can be downloaded from the site.

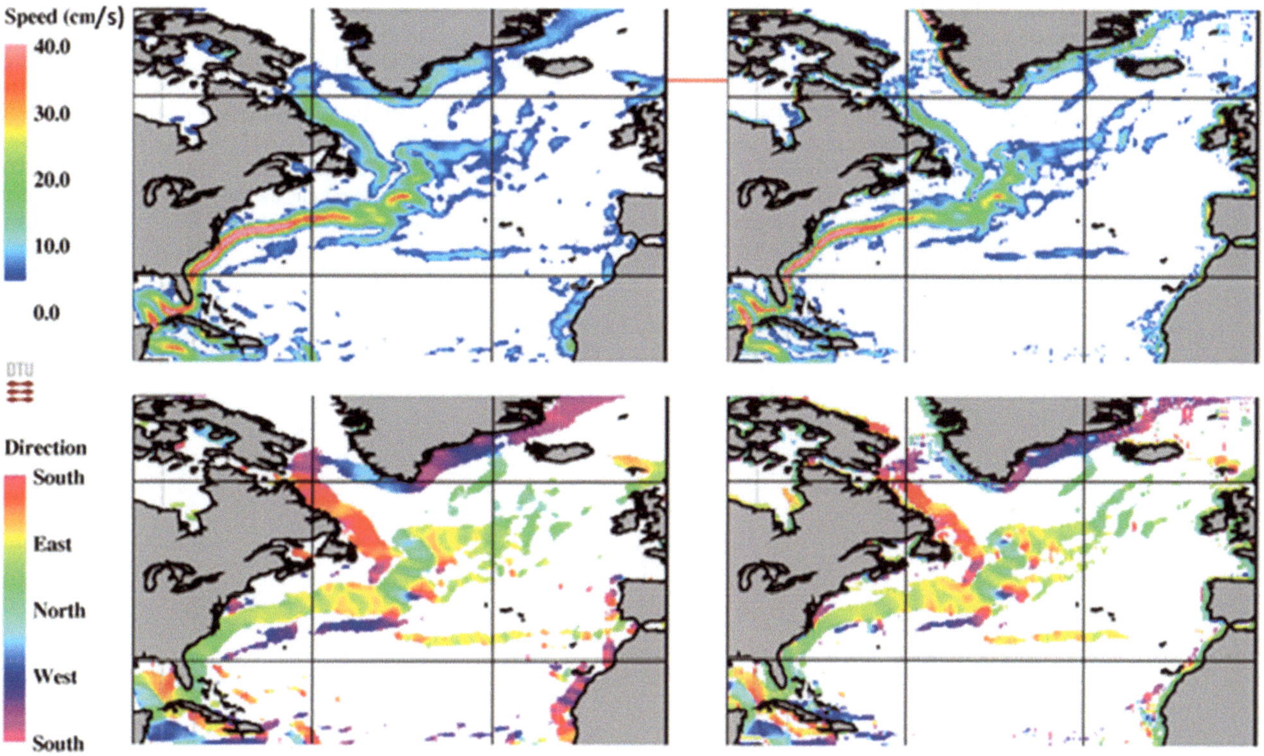


**Fig. 7** Surface geostrophic currents in the North Atlantic Ocean. Magnitudes from DTU13MDT (*left*) and Maximenko (*right*) are shown in the *upper panels*. Directions of the surface geostrophic currents are shown in the lower figures. Currents weaker than 5 cm/s are not shown and the same color scale as in Figs. 5 and 6 has been used

## References

Andersen OB (2010) The DTU10 global gravity field and mean sea surface – improvements in the Arctic. Paper presented at the 2nd IGFS symposium, Fairbanks, Alaska. http://space.dtu.dk/

Andersen OB, Knudsen P (2009) The DNSC08 mean sea surface and mean dynamic topography. J Geophys Res 114:C11. doi:10.1029/2008JC005179

Andersen OB, Scharroo R (2011) Range and geophysical corrections in coastal regions: and implications for mean sea surface determination. In: Vignudelli S et al (eds) Coastal altimetry. Springer, New York, pp 103–145. doi:10.1007/978-3-642-12796-0_5

Benveniste J, Knudsen P, the GUTS Team (2007) The GOCE user toolbox. In: Fletcher K (ed) Proceedings of the 3rd international GOCE User Workshop, 6–8 November 2006, Frascati, Italy. European Space Agency, Noordwijk

Bingham RJ, Haines K (2006) Mean dynamic topography: intercomparison and errors. Phil Trans R Soc A 364:903–916

Bingham RJ, Haines K, Hughes CW (2008) Calculating the Ocean's Mean dynamic topography from a Mean sea surface and a Geoid. J Atmos Ocean Tech 25:1808–1822. doi:10.1175/2008JTECHO568.1

Bingham RJ, Knudsen P, Andersen O, Pail R (2011) A new view of the mean North Atlantic circulation provided by GOCE. Geophys Res Let 38(1). doi:10.1029/2010GL045633

ESA and UCL (2013) CryoSat product handbook, April 2013. ESRIN – ESA and Mullard Space Science Laboratory – University College London. www.emits.esa.int/emits-doc/.../CryoSat-PHB-17apr2012.pdf

ESA, Parrinello T (2013) Known biases in Cryosat-2. https://wiki.services.eoportal.org/tiki-download_wiki_attachment.php?attId=2699

Forsberg R, Skourup H, Andersen O, Laxon SW, Ridout A, Braun A, Johannessen J, Siegismund F, Tscherning CC, Knudsen P (2007) Combination of spaceborne, airborne and surface gravity in support of Arctic ocean sea-ice and MDT mapping. Poster presented at IUGG 24th General Assembly, Perugia, Italy

Förste C, Bruinsma S, Shako R, Marty J-C, Flechtner F, Abrikosov O, Dahle C, Lemoine J.-M, Neumayer KH, Biancale R, Barthelmes F, König R, Balmino G (2011) EIGEN-6 – a new combined global gravity field model including GOCE data from the collaboration of GFZ-Potsdam and GRGS-Toulouse. Geophys Res Abstr 13:EGU2011-3242-2, EGU General Assembly

Giles K, Laxon S, Ridout S (2008) Circumpolar thinning of Arctic sea ice following the 2007 record ice extent minimum. Geop Res Lett 35, L22502. doi:10.1029/2008GL035710

Giles KA, Laxon SW, Ridout AL, Wingham DJ, Bacon S (2012) Western Arctic Ocean freshwater storage increase by wind-driven spin-up of the Beaufort Gyre. Nature 5:194–197

Hughes CW, Bingham RJ (2008) An oceanographer's guide to GOCE and the geoid. Ocean Sci 4(1):15–29

Jain, M, O. Andersen, L. Stenseng and J. Dall (2015) The importance of empirical retracking in the Arctic for sea surface and marine gravity recovery. Advances in Space Research 55(1):40–50

Johannessen JA, Balmino G, Le Provost C, Rummel R, Sabadini R, Sünkel H, Tscherning CC, Visser P, Woodworth P, Hughes CW, LeGrand P, Sneeuw N, Perosanz F, Aguirre-Martinez M, Rebhan H, Drinkwater M (2003) The European gravity field and steady-state ocean circulation explorer satellite mission: impact in geophysics. Surv Geophys 24:339–386

Knudsen P, Andersen O (2013) The DTU12MDT Global mean dynamic topography and ocean circulation model. In: Proceeding of the living planet symposium, Edinburgh, 2013, ESA SP-772. ESA ESTEC. Nordweik

Knudsen P, Bingham R, Andersen O, Rio MH (2011) Enhanced mean dynamic topography and ocean circulation estimation using GOCE preliminary models. J Geodesy. doi:10.1007/s00190/011/04858

Maximenko N, Niiler P, Rio M-H, Melnichenko O, Centurioni L, Chambers D, Zlotnicki V, Galperin B (2009) Mean dynamic topography of the ocean derived from satellite and drifting buoy data using three different techniques. J Atmos Oceanic Tech 26(9):1910–1919

Raney RK (1998) The delay Doppler radar altimeter. IEEE Trans Geosci Remote Sens 36:1578–1588

Stenseng L, Andersen O (2012) Preliminary gravity recovery from CryoSat-2 data in the Baffin Bay. Adv Space Res 50(8):1158–1163

Wingham D, Francis CR, Baker S, Bouzinac C, Brockley D, Cullen R, de Chateau-Thierry P, Laxon SW, Mallow U, Mavrocordatos C, Phalippou L, Ratier G, Rey L, Rostan F, Viau P, Wallis DW (2006) CryoSat-2: a mission to determine the fluctuations in Earths land and marine ice fields. Adv Space Res 37(4):841–871

# Part III

# Local Geoid/Gravity Modeling

# A New Gravimetric Geoid Model for the Area of Sudan Using the Least Squares Collocation and a GOCE-Based GGM

Walyeldeen Godah and Jan Krynski

**Abstract**

The determination of an accurate geoid model remains an important challenge for geodetic research in Sudan. The presented contribution concerns the determination of a new gravimetric geoid model (SUD-GM2014) for the area of Sudan using recent released GOCE-based Global Geopotential Models (GGMs), available terrestrial mean free-air gravity anomalies and the high-resolution SRTM30_PLUS global digital elevation model. The computations of the SUD-GM2014 were performed using remove-compute-restore (RCR) procedure and the least squares collocation method. The residual terrain modelling (RTM) reduction method was applied to estimate the topography effect on the geoid. The resulting gravimetric geoid model has been evaluated using geoid heights at 19 GNSS/levelling points distributed over the country. The evaluation results and the expected quality of the SUD-GM2014 geoid model were discussed considering the quality of GNSS/levelling data in Sudan. The SUD-GM2014 or the geoid model computed from GOCE-based GGMs only has been recommended as reference for GNSS heighting in Sudan.

**Keywords**

Geoid • Global geopotential model • GNSS/levelling • GOCE • Least squares collocation

## 1 Introduction

The dedicated gravity satellite missions, CHAllenging Minisatellite Payload, CHAMP (Reigber et al. 2002), Gravity Recovery And Climate Experiment, GRACE (Tapley et al. 2004) and Gravity Field and Steady-State Ocean Circulation Explorer, GOCE (Floberghagen et al. 2011), have considerably revolutionized the knowledge of the Earth's gravity field and its temporal variations by several orders of magnitude. In particular, a substantial improvement in the modelling of the long/medium wavelength component of the Earth's gravity field, e.g. up to degree/order (d/o) 260 has been achieved. For example, geoid heights with an accuracy of 2–4 cm could be determined using global geopotential models (GGMs) based on about 27 months of GOCE mission data in combination with high quality of terrestrial gravity data or with the EGM08 (Pavlis et al. 2012) in well gravity surveyed areas, e.g. in Europe, Australia. Moreover, geoid heights accuracy of 12–20 cm could be obtained from satellite only GGMs based on about 27 months of GOCE mission data (e.g. Gruber and Rummel 2013; Godah et al. 2014; Rexer et al. 2014; Yi and Rummel 2014). On the other hand, in the recent decades, a significant progress in modelling methods of a regional/local geoid as well as in the acquisition of gravity data from terrestrial and airborne measurements has also been achieved. Thus, modelling a regional/local geoid of the accuracy of 1 cm becomes one of the main aims of numerous research groups as well as surveying and mapping agencies worldwide. Unfortunately, so far this accuracy level has not been reached in many parts of the Earth, e.g. in Africa. However, recent activities in the

W. Godah (✉) • J. Krynski
Centre of Geodesy and Geodynamics, Institute of Geodesy and Cartography, 27 Modzelewskiego St., 02-679 Warsaw, Poland
e-mail: walyeldeen.godah@igik.edu.pl



S. Jin, R. Barzaghi (eds.), *IGFS 2014*, International Association of Geodesy Symposia 144,
DOI 10.1007/1345_2015_196

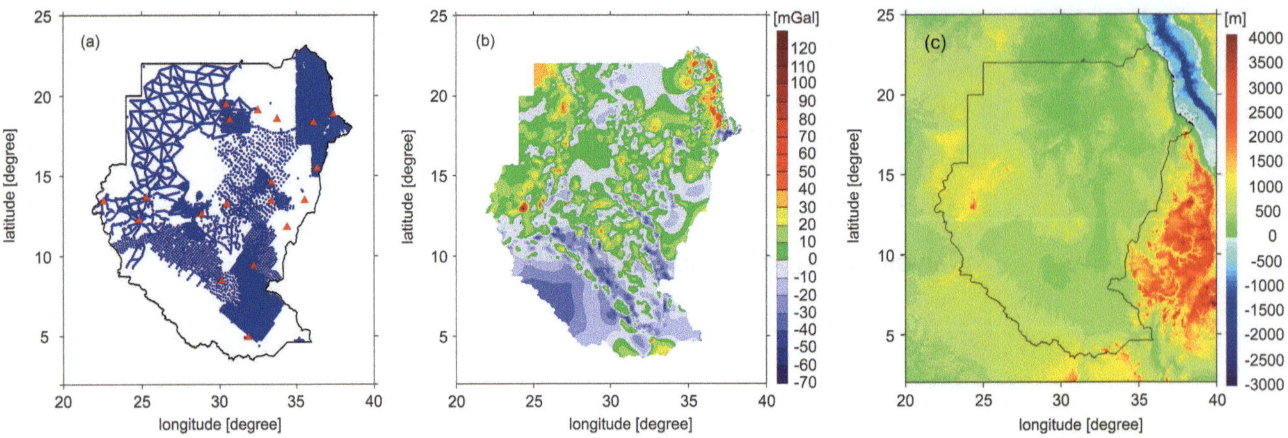


**Fig. 1** (**a**) Distribution of original GETECH point gravity data (*blue dots*), and GNSS/levelling stations (*red triangles*) in the area of Sudan, (**b**) free-air gravity anomalies provided by the GETECH on $5' \times 5'$ grid, and (**c**) the topography of the study area from the SRTM30_PLUS

framework of the African Geoid Project (e.g. Abd-Elmotaal et al. 2014) lead to the improvement of geoid models in the region.

In the area of Sudan, the determination of an accurate geoid model remains an important challenge for geodetic research. The first geoid model for the area of Sudan was developed by Adam (1967) using deflections of the vertical at 46 stations. Next, a gravimetric geoid model based on terrestrial gravity data, topography data and the GEM-T1 was developed by Fashir (1991) using modified Stokes' kernel. The accuracy of the following geoid models developed by Fashir et al. (1997) using the same method, the same terrestrial and topography data and the EGM96 was estimated to 1.43 m using Doppler/levelling data. The latest geoid model for the area of Sudan was developed using the least squares modification of Stokes integration formula (Sjöberg 2003) by Abdalla (2009). It is based on terrestrial gravity data and the GRACE-based GGM (i.e. EIGEN-GRACE02S) up to d/o 120. The fit of this model to the available GNSS/levelling data is 30 cm in terms of the standard deviation of the differences. The aim of this study is to develop a new gravimetric geoid model for the area of Sudan using the available terrestrial gravity data, a high resolution digital elevation model, and the recent GOCE-based GGM.

## 2 Data Used

### 2.1 Terrestrial Free-Air Gravity Data

A regular $5' \times 5'$ grid of mean terrestrial free-air gravity anomalies $\Delta g_{\mathrm{Faa}}$ covering the area of Sudan (Fig. 1b) provided by the GETECH (Geophysical Exploration Technology), a division of the University of Leeds Industrial Services Ltd. was developed using a simple interpolation (i.e. a minimum curvature interpolation algorithm) of 28 156 point gravity data covering about 33% of the total area investigated (Fig. 1a), acquired in the framework of the African Gravity Project (Fairhead et al. 1988). For most surveys the error details are not available. An accuracy of 1 mGal has been suggested for $\Delta g_{\mathrm{Faa}}$ by the provider of gravity data. It is expected, however, that the actual accuracy of $\Delta g_{\mathrm{Faa}}$ based on the highly inhomogeneous data is substantially lower for the majority of the country. Moreover, in order to minimize the error around the boundary due to lack of data from the neighbouring countries, the adjacent areas from the border of Sudan to parallels of 2°N and 25°N and meridians of 20°E and 40°E were filled with $5' \times 5'$ free-air gravity anomalies obtained from the EGM08.

In addition, a high-resolution $30'' \times 30''$ digital elevation model (Fig. 1c), the SRTM30_PLUS (Becker et al. 2009), has been used to compute the topography effect on the geoid.

### 2.2 GOCE-Based GGM

The recent satellite-only GOCE-based GGM, GO_CONS_GCF_2_TIM_R4 (TIM-R4) (Pail et al. 2011), developed and released for public use by ESA, has been used in this study. This model, distinguished as a GOCE-only model in a rigorous sense, i.e. no external gravity field information is used, neither as a reference model, nor for constraining the solution (Pail et al. 2011), has shown slightly better performance than the other GGMs based on the same GOCE data (e.g. Gruber and Rummel 2013; Godah et al. 2014). In the spectral band up to d/o 200 its estimated accuracy in terms of geoid heights is below 4 cm; it is representative for any area on the Earth, except polar regions that were not flown over by the GOCE satellite. TIM-R4 also shows an improvement of about 60–70% with respect to its previous 3rd release (Gruber and Rummel 2013; Godah et al. 2014;

**Table 1** The main characteristics of the TIM-R4 GGM used (Pail et al. 2011)

| | |
|---|---|
| ESA's name | GO_CONS_2_TIM_R4 |
| Maximum d/o | 250 |
| Semi-major axis $a$ (m) | 6,378,136.30 |
| GOCE GPS-SST data | d/o 130 (~26.5 months) |
| GOCE SGG data | d/o 250 (~26.5 months) |
| Kaula regulation constraint beyond d/o | 180 |
| Time of releasing | March 2013 |

Yi and Rummel 2014). The main characteristics of the TIM-R4 GGM are summarized in Table 1.

## 2.3 GNSS/Levelling Data

GNSS/levelling data at 19 stations from the area of Sudan (Fig. 1a) were used. The normal orthometric heights (orthometric heights based on normal gravity) of those stations have been determined by spirit levelling referred to 1st, 2nd and 3rd order vertical control. They are referred in this study as orthometric heights. It should be noted that the error due to the use of the normal orthometric height instead of the orthometric height has not been considered. Ellipsoidal heights of those stations were obtained from GNSS survey conducted in 12 h observing sessions in the framework of several commercial geodetic projects conducted between the years 2005 and 2008. The estimated accuracy of GNSS/levelling geoid heights at these stations ranges from 0.1 to 0.5 m (Abdalla 2009).

# 3 Methodology

The gravimetric geoid model was computed using the remove-compute-restore (RCR) procedure (e.g. Forsberg and Tscherning 1981; Torge and Müller 2012). Long- and very short wavelength components of the disturbing potential functionals were computed from the TIM-R4 GGM truncated to d/o 200 and the SRTM30_PLUS, respectively. Thereafter, those effects were removed from $\Delta g_{\mathrm{Faa}}$ providing residual gravity anomalies $\Delta g_{\mathrm{res}}$

$$\Delta g_{\mathrm{res}} = \Delta g_{\mathrm{Faa}} - \Delta g_{\mathrm{GGM}} - \Delta g_{\mathrm{RTM}} \tag{1}$$

where the long wavelength component of gravity anomaly $\Delta g_{\mathrm{GGM}}$ is computed as follows

$$\Delta g_{\mathrm{GGM}} = \frac{GM}{r^2}\sum_{n=2}^{N_{\max}}(n-1)\left(\frac{a}{r}\right)^n\sum_{m=0}^{n}\left(\Delta\overline{C}_{nm}\sin m\lambda + \Delta\overline{S}_{nm}\cos m\lambda\right)\overline{P}_{nm}(\sin\varphi) \tag{2}$$

where $GM$ is the geocentric gravitational constant, $a$ is the semi-major axis of the reference ellipsoid, $\overline{P}_{nm}(\sin\varphi)$ are fully normalized associated Legendre functions of degree $n$ and order $m$, $N_{\max}$ is the maximum applied degree of the GGM, $\Delta\overline{C}_{nm}$ and $\Delta\overline{S}_{nm}$ are differences between fully normalised spherical harmonic coefficients of actual and normal gravity field, $r$, $\varphi$, $\lambda$ are the geocentric coordinates of the computation point.

The $\Delta g_{\mathrm{RTM}}$ presents the very short wavelength component of gravity anomaly induced by the local topography, which is determined with the use of the residual terrain modelling (RTM) method (Forsberg 1984) as follows

$$\Delta g_{\mathrm{RTM}} = 2\pi G\rho\left(H - H_{\mathrm{ref}}\right) - c \tag{4}$$

where $c$ is the classical terrain correction (see Heiskanen and Moritz 1967), $H$ presents the topographic height of the gravity point, $H_{\mathrm{ref}}$ is the height of the mean elevation surface, and $\rho$ is the density of the Earth's crust. In order to avoid the inconsistency between the Earth's crust density and the water density ($\rho_w = 1{,}030$ kg m$^{-3}$), the so-called rock-equivalent topography approach (e.g. Balmino et al. 1973; Rummel et al. 1988) has been implemented. The mean elevation surface has been constructed by averaging the SRTM30_PLUS (Fig. 1c) using a spatial resolution of 100 km, that almost corresponds to the applied maximum d/o (i.e. d/o 200) of the GGM used.

The gravimetric geoid height $N$ computed using the RCR procedure is expressed as

$$N = N_{\mathrm{GGM}} + N_{\mathrm{res}} + N_{\mathrm{RTM}} \tag{5}$$

where $N_{\mathrm{GGM}}$ presents the long wavelength component of the geoid height determined as follows

$$N_{\mathrm{GGM}} = \frac{GM}{r\gamma}\sum_{n=2}^{N_{\max}}\left(\frac{a}{r}\right)^n\sum_{m=0}^{n}\left(\Delta\overline{C}_{nm}\sin m\lambda + \Delta\overline{S}_{nm}\cos m\lambda\right)\overline{P}_{nm}(\sin\varphi) \tag{6}$$

where $\gamma$ is the normal gravity at the computation point. The $N_{\mathrm{RTM}}$ presents the very short wavelength component of the geoid height. The RTM effect on the geoid is expressed in planar approximation as

$$N_{\mathrm{RTM}} = \frac{G\rho}{\gamma}\iint_E\int_{H_{\mathrm{ref}}}^{H}\frac{1}{r}\,dx\,dy\,dH \approx \frac{G\rho\left(H - H_{\mathrm{ref}}\right)}{\gamma}\cdot\frac{1}{r_0} \tag{7}$$

where $E$ denotes the area of the computation and $r_0$ is the planar distance between the computation point and the current point integrated.

The residual geoid heights $N_{res}$ were determined from $\Delta g_{res}$ using the least squares collocation (LSC) as follows (Moritz 1980)

$$N_{res} = \boldsymbol{C}_{N_{res}\Delta g_{res}} \boldsymbol{C}^{-1}_{\Delta g_{res}\Delta g_{res}} \Delta g_{res} \tag{8}$$

where $\boldsymbol{C}_{\Delta g_{res}\Delta g_{res}}$ is the auto-covariance matrix of "observations" $\Delta g_{res}$, $\boldsymbol{C}_{N_{res}}\Delta g_{res}$ is the matrix of the cross-covariance between $N_{res}$ and $\Delta g_{res}$.

In order to evaluate $N_{res}$ in Eq. (8), the empirical covariance function

$$\text{cov}_{\Delta g_{res}\Delta g_{res}}\left(\psi_{jk}\right) = \frac{1}{N_{jk}} \sum_{j,k}^{N_{jk}} \Delta g_{res}\left(\varphi_j, \lambda_j\right) \Delta g_{res}\left(\varphi_k, \lambda_k\right) \tag{9}$$

for $\Delta g_{res}$ has to be estimated first. $N_{jk}$ is the number of pairs of gravity anomalies for each interval

$$\psi_{jk} - \frac{\Delta\psi}{2} < \psi_{jk} < \psi_{jk} + \frac{\Delta\psi}{2} \tag{10}$$

and $\psi_{jk}$ is the spherical distance between gravity anomalies in a pair (Heiskanen and Moritz 1967)

$$\cos\psi_{jk} = \sin\varphi_k \sin\varphi_j + \cos\varphi_k \cos\varphi_j \cos\left(\lambda_j - \lambda_k\right) \tag{11}$$

while $\frac{\Delta\psi}{2}$ denotes the sampling interval size (Tscherning 2013).

In practice, modelling covariance function means the fitting of the empirical covariance function to the analytical covariance function model and determining its parameters. The well-known analytical Tscherning/Rapp model (Tscherning and Rapp 1974) of the covariance function for the anomalous potential $T$ has been used

$$\text{cov}_{T(P)T(Q)} = \alpha \sum_{n=2}^{N_{max}} \sigma_n^2 \left(\frac{R_E^2}{r_P r_Q}\right)^{n+1} P_n(t) + \sum_{n=N_{max}+1}^{\infty} \frac{A}{(n-1)(n-2)(n+4)} \left(\frac{R_B^2}{r_P r_Q}\right)^{n+1} P_n(t) \tag{12}$$

where $R_E$ is the mean Earth's radius, $R_B$ is the radius of the Bjerhammar sphere, $P_n(t) = P_n\left(\cos\psi_{PQ}\right)$ is the Legendre polynomial of degree $n$ with the spherical distance $\psi_{PQ}$ between the points $P$ and $Q$, $r_P$ and $r_Q$ are geocentric distances of the points $P$ and $Q$, $\sigma_n^2$ is the error degree variance for the anomalous potential, $A$ is a constant in units of $(m/s)^4$ and $\alpha$ is the scale factor of the error degree variance.

**Table 2** Statistics of gravity anomalies and residual gravity anomalies (mGal)

| Anomaly | Min | Max | Mean | Std dev. |
|---|---|---|---|---|
| $\Delta g_{Faa}$ | –63.40 | 116.60 | –0.91 | 15.15 |
| $\Delta g_{Faa} - \Delta g_{GGM}$ | –75.91 | 84.20 | –3.03 | 12.22 |
| $\Delta g_{res} = \Delta g_{Faa} - \Delta g_{GGM} - \Delta g_{RTM}$ | –95.18 | 80.79 | 0.47 | 12.64 |

## 4 Results and Analysis

The gravimetric geoid model SUD-GM2014 for the area of Sudan has been developed following the described methodology with the use of the GRAVSOFT package (Tscherning et al. 1992). Table 2 provides the statistics of free-air gravity anomalies as well as residual gravity anomalies.

The results presented in Table 2 show that removing $\Delta g_{GGM}$ from $\Delta g_{Faa}$ results in substantial reduction in terms of dispersion and the standard deviation. Due to the poor quality of available terrestrial free-air gravity anomalies in Sudan as well as since the topography is very smooth (i.e. flat areas) in the majority of the area of Sudan, such clear reduction of dispersion and the standard deviation after subtracting $\Delta g_{RTM}$ is not observed. On the other hand, the removal of the topography effect ($\Delta g_{RTM}$) results in the substantial decrease of the mean value, which may indicate the systematic influence of the terrestrial gravity measurements which are mainly located in easily accessible areas, e.g. in valleys.

The covariance function parameters estimated for the whole area of Sudan using residual gravity anomalies $\Delta g_{res}$ are $R_B - R_E$ = –0.52043 km, the variance of gravity anomalies at zero altitude of 159.77 mGal$^2$, and the error degree variance scale factor of 67.9199. In order to avoid artefacts in determining the covariance function that may result from inhomogeneous distribution of terrestrial gravity data, limited area has been chosen to estimate the representative covariance function parameters. After numerous trials the area bounded by parallels of 11°N and 21°N and meridians of 25°E and 35°E was selected. Graphical representation of the empirical and analytical fitted covariance functions for $\Delta g_{res}$ from the selected area is given in Fig. 2.

The empirical and analytical fitted covariance functions depicted in Fig. 2 exhibit good agreement up to ~0.65° spherical distance, where the covariance value reaches zero. For those covariance functions the following parameters were obtained: $R_B - R_E$ = –4.58918 km, the variance of gravity anomalies at zero altitude of 89.9 mGal$^2$ and the error degree variance scale factor of 15.6598. With the use of these parameters and the LSC (Eq. (8)), residual geoid heights (Fig. 3b) were determined.

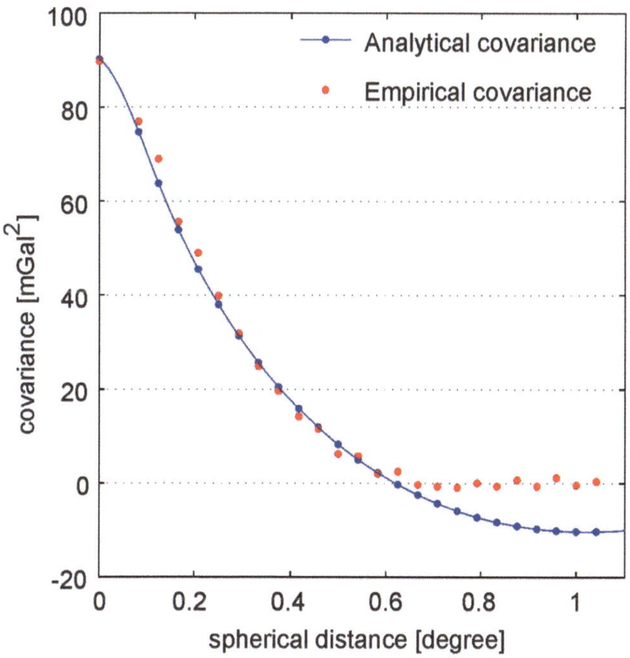


**Fig. 2** Empirical and analytical fitted covariance functions for the residual gravity anomalies from the area bounded by parallels of 11°N and 21°N and meridians 25°E and 35°E

The topography effect on the geoid height (Fig. 3c) was computed from the SRTM30_PLUS using Eq. (7), whereas the reference geoid heights (Fig. 3a) were obtained from the TIM-R4 GGM truncated to d/o 200 using Eq. (6). Finally, the SUD-GM2014 was obtained using Eq. (5).

The major contribution to the SUD-GM2014 comes obviously from the reference geoid heights (Fig. 3a). Large residual geoid heights (Fig. 3b) are correlated with the spatial gaps in gravity measurements (c.f. Fig. 1). The RTM effects (Fig. 3c) on geoid heights range from −25 to 50 cm with an average of −2.4 cm.

The geoid heights from the SUD-GM2014 were interpolated at the sites of GNSS/levelling data using a simple bilinear interpolation method. Thereafter, they have been compared with the corresponding ones obtained from GNSS/levelling data. In order to determine better fit of geoid heights from the SUD-GM2014 model to GNSS/levelling data, 4- (Heiskanen and Moritz 1967), 5- (Kotsakis and Sideris 1999), and 7-parameter (e.g. Fotopoulos 2003) transformation models of corrector surface have been applied (for more details see Godah and Krynski 2015). The statistics of geoid heights differences are given in Table 3.

The results presented in Table 3 show that the SUD-GM2014 fits to GNSS/levelling data before applying the 4-, 5-, and 7-parameter transformation models are at the level of 92 cm and 64 cm in terms of the mean and the standard deviation of differences, respectively. It indicates considerable reduction (almost 43%) in terms of the mean of geoid heights differences compared to the corresponding ones obtained from the KTH-SDG08 presented in Abdalla (2009) and Abdalla and Fairhead (2011). This may indicate the usefulness of using GOCE satellite data for modelling the long wavelength of the gravity signal (e.g. up to d/o 200) in the geoid determination process for the area of Sudan. On the other hand, no improvement in terms of the standard deviations of geoid heights differences was observed when using GOCE-based GGMs. This might be due to the accuracy of GNSS/levelling data used to evaluate the SUD-GM2014 and their number which are insufficient for reliable accuracy assessment of the model investigated. After applying 4-, 5- and 7-parameter transformation for fitting the SUD-GM2014 to GNSS/levelling data, the systematic error corresponding to the mean is totally removed, and standard deviations of differences are clearly reduced. When applying the 7-parametr transformation the standard deviation of differences becomes about 30 cm, which is at the same level as in case of the KTH-SDG08. This may also emphasize that low accuracy, low number, and inhomogeneous distribution of the available GNSS/levelling data do not allow a reliable evaluation of gravimetric geoid models. For reliable evaluation of gravimetric geoid models in Sudan larger number of higher accuracy GNSS/levelling data would be needed.

## 5 Summary and Conclusions

In this paper, a new gravimetric geoid model SUD-GM2014 has been developed for the area of Sudan using the available terrestrial free-air gravity anomalies, the TIM-R4 GGM truncated to d/o 200, and the SRTM30_PLUS. The RCR procedure and LSC have been implemented to develop this model.

The fit of the SUD-GM2014 to the available GNSS/levelling data in terms of the standard deviation of geoid heights differences is 64 cm. After applying the 7-parameter transformation, this fit gets reduced to 30 cm, which almost equals to the one of the latest gravimetric geoid model over the area of Sudan. However, this estimate of geoid model quality might not be fully representative due to low accuracy, very limited number and inhomogeneous distribution of GNSS/levelling heights used. The results have also exhibited that the use of TIM-R4 provides a substantial reduction in terms of the mean of differences, although its value is still large (up to 90 cm), which might imply serious problems within the vertical datum in Sudan. Therefore, re-establishment of the levelling networks and unification of the vertical datum for the area of Sudan is essentially needed.

Since GOCE-based GGMs provide geoid heights with the accuracy of 10–20 cm (Godah et al. 2014) they can play a significant role in the unification of the height system

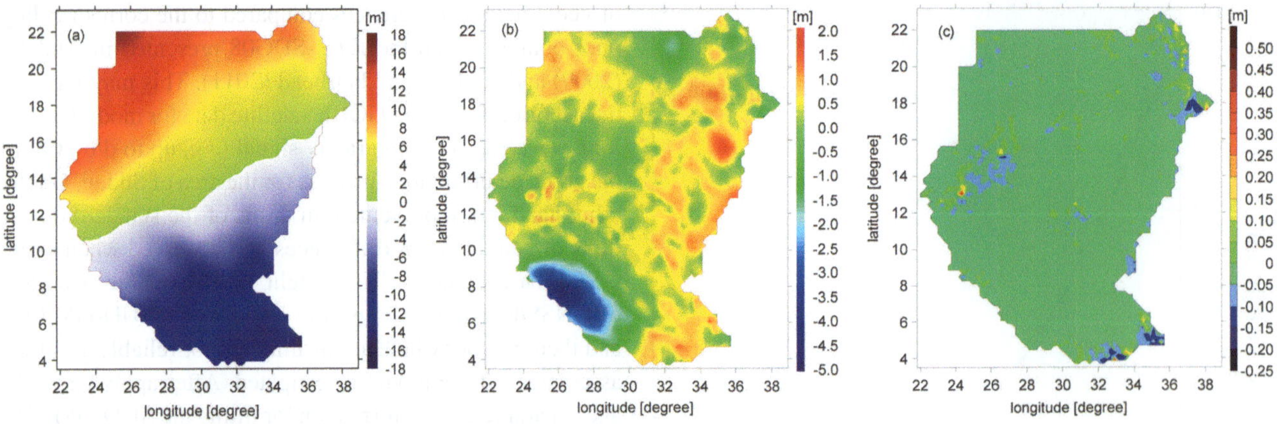


**Fig. 3** (**a**) The reference geoid heights, (**b**) the residual geoid heights, and (**c**) the RTM effects on geoid heights

**Table 3** Statistics of geoid heights differences between obtained from GNSS/levelling data and the SUD-GM2014 before and after applying 4-, 5-, and 7-parameter transformation models (m)

| Statistics | Before fitting | 4-Parameter | 5-Parameter | 7-Parameter |
|---|---|---|---|---|
| Min | 0.096 | –0.715 | –0.708 | –0.513 |
| Max | 2.006 | 0.834 | 0.887 | 0.779 |
| Mean | 0.920 | 0.000 | 0.000 | 0.000 |
| Std dev. | 0.644 | 0.467 | 0.451 | 0.306 |

for the area of Sudan. Moreover, the SUD-GM2014 or the geoid model computed from GOCE-based GGMs only are recommended as reference for GNSS heighting in Sudan.

**Acknowledgements** The authors would like to thank Prof. Derek Fairhead from GETECH, Leeds University for providing terrestrial gravity data for the area of Sudan. This study is a part of the first author PhD research. The University of Khartoum, Sudan, as well as the Institute of Geodesy and Cartography (IGiK), Warsaw, Poland, are kindly acknowledged for their partial financial support. The discussions concerning the research and the preparation of the manuscript with the team of the Centre of Geodesy and Geodynamics of the IGiK are kindly acknowledged.

## References

Abdalla A (2009) Determination of a gravimetric geoid model of Sudan using the KTH Method. MSc Thesis in Geodesy No 3109, Royal Institute of Technology (KTH), Stockholm

Abdalla A, Fairhead D (2011) A new Geoid model for Sudan using the KTH method. J African Earth Sci 60(4):213–221

Abd-Elmotaal A, Seitz K, Kühtreiber N, Heck B (2014) Establishment of the gravity database for the African Geoid. In: 3rd Int Gravity Field Service (IGFS) Gen Ass, 30 June–6 July 2014, Shanghai, China

Adam MO (1967) A geodetic datum for the Sudan. Master Thesis, Cornell Univ., USA

Balmino G, Lambeck K, Kaula WM (1973) A spherical harmonic analysis of Earth's topography. J Geoph Res 78(2):478–481

Becker J, Sandwell DT, Smith WHF, Braud J et al (2009) Global bathymetry and elevation data at 30 arc seconds resolution: SRTM30_PLUS. Mar Geod 32(4):355–371

Fairhead JD, Watts AB, Chevalier P et al (1988) African Gravity Project. Tech Rep, Univ. of Leeds Industrial Services Ltd., Leeds, UK

Fashir HH (1991) The gravimetric geoid of Sudan. J Geodyn 14(1–4):19–36

Fashir HH, Majid A, Kadir A (1997) The Sudanese gravimetric geoid 1998. Int Geoid Service Bull 10:59–77

Floberghagen R, Fehringer M, Lamarre D et al (2011) Mission design, operation and exploitation of the gravity field and steady-state ocean circulation explorer mission. J Geod 85(11):749–758

Forsberg R (1984) A study of terrain reductions, density anomalies and geophysical inversion methods in gravity field modelling. Report No 355, Department of Geodetic Science and Surveying, OSU, Columbus, USA

Forsberg R, Tscherning CC (1981) The use of heights data in gravity field approximation. J Geophys Res 86:7843–7854

Fotopoulos G (2003) An analysis on the optimal combination of geoid, orthometric and ellipsoidal height data. Dissertation No 20185, Department of Geomatics Engineering, Univ. of Calgary, Canada

Godah W, Krynski J (2015) Comparison of GGMs based on one year GOCE observations with the EGM08 and terrestrial data over the area of Sudan. Int J Appl Earth Obs Geoinform 35:128–135

Godah W, Szelachowska M, Krynski J (2014) Accuracy assessment of GOCE-based geopotential models and their use for modelling the gravimetric quasigeoid – a case study for Poland. Geod Cartogr Walter de Gruyter 63(1):3–24

Gruber T, Rummel R (2013) The 4th release of GOCE gravity field models-overview and performance analysis. In: ESA living planet symposium, Edinburgh, 9 Sept 2013

Heiskanen WA, Moritz H (1967) Physical geodesy. W.H. Freeman and Company, San Francisco

Kotsakis C, Sideris MG (1999) On the adjustment of combined GPS/levelling/geoid networks. J Geod 73(8):412–421

Moritz H (1980) Advanced physical geodesy. Wichmann, Karlsruhe

Pail R, Bruinsma S, Migliaccio F et al (2011) First GOCE gravity field models derived by three different approaches. J Geod 85(11):819–843

Pavlis NK, Holmes SA, Kenyon SC, Factor JK (2012) The development and evaluation of the earth gravitational model 2008 (EGM2008). J Geophys Res 117(B04406):1–38

Reigber C, Luhr H, Schwintzer P (2002) CHAMP mission status. Adv Space Res 30:129–134

Rexer M, Hirt C, Pail R, Claessens S (2014) Evaluation of the third- and fourth-generation GOCE Earth gravity field models with Australian terrestrial gravity data in spherical harmonics. J Geod 88(4):319–333

Rummel R, Rapp RH, Sünkel H, Tscherning CC (1988) Comparisons of global topographic/isostatic models to the Earth's observed gravity field. Report No 388, Department of Geod. Sci. and Surv., Ohio State Uni., Columbus, USA

Sjöberg LE (2003) A general model of modifying Stokes' formula and its least squares solution. J Geod 77(7–8):459–464

Tapley BD, Bettadpur S, Watkins M, Reigber C (2004) The gravity recovery and climate experiment: mission overview and early results. Geophys Res Lett 31, L09607

Torge W, Müller J (2012) Geodesy, 4th edn. Walter de Gruyter, Berlin-Boston

Tscherning CC (2013) Geoid determination by 3D least-squares collocation. In: Sanso F, Sideris MG (eds) Geoid determination; theory and methods, lecture notes in Earth system sciences 110. Springer-Verlag, Berlin Heidelberg pp 311–329

Tscherning CC, Rapp RH (1974) Closed covariance expressions for gravity anomalies geoid undulations, and the deflections of the vertical implied by anomaly degree-variance models. Reports No 208, Dept. of Geodetic Science, OSU, Columbus, USA

Tscherning CC, Forsberg R, Knudsen P (1992) The GRAVSOFT package for geoid determination. In: Proceedings of the 1st continental workshop on the Geoid in Europe, Prague, May 1992, Research Institute of Geodesy, Topography and Cartography, Prague, pp 327–334

Yi W, Rummel R (2014) A comparison of GOCE gravitational models with EGM2008. J Geodyn 73:14–22

# Establishment of the Gravity Database AFRGDB_V1.0 for the African Geoid

Hussein A. Abd-Elmotaal, Kurt Seitz, Norbert Kühtreiber, and Bernhard Heck

**Abstract**

In the framework of the IAG African Geoid Project it is needed to have a uniform gridded gravity data set to compute the geoid using Stokes' integral in the frequency domain by 1-D FFT technique. The available gravity data set consists of land point gravity data as well as shipborne and altimetry derived gravity anomaly data. The available gravity data set has a lot of significant gaps allover the continent. The establishment of the gravity database AFRGDB_V1.0 for the African geoid has been carried out using an iterative process employing a tailored reference model and weighted least-squares prediction technique. The point gravity data on land has got the highest precision, while the shipborne and altimetry gravity data got a moderate precision. In each iteration step, the data gaps are filled with the tailored reference model computed at the previous iteration step, getting the lowest precision within the prediction technique. The weighted least-squares prediction technique is thus carried out to estimate gridded gravity anomalies, which are used to estimate a new tailored reference model employing two harmonic analysis techniques. The gravity database on a uniform grid size of $5' \times 5'$ has been established by the developed process. It has been validated with real data. A standard deviation of the residuals is about 9 mgal. 68% of the data points have residuals less than 4 mgals.

**Keywords**

Africa • Gravity • Geoid determination • Harmonic analysis • Tailored geopotential models

## 1 Introduction and Basic Idea

The geoid for Africa is going to be computed using Stokes' integral in the frequency domain by 1-D FFT technique. This needs the gravity data to be known (interpolated) on a regular grid. This paper presents the current status of the ongoing research to establish such a grid from all currently available gravity data sets.

The currently available gravity data set for Africa is full of significantly large gaps. In order to avoid the random freedom of the interpolation solution at those areas, an underlying grid is proposed there. The idea is based on the fact, that the smoother the field, the smaller are the interpolation errors. Thus, the used approach aims to obtain

H.A. Abd-Elmotaal (✉)
Faculty of Engineering, Civil Engineering Department, Minia University, Minia, Egypt
e-mail: abdelmotaal@lycos.com

K. Seitz • B. Heck
Geodetic Institute, Karlsruhe Institute of Technology, Englerstrasse 7, 76128 Karlsruhe, Germany
e-mail: kurt.seitz@kit.edu; bernhard.heck@kit.edu

N. Kühtreiber
Institute of Navigation, Graz University of Technology, Steyrergasse 30, 8010 Graz, Austria
e-mail: norbert.kuehtreiber@tugraz.at


S. Jin, R. Barzaghi (eds.), *IGFS 2014*, International Association of Geodesy Symposia 144,
DOI 10.1007/1345_2015_51

as much smoother residual field as possible. This underlying grid is computed using a tailored geopotential model for Africa, as none of the existing global models fits the African gravity field with the desired accuracy (cf. Abd-Elmotaal 2015).

The tailored geopotential model for Africa is generated employing an iterative process. In each iteration step, the data gaps are filled with the tailored reference model values computed at the previous iteration step. The weighted least-squares prediction technique is thus carried out to estimate gridded gravity anomalies, which are used to compute a new tailored reference model. This iterative process is terminated when two successive tailored models give practically the same results. The latest tailored geopotenial model is thus used to generate the final underlying grid, employing all available gravity data within an unequal weight least-squares prediction technique, to generate the gravity database for Africa.

The available data sets are described and the methodology to create the tailored geopotential model within the window remove-restore technique is explained. The reduced local gravity anomalies for the African data window are gridded in a $30' \times 30'$ grid using the unequal weight least-squares prediction interpolation technique. The local gridded data are merged with the global $30' \times 30'$ gravity anomalies, computed using EGM2008 (Pavlis et al. 2012) till degree and order 360, to establish the data set for computing the tailored geopotential models. The merged $30' \times 30'$ global field is then used to estimate the harmonic coefficients of the tailored reference model by two different harmonic analysis techniques, namely the FFT technique and the least-squares technique.

It should be noted that many researchers have computed tailored geopotential models to best suit their specific areas of interest. The reader may refer, e.g., to Weber and Zomorrodian (1988), Wenzel (1998), Abd-Elmotaal (2007), Abd-Elmotaal et al. (2014).

It is worth mentioning that the first attempt to compute a geoid model for Africa has been made by Merry et al. (2005). In this paper, a $5' \times 5'$ gravity anomaly grid values developed at Leeds University has been used. The remove-restore method, based on the EGM96 geopotential model, was employed to compute the geoid.

## 2 Available Data

### 2.1 Land Data

A number of 96,472 gravity data points on land, which are collected over the past years by the first author from different sources, are available for this investigation. Figure 1a shows their distribution, containing very large data gaps. A gross-error detection has been implemented on the land data set using a smart gross-error detection technique (Abd-Elmotaal and Kühtreiber 2014), based on the least-squares prediction algorithm. The technique works first to estimate the gravity anomaly value at the data station using values of points different from the current data point. It thus compares the estimated value to the data value for possible blunder detection. Hence the technique measures the influence of removing the data value of a current point on the neighbourhood stations. Only if the value of a certain station proves to be a blunder, it is then removed from the database. The free-air gravity anomalies on land range between −624.5 and 452.8 mgal with an average of about 1.9 mgal and a standard deviation of 60.7 mgal.

### 2.2 Shipborne Data

A number of 971,945 shipborne gravity data points are available. Figure 1b shows their distribution, revealing better distribution than that of the land data. Still some data gaps exist, which are filled by altimetry gravity anomalies. A gross-error detection has been implemented on the shipborne data set using an approach described in (Abd-Elmotaal and Makhloof 2013) based on the least-squares prediction algorithm. It estimates the gravity anomaly value at the data station using values of points different from the current data point and then compares the estimated value to the data value for possible blunder detection. The technique works in an iterative scheme till the standard deviation of the residuals (data minus estimated) becomes less than 1.5 mgal. The shipborne free-air gravity anomalies range between −238.3 and 364.8 mgal with an average of about −6.5 mgal and a standard deviation of 40.4 mgal.

### 2.3 Altimetry Data

119,249 altimetry gravity anomalies are available. Figure 1c illustrates the distribution of the altimetry data, showing a regular distribution. A gross-error detection technique, similar to that implemented on the shipborne data, was applied. A study of combining the shipborne and altimetry data (cf. Abd-Elmotaal and Makhloof 2014) has resulted in some gaps along altimetry tracks when they don't match with the shipborne data. The altimetry free-air gravity anomalies range between −172.2 and 172.7 mgal with an average of 4.0 mgal and a standard deviation of 18.6 mgal.

### 2.4 Digital Height Models

For the terrain reduction computation, a set of fine and coarse Digital Height Models (DHM's) is needed. The $30'' \times 30''$ SRTM30+ (Shuttle Radar Topography Mission) (Farr et al.

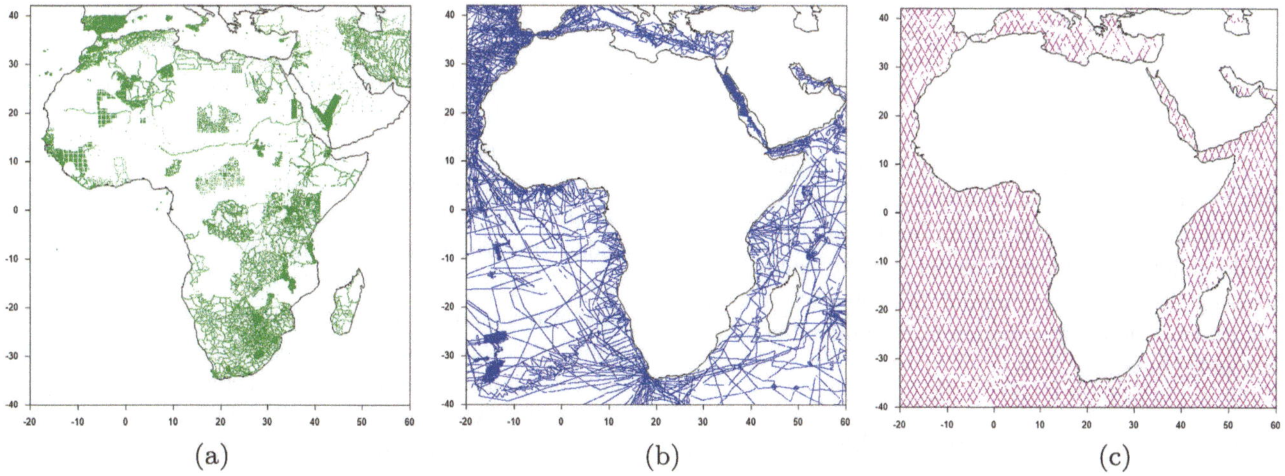


**Fig. 1** Distribution of the (**a**) land, (**b**) shipborne and (**c**) altimetry African free-air gravity anomalies

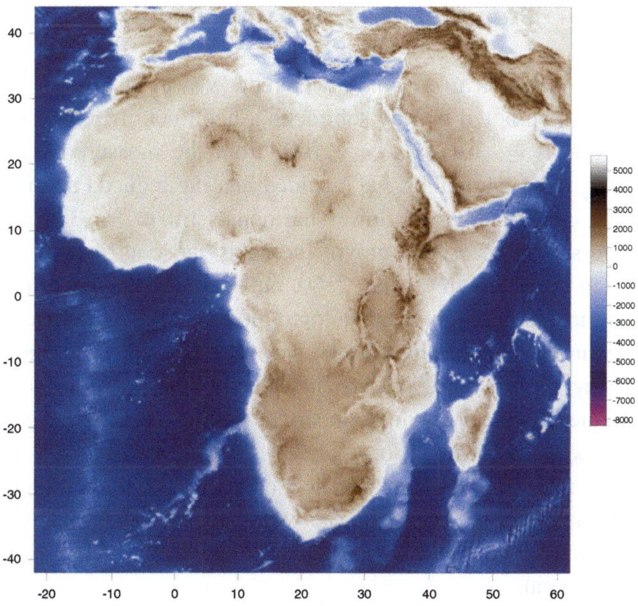

**Fig. 2** The fine 30″ × 30″ SRTM30+ DHM. Units in [m]

2007) is employed as fine DHM, and the 3′×3′ SRTM is used as coarse DHM. Figure 2 illustrates the 30″ × 30″ SRTM30+ fine DHM.

# 3 Methodology

## 3.1 Window Technique

The ultra high-degree tailored geopotential model, used to create the underlying grid, is going to be created in the framework of the window remove-restore technique. The remove step of the window remove-restore technique can mathematically be written as (Abd-Elmotaal and Kuhtreiber 2003) (cf. Fig. 3)

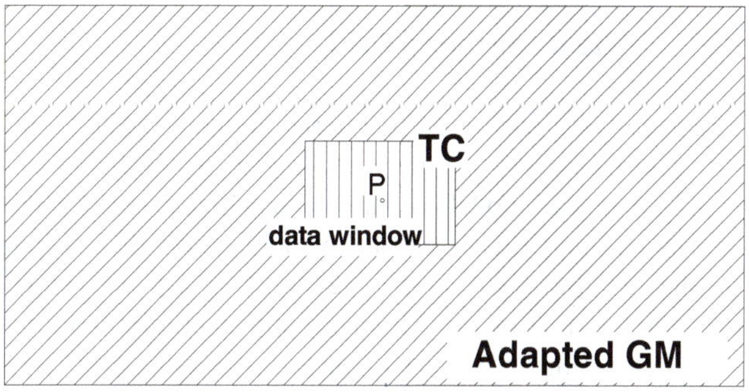


**Fig. 3** The window remove-restore technique

$$\begin{aligned}\Delta g_{red} &= \Delta g_F - \Delta g_{TI\,win} - \Delta g_{GM\,Adapt} \\ &= \Delta g_F - \Delta g_{TI\,win} - (\Delta g_{GM} - \Delta g_{wincof}) \,, \qquad (1)\end{aligned}$$

where $\Delta g_F$ stands for the measured free-air gravity anomalies, $\Delta g_{GM\,Adapt}$ is the contribution of the adapted reference field, $\Delta g_{TI\,win}$ stands for the contribution of the topographic-isostatic masses for a fixed data window, and $\Delta g_{wincof}$ stands for the contribution of the harmonic coefficients of the topographic-isostatic masses of the data window. The contribution of the global geopotential model $\Delta g_{GM}$ (if the maximum available degree of the model is, e.g., 2160) can be decomposed as

$$\Delta g_{GM} = \Delta g_{GM}\Big|_{2\leq n\leq n_{max}} + \Delta g_{GM}\Big|_{n_{max}+1\leq n\leq 2160} \,, \qquad (2)$$

where $n_{max}$ is an arbitrary degree to be used to estimate the harmonic coefficients of the tailored model. In this investigation, a value of $n_{max} = 360$ has been used. Thus (1) can finally be written as

$$\Delta g_{red} = \Delta g_F - \Delta g_{TI\,win} - \Delta g_{GM}\Big|_{2\le n\le n_{max}} - \Delta g_{GM}\Big|_{n_{max}+1\le n\le 2160} + \Delta g_{wincof}\Big|_{2\le n\le n_{max}} . \qquad (3)$$

If we aim to produce, theoretically, zero reduced anomalies by tailoring the geopotential model, then the left hand-side of (3) is put to zero. Keeping the higher harmonics of the tailored geopotential model to their values as of the existing global model (e.g., EGM2008), thus we can compute the gravity anomalies referring to the tailored geopotential model $\Delta g_{GM_T}$ as:

$$\Delta g_{GM_T}\Big|_{2\le n\le n_{max}} = \Delta g_F - \Delta g_{TI\,win} - \Delta g_{GM}\Big|_{n_{max}+1\le n\le 2160} + \Delta g_{wincof}\Big|_{2\le n\le n_{max}} . \qquad (4)$$

## 3.2 The Underlying Grid

The underlying grid for the first iteration step is created using the tailored model of Africa computed in (Abd-Elmotaal et al. 2014). This model is complete to degree and order 360. The underlying grid is generated on a $30' \times 30'$ grid covering the data window ($-40° \le \phi \le 42°; -20° \le \lambda \le 60°$) which exceeds the area of the African continent to eliminate the well-known edge effect. Thus for the underlying grid, $\Delta g_F$ is defined for the first iteration as

$$\Delta g_F = \Delta g_{GM_{AFR2013}}\Big|_{2\le n\le 360} , \qquad (5)$$

where $\Delta g_{GM_{AFR2013}}$ refers to the contribution of the tailored geopotential model computed in (Abd-Elmotaal et al. 2014). Equation (4) then holds true for the reduction step of the underlying grid.

For the successive iteration steps, the free-air anomalies at the underlying grid for the iteration number $i$ are computed using the previous $(i-1)^{th}$ tailored geopotential model $GM_{T_{i-1}}$ as follows

$$\Delta g_F = \Delta g_{GM_{T_{i-1}}}\Big|_{2\le n\le n_{max}} + \Delta g_{GM}\Big|_{n_{max}+1\le n\le 2160} . \qquad (6)$$

where $\Delta g_{GM_{T_{i-1}}}$ provides the contribution of the $(i-1)^{th}$ iteration step of the tailored geopotential model. For that case, the reduction step (4) for computing the $i^{th}$ tailored geopotential model becomes

$$\Delta g_{GM_{T_i}}\Big|_{2\le n\le n_{max}} = \Delta g_{GM_{T_{i-1}}}\Big|_{2\le n\le n_{max}} - \Delta g_{TI\,win} + \Delta g_{wincof}\Big|_{2\le n\le n_{max}} . \qquad (7)$$

This means that the difference between the previous and current gravity anomaly values for the underlying grid equals the difference between the effect of the topographic-isostatic masses computed using TC program (Forsberg 1984; Abd-Elmotaal and Kühtreiber 2003) from the detailed local DHM and their effect in terms of dimensionless potential coefficients computed using Eq. (6) of Abd-Elmotaal and Kühtreiber (2003). In other words, the difference between the previous and current gravity anomaly values for the underlying grid equals the difference between the effect of using the fine and coarse DHM for the data window (mainly short wavelength component). Equation (7) thus assures the stabilization of the solution and guarantees its convergence, since the sum of the second and third components of the right hand side slowly converges to zero, which then comes to our final solution of the tailored geopotential model applied for generating the underlying grid to fill the data gaps.

## 3.3 Preparing the Gravity Anomalies

The local African data have been reduced according to (4). It should be noted that by subtracting the effect of EGM2008 (from $n = 361$ to $n = 2160$) from the measured free-air gravity anomalies, we assumed that the reduced anomalies contain only the lower harmonics till $n = 360$ (of course the measured gravity contains the contribution of even higher degrees than 2160, but we neglect their effect in the computation of the tailored model). The local African reduced anomalies have been interpolated on a $30' \times 30'$ grid using the unequal weight least-squares interpolation technique with the following standard deviations: $\sigma_{land} = 1$ mgal, $\sigma_{shipborne} = 3$ mgal, $\sigma_{altimetry} = 5$ mgal, $\sigma_{underlying\ grid} = 10$ mgal. The African interpolated reduced anomalies are shown in Fig. 4. They range between −268.4 and 323.9 mgal with an average of 0.7 mgal and a standard deviation of 27.0 mgal.

The EGM2008 global geopotential model (Pavlis et al. 2012) has been used, till degree and order 360, to create a $30' \times 30'$ global grid of gravity anomalies. The local African $30' \times 30'$ interpolated gravity anomalies have been merged with the created $30' \times 30'$ global gravity anomalies forming the data set for computing the tailored geopotential model for Africa. Figure 5 shows that merged field, where the boundaries of the African window (local field) are indicated by a solid rectangle.

## 3.4 Harmonic Analysis – Computing the Tailored Geopotential Model for Africa

The lower harmonic coefficients (complete up to degree and order $n_{max} = 360$) of the tailored geopotential model of

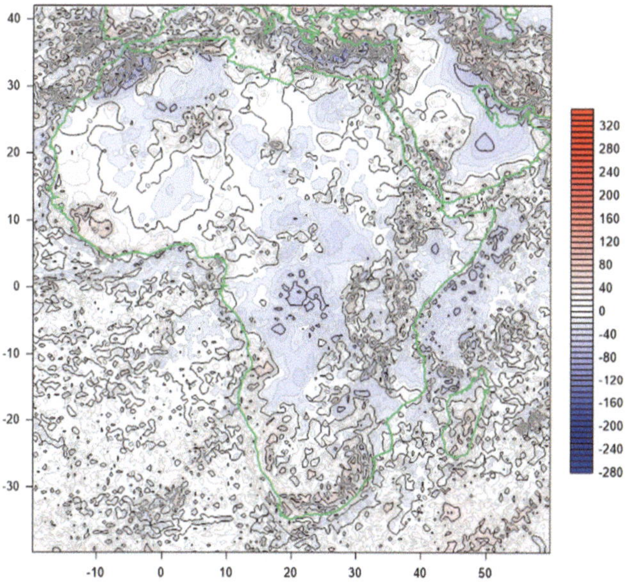


**Fig. 4** The $30' \times 30'$ interpolated reduced gravity anomalies for Africa. Contour interval: 10 mgal

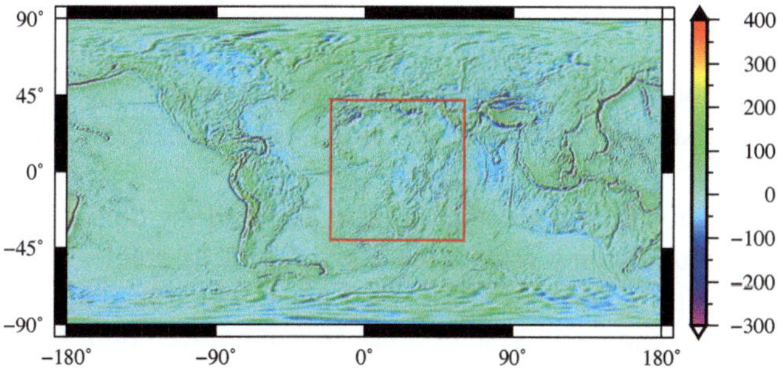


**Fig. 5** The global $30' \times 30'$ merged field including the gravity anomalies for Africa (*solid rectangle*). Units in [mgal]

Africa have been computed using the following harmonic analysis techniques:

- least-squares harmonic analysis technique (Heck and Seitz 1991).
- FFT harmonic analysis technique (Abd-Elmotaal 2004).

The higher harmonic coefficients ($361 \leq n \leq 2160$) are kept as their values of EGM2008, as stated in Sects. 3.1 and 3.3 and explained in Eq. (4).

### 3.5 Iteration Process

An iteration process has been carried out to generate the best tailored geopotential model for Africa. The tailored models converge as already proved mathematically (cf. Eq. (7)). The iteration process has been terminated after two steps when the tailored model doesn't practically improve the residual field (see below).

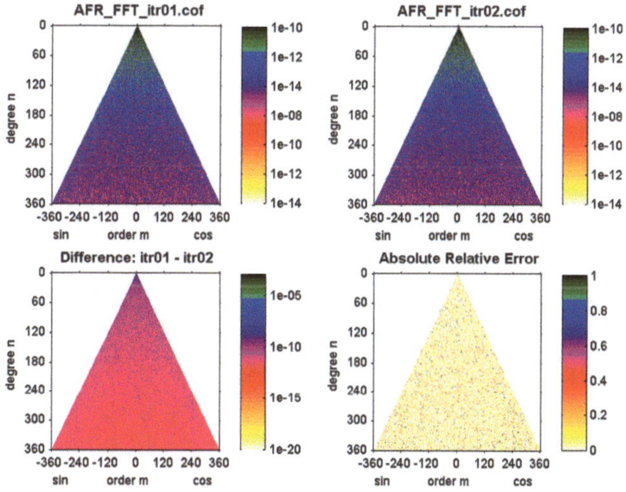


**Fig. 6** The lower harmonic coefficients (complete up to degree and order $n_{max} = 360$) of the tailored geopotential model of Africa, related to the first two iterations, and their differences

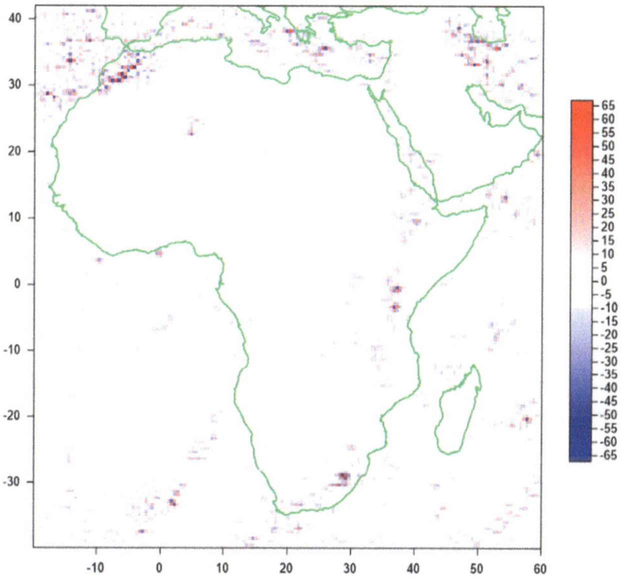


**Fig. 7** Validation in the space domain of the tailored model for Africa. Units in [mgal]

Figure 6 illustrates the lower harmonic coefficients (complete up to degree and order $n_{max} = 360$) of the tailored geopotential model of Africa for the first two iterations and their differences. It shows a small change of the harmonic coefficients of the second iteration step. It is worth mentioning that the ultra-high degree tailored geopotential model for Africa is complete to degree and order 2160 as stated in Sects. 3.1 and 3.3.

Figure 7 shows the validation in the space domain of the tailored model for Africa comparing the gridded data with the synthesized values. The residuals are perfectly unbiased with a standard deviation of about 5.9 mgal. Figure 7 shows

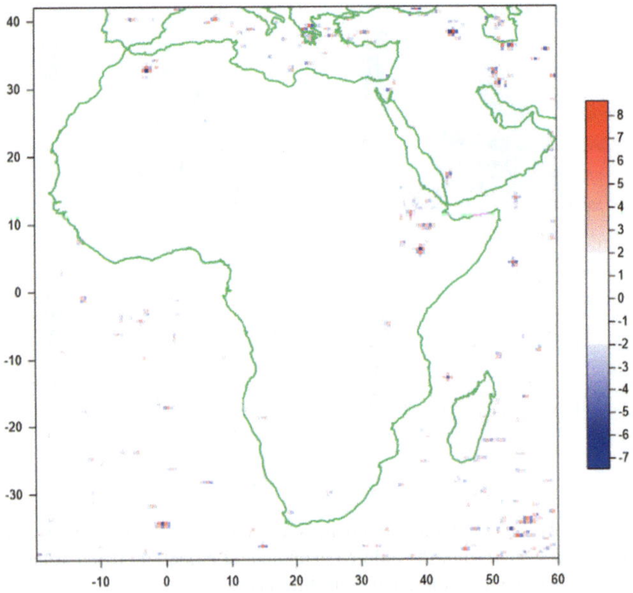


**Fig. 8** Difference between the first and second iteration in the space domain. Units in [mgal]

that most of the area, especially on land, has residuals below 10 mgals (the white pattern).

The difference between the first and second iteration in the space domain is provided in Fig. 8. It shows that the differences are practically very small having a standard deviation of only 0.7 mgal. Therefore it has been decided to terminate the iteration process after the second iteration.

It should be mentioned that both harmonic analysis techniques used in the current investigation give comparable results. We show here only the results of the FFT harmonic analysis technique because of the limited space.

## 4 Establishment of the Gravity Database of Africa

### 4.1 Reduction

In the framework of the window remove-restore technique, and after computing the ultra-high tailored geopotential model for Africa, the reduced gravity anomalies $\Delta g_{red}$ for all actual data types (land, shipborne and altimetry) within the African window can be computed by

$$\begin{aligned}\Delta g_{red} = \Delta g_F - \Delta g_{TI\,win} - \Delta g_{GM_T}\Big|_{2\le n\le n_{max}} \\ - \Delta g_{GM}\Big|_{n_{max}+1\le n\le 2160} + \Delta g_{wincof}\Big|_{2\le n\le n_{max}} .\end{aligned} \quad (8)$$

For the underlying grid, the gravity anomaly $\Delta g_F$ is given by

$$\begin{aligned}\Delta g_F &= \Delta g_{GM_T}\Big|_{2\le n\le 2160} \\ &= \Delta g_{GM_T}\Big|_{2\le n\le n_{max}} + \Delta g_{GM}\Big|_{n_{max}+1\le n\le 2160} .\end{aligned} \quad (9)$$

Thus by inserting (9) into (8), the reduced gravity anomalies $\Delta g_{red}$ for the underlying grid is given by

$$\Delta g_{red} = \Delta g_{wincof}\Big|_{2\le n\le n_{max}} - \Delta g_{TI\,win} . \quad (10)$$

### 4.2 Interpolation

An unequal weight least-squares interpolation process was applied for the African window ($-40° \le \phi \le 42°$; $-20° \le \lambda \le 60°$) on a $5' \times 5'$ grid yielding the interpolated gridded reduced anomalies $\Delta g^G_{red}$. The relative standard deviations for the different gravity anomaly types are set as follows: $\sigma_{land} = 1$ mgal, $\sigma_{shipborne} = 3$ mgal, $\sigma_{altimetry} = 5$ mgal, $\sigma_{underlying\ grid} = 20$ mgal. The underlying grid is computed using Eq. (10) on a $15' \times 15'$ unregistered (non-identical) grid with the final output grid.

It should be noted that in order to avoid the domination of the underlying grid on the gravity field, its precision has been downgraded from 10 to 20 mgal, which matches the fact that the underlying grid has been employed on a $30' \times 30'$ resolution for computing the tailored model, and on a $15' \times 15'$ resolution for computing the interpolated gridded reduced anomalies $\Delta g^G_{red}$.

### 4.3 Restore

The restore step, which takes place on the $5' \times 5'$ grid points, reads

$$\begin{aligned}\Delta g^G_F = \Delta g^G_{red} + \Delta g^G_{TI\,win} + \Delta g^G_{GM_T}\Big|_{2\le n\le n_{max}} \\ + \Delta g^G_{GM}\Big|_{n_{max}+1\le n\le 2160} - \Delta g^G_{wincof}\Big|_{2\le n\le n_{max}} ,\end{aligned} \quad (11)$$

where the superscript $G$ stands for values computed at the grid points. Thus, the established grid database $\Delta g^G_F$, computed by (11), is already on the free-air scale.

Figure 9 shows the $5' \times 5'$ African free-air gravity anomaly database, ver. 1.0 (AFRGDB_V1.0). The anomalies range between −321.3 and 759.5 mgal with an average of 1.2 mgal and a standard deviation of 35.4 mgal.

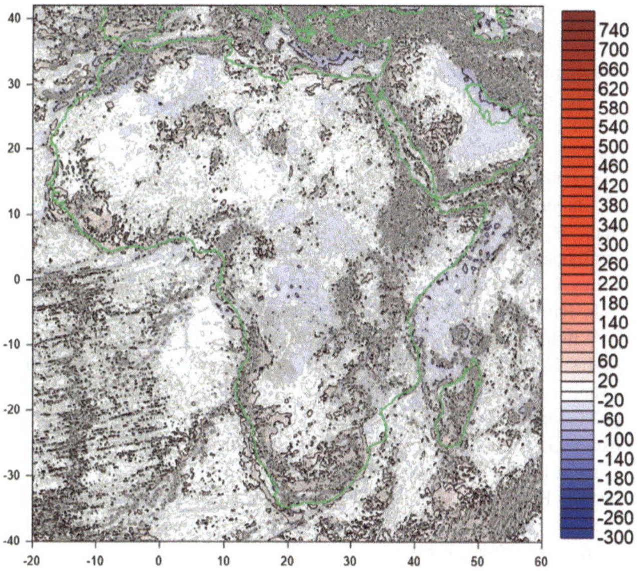


**Fig. 9** The $5' \times 5'$ African free-air gravity anomaly database AFRGDB_V1.0. Contour interval: 20 mgal

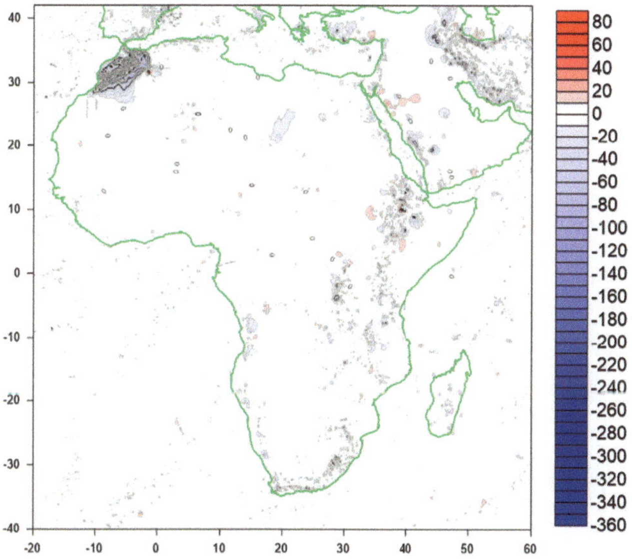


**Fig. 10** Residuals at the data points between measured values and the $5' \times 5'$ African free-air gravity anomaly database AFRGDB_V1.0. Contour interval: 10 mgal

## 4.4 Validation

To estimate the quality of the established free-air gravity anomaly database of Africa AFRGDB_V1.0, the residuals at the data points have been computed. Figure 10 shows the residuals between the measured and the database values at the data points. These residuals range between −371.7 and 99.0 mgal with an average of −0.6 mgal and a standard

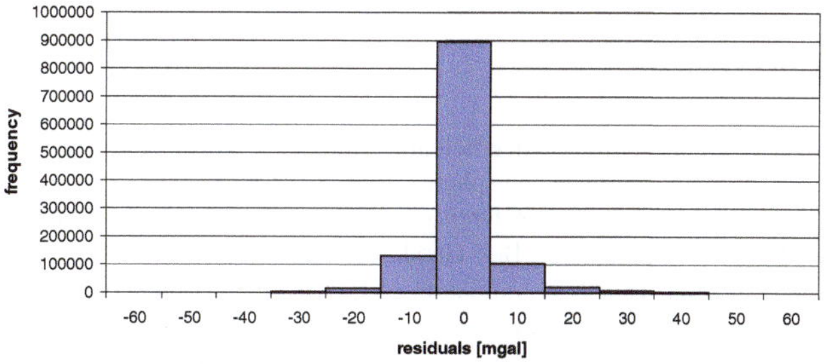


**Fig. 11** Histogram of comparing the $5' \times 5'$ African free-air gravity anomaly database AFRGDB_V1.0 with real data

deviation of 9.8 mgal. Figure 10 shows that most of the area has residuals below 10 mgal (the white pattern).

It is worth mentioning that Fig. 10 shows significantly large residuals at the high mountainous area of Morocco, which is likely due to vertical datum inconsistency. Deeper investigation for this area is essentially needed.

Figure 11 illustrates a histogram of comparing the established free-air gravity anomaly database of Africa AFRGDB_V1.0 with the real point data. The differences show Gaussian normal distribution with high precision index, which indicates a high precision of the established gravity database. 90.4% of the data points have differences less than 10 mgals, and 68.4% of the data points have differences less than 3.9 mgals.

## 5 Summary and Conclusion

An ultra high-degree tailored geopotential model for Africa, complete to degree and order 2160, has been developed. This model, created within a successful iterative process, has been used to fill in the gaps within the African gravity data set. An unequal weight least-squares prediction technique is employed to generate the gridded $5' \times 5'$ gravity anomalies, defining version 1.0 of the gravity database of Africa AFRGDB_V1.0.

The established $5' \times 5'$ African free-air gravity anomaly database AFRGDB_V1.0 has been validated using real point data. The created AFRGDB_V1.0 has been compared with the available point data at land and sea areas for a set of about 1.2 million points. Differences follow perfectly a Gaussian normal distribution. 90.4% of the data points have differences less than 10 mgals, and 68.4% of the data points have differences less than 3.9 mgals.

It should be noted that both the developed ultra high-degree tailored model for Africa (complete to d/o 2160) and the gravity database of Africa AFRGDB_V1.0 will be available from the IAG African Geoid Project webpage (mu.minia.edu.eg/Geodesy/AFRgeo).

## References

Abd-Elmotaal H (2004) An efficient technique for harmonic analysis on a spheroid (ellipsoid and sphere). VGI 3(4):126–135

Abd-Elmotaal H (2007) Reference geopotential models tailored to the Egyptian gravity field. Boll Geod Sci Affini 66(3):129–144

Abd-Elmotaal H (2015) Validation of GOCE models in Africa. Newton's Bulletin 5 (in press)

Abd-Elmotaal H, Kühtreiber N (2003) Geoid determination using adapted reference field, seismic Moho depths and variable density contrast. J Geod 77(1–2):77–85

Abd-Elmotaal H, Kühtreiber N (2014) Automated gross error detection technique applied to the gravity database of Africa. In: General Assembly of the European Geosciences Union, Vienna, Austria, April 27 to May 2, 2014

Abd-Elmotaal H, Makhloof A (2013) Gross-errors detection in the shipborne gravity data set for Africa. In: Geodetic Week, Essen, Germany, October 8–10, 2013

Abd-Elmotaal H, Makhloof A (2014) Combination between altimetry and shipborne gravity data for Africa. In: 3rd International Gravity Field Service (IGFS) General Assembly, Shanghai, China, June 30 to July 6, 2014

Abd-Elmotaal H, Seitz K, Abd-Elbaky M, Heck B (2014) Tailored reference geopotential model for Africa. In: International Association of Geodesy Symposia 143

Farr T, Rosen P, Caro E, Crippen R, Duren R, Hensley S, Kobrick M, Paller M, Rodriguez E, Roth L, Seal D, Shaffer S, Shimada J, Umland J, Werner M, Oskin M, Burbank D, Alsdorf D (2007) The shuttle radar topography mission. Rev Geophys 45(RG2004). doi:110.1029/2005RG000183

Forsberg R (1984) A study of terrain reductions, density anomalies and geophysical inversion methods in gravity field modelling. Ohio State University, Department of Geodetic Science and Surveying, Rep 355

Heck B, Seitz K (1991) Harmonische Analyse. Technical Report, Geodetic Institute, University of Karlsruhe

Merry C, Blitzkow D, Abd-Elmotaal H, Fashir H, John S, Podmore F, Fairhead J (2005) A preliminary geoid model for Africa. In: A Window on the Future of Geodesy, pp 374–379

Pavlis N, Holmes S, Kenyon S, Factor J (2012) The development and evaluation of the earth gravitational model 2008 (EGM2008). J Geophys Res 117(B04406). doi:10.1029/2011JB008916

Weber G, Zomorrodian H (1988) Regional geopotential model improvement for the Iranian geoid determination. Bull Geod 62:125–141

Wenzel H (1998) Ultra high degree geopotential models GPM98A, B and C to degree 1800 tailored to Europe. Reports Finnish Geodetic Institute 98(4):71–80

# New Geoid Model in the State of São Paulo

G.N. Guimarães, A.C.O.C. de Matos, and D. Blitzkow

**Abstract**

The purpose of this manuscript is to present some efforts in terms of gravity measurements in the State of São Paulo with the aim to improve the geoid model, and to show the establishment of an absolute gravity network in this state. The efforts resulted in a geoid model called GEOIDSP, limited by 19°S and 26°S in latitude and 44°W and 54°W in longitude, which has been derived using the modified Stokes' integral through Fast Fourier Transform (FFT). Another objective of this study is to verify the potentiality of GOCE-based models. The spectral decomposition was employed in the geoid models computation and the long wavelength component was represented by GOCE-based models up to degree and order 200. The models were compared in terms of absolute comparisons from GPS/leveling and the results show consistency between them. EIGEN6C3STAT model and GO_CONS_GCF_2_DIR_R4 presented the same results (0.18 m), in terms of RMS, while GO_CONS_GCF_2_TIM_R4, 0.20 m. In the case of absolute gravity measures, 18 stations are being measured (15 news and 3 re-occupations).

**Keywords**

Recent GGM • São Paulo geoid modeling • Spectral decomposition

## 1 Introduction

A new version of the geoid model for the State of São Paulo was computed and it is limited by 26°S and 19°S in latitude and 54°W and 44°W in longitude. The model was computed by using three Global Gravitational Models (GGMs) as reference field: GO_CONS_GCF_2_DIR_R4 (Bruinsma et al. 2013), GO_CONS_GCF_2_TIM_R4 (Pail et al. 2011) and EIGEN6C3stat (Förste et al. 2012).

G.N. Guimarães (✉)
Institute of Geography, University Federal of Uberlândia, IGUFU, 38500-000 Monte Carmelo, Minas Gerais, Brazil
e-mail: gabriel@ig.ufu.br

A.C.O.C. de Matos • D. Blitzkow
Laboratory of Topography and Geodesy, Department of Transportation, University of São Paulo, EPUSP-PTR, P.O. Box 61548, 05424-970 Butanta, Sao Paulo, Brazil
e-mail: acocmatos@gmail.com; dblitzko@usp.br

The reduced Helmert mean gravity anomalies were estimated in blocks of 5′ for continental areas. The oceanic region was completed with the mean free-air anomalies derived from a satellite altimetry model by the Danish National Space Center called DTU10 (Andersen 2010). The digital terrain model SAM3s_v2, with a grid size of $3'' \times 3''$ ($\sim$90 × 90 m), was chosen for computing the new geoid. The complete Bouguer, Helmert gravity anomalies and primary indirect topographical effect have been derived through the Canadian package SHGEO. The short wavelength component (residual co-geoid) was estimated via FFT and the modified Stokes kernel proposed by Featherstone (2003) was used. A set of 170 GPS/leveling stations were selected in the comparisons with the geoid model and the recent GGMs.

In this manuscript, Sect. 2 is about recent gravity measurements in the State of São Paulo, while Sect. 3 presents the geoid model estimation procedures. Section 4 shows the data set used in the computation. Finally, Sect. 5 are the results and discussion and Sect. 6 the conclusions.


S. Jin, R. Barzaghi (eds.), *IGFS 2014*, International Association of Geodesy Symposia 144,
DOI 10.1007/1345_2015_131

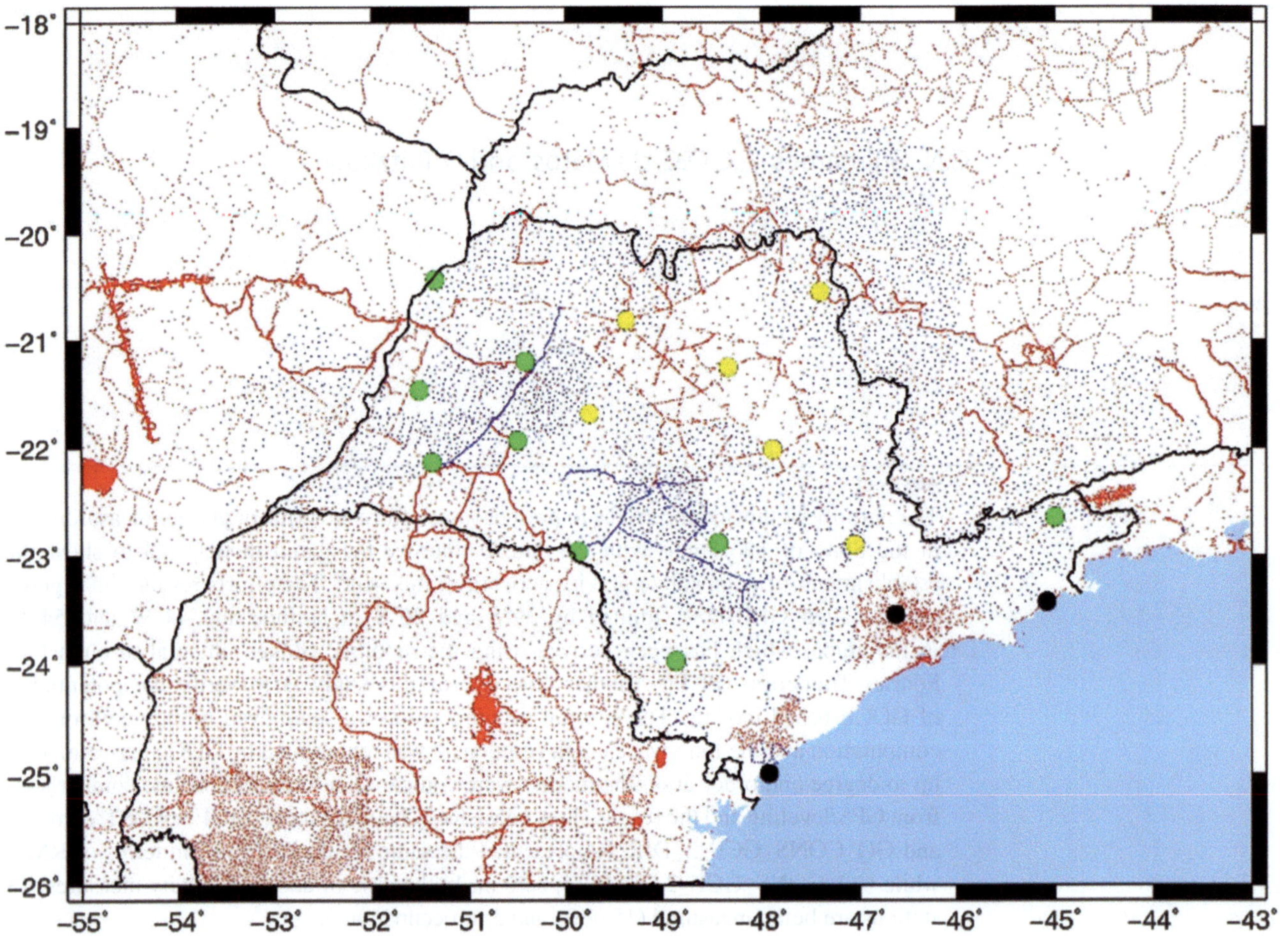


**Fig. 1** Distribution of gravity measurements in the State of São Paulo

## 2 Gravity Coverage in the State of São Paulo

September 2008 saw the start of the thematic project, and since then gravimetric campaigns have been carried out. A whole of 5,822 stations were measured during this time. Figure 1 shows the distribution of the gravimetric stations in the state of São Paulo and its surroundings. The red points are the stations surveyed up to 2008 and the blue points from 2008 up to 2014. It is important to mention that several institutions collaborated to collect the data: the National Observatory, PETROBRAS, National Petroleum Agency and Brazilian Institute of Geography and Statistics. In the South and North of the state, there are some lacks due the mountains area.

Since the beginning of 2014, some efforts have been carried out to establish an absolute gravity network. The absolute gravimeter A-10 is portable and can determine the value of the gravity acceleration with an accuracy of 10 μGal. Its efficiency and accuracy is higher than that of relative gravimeters. The establishment of 18 stations with absolute values of gravity will allow a greater efficiency and accuracy in traditional gravity surveys. Black dots represent existing stations measured in the past, yellow dots are new stations measured in 2014 and green dots are future stations.

The measurements have been conducted in the universities or agencies that already have a stable pillars or a surface with minimal disturbance. In this sense, external noises does not influence the observations.

## 3 Geoid Model Computation

The schedule to determine the geoid model using Fast Fourier Transform (FFT) can be described in five steps (Blitzkow et al. 2008):

1. Calculation of point free air gravity anomalies (FA) through terrestrial gravimetric data (coordinates, orthometric height and gravity acceleration);
2. Calculation of complete Bouguer anomalies in order to derive mean free air gravity anomalies. The $5' \times 5'$ grid of these anomalies was computed from point gravity data. Over the ocean, DTU10 (Andersen 2010) was used;

**Table 1** Statistics of the anomalies computation

| mGal | FA | HGES |
|---|---|---|
| Mean | −4.86 | −4.91 |
| Stdev | 24.68 | 24.97 |
| Máx. | 134.29 | 134.11 |
| Min. | −96.59 | −97.93 |

3. Calculation of Helmert gravity anomalies referred to the surface of the Earth (HGES), which are obtained from the mean free air anomaly by adding Direct Topographical Effect, Direct Atmospheric Effect and Secondary Indirect Topographical Effect (Ellmann and Vaníček 2007);
4. Stokes' integration with the use of the spectral decomposition to calculate the co-geoid. The modified Stokes' kernel was computed according to Featherstone (2003). It was used the remove-restore technique to remove the long-wavelength component from the geopotential model and the residual terrain correction;
5. Primary Indirect Topographical Effect was added to co-geoid heights to obtain geoid heights (Martinec and Vaníček 1994) and (Martinec 1998).

The statistics of FA and HGES computation can be seen in Table 1.

## 4 Data Set

Figure 2 presents the study area delimited by the smallest square. This area includes the State of São Paulo, as well as some of its and surroundings, and extends from 26°–19° South in latitude and 54°–44° West in longitude. The medium square represents the gravity data area and it is limited by 28°–17° South in latitude and 56°–42° West in longitude. The largest square is where the Digital Terrain Model and the Digital Bathymetric Model exists. It is one degree larger than the gravity area.

In this paper, the Brazilian national gravity data set was used. The study area consists of 46,290 stations (Fig. 1). The accuracy of the Brazilian terrestrial gravity data is at the 0.1 mGal level or better (Blitzkow et al. 2010). The gravity information was validated by a package called DIVA developed by *Bureau Gravimétrique International* (BGI) (Toustou 1991).

In terms of digital terrain model, for the present study, a suitable grid model for the topography with a grid size of $3'' \times 3''$ (approximately $90 \times 90$ m) from SAM3s_v2 (Matos and Blitzkow 2008) was used. This model consists of SRTM3 data (Farr et al. 2007), but EGM96 (Lemoine et al. 1998a, 1998b) geoid heights used in the SRTM3 were replaced by EIGEN-GL04C (Förste et al. 2006) geoid heighs, in order to derive the orthometric height. For the ocean, the model DTU10 was used.

In this manuscript a set of 170 GPS/leveling stations were selected in the comparisons with the geoid model and the recent GGMs. The spirit leveling was carried out by the Brazilian surveying institute and IGC (*Instituto Geográfico e Cartográfico*). The orthometric heights are referred to a local height datum (Imbituba tide gauge).

The geoid model of the State of São Paulo was computed by using three GGMs as reference field: GO_CONS_GCF_2_DIR_R4 (Bruinsma et al. 2013), GO_CONS_GCF_2_TIM_R4 (Pail et al. 2011) and EIGEN6C3stat (Förste et al. 2012).

In the spectral decomposition, the long wavelength component was represented up to degree and order 200. The reason is that the models lost accuracy for resolutions above degree and order 210. This suggests that GOCE-based models do not present the same performance for the spectral band of spherical harmonics above degree 210. This agrees with demonstrations of similar results in other regions (Gruber et al. 2011).

## 5 Results and Discussions

The absolute comparison allows the analysis on how consistent is the geoid model $N$ (GGM + terrestrial gravity) and the $N$ from GPS/leveling stations in relation to geoid height. The comparison between these two quantities has been performed in terms of root mean square errors and it is presented in Table 2.

Figure 3 shows the RMS between GPS/leveling and geoid models using GGMs CONS_GCF_2_DIR_R4, CONS_GCF_2_TIM_R4 and EIGEN6C3STAT, respectively.

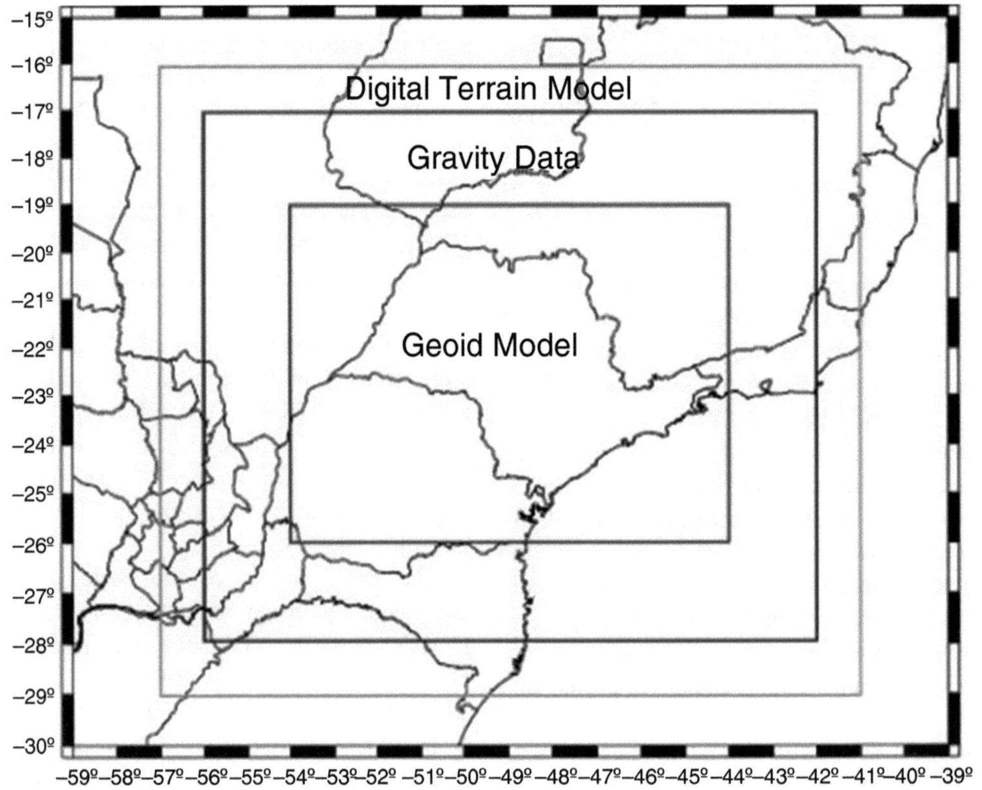


**Fig. 2** Data area

**Table 2** Statistics of the geoidal models computed in the State of São Paulo

| Geoid model | Mean (m) | RMS (m) | Max. (m) | Min. (m) |
|---|---|---|---|---|
| GO_CONS_GCF_2_DIR_R4 | 0.12 | 0.18 | 0.59 | −0.37 |
| GO_CONS_GCF_2_TIM_R4 | 0.11 | 0.20 | 0.51 | −0.43 |
| EIGEN6C3STAT | 0.11 | 0.18 | 0.59 | −0.37 |

Figure 4 shows the comparisons between GPS/leveling geoid undulation and GGM height anomaly. The horizontal spatial variations of TIM_R4 and DIR_R4 residuals $N_{GPS/BM} - N_{GGM}$ did not reveal any particular systematic pattern within the test network. Both their latitude-dependent and longitude-dependent scatter plots are free of any sizeable north/south or east/west tilts.

## 6 Conclusions

The geoid model in the State of São Paulo was computed with basis in Stokes' integral applying Fast Fourier Transform. The computation was performed by using, DIR_R4, TIM_R4, EIGEN-6C3stat as the reference field for the long wavelength component. The comparisons were carried out to verify the quality and consistency of the models. In the first comparison, EIGEN6C3STAT and DIR_R4 show the same results (0.18 m) in terms of RMS, while TIM_R4 is in the level of 0.20 m. The second evaluation shows the comparisons between GPS/leveling geoid undulation and GGM height anomaly in terms of latitude, longitude and height.

**Fig. 3** Difference between GPS/leveling and the geoidal models using the GGMs GO_CONS_GCF_2_DIR_R4, GO_CONS_GCF_2_TIM_R4 and EIGEN6C3STAT, respectively

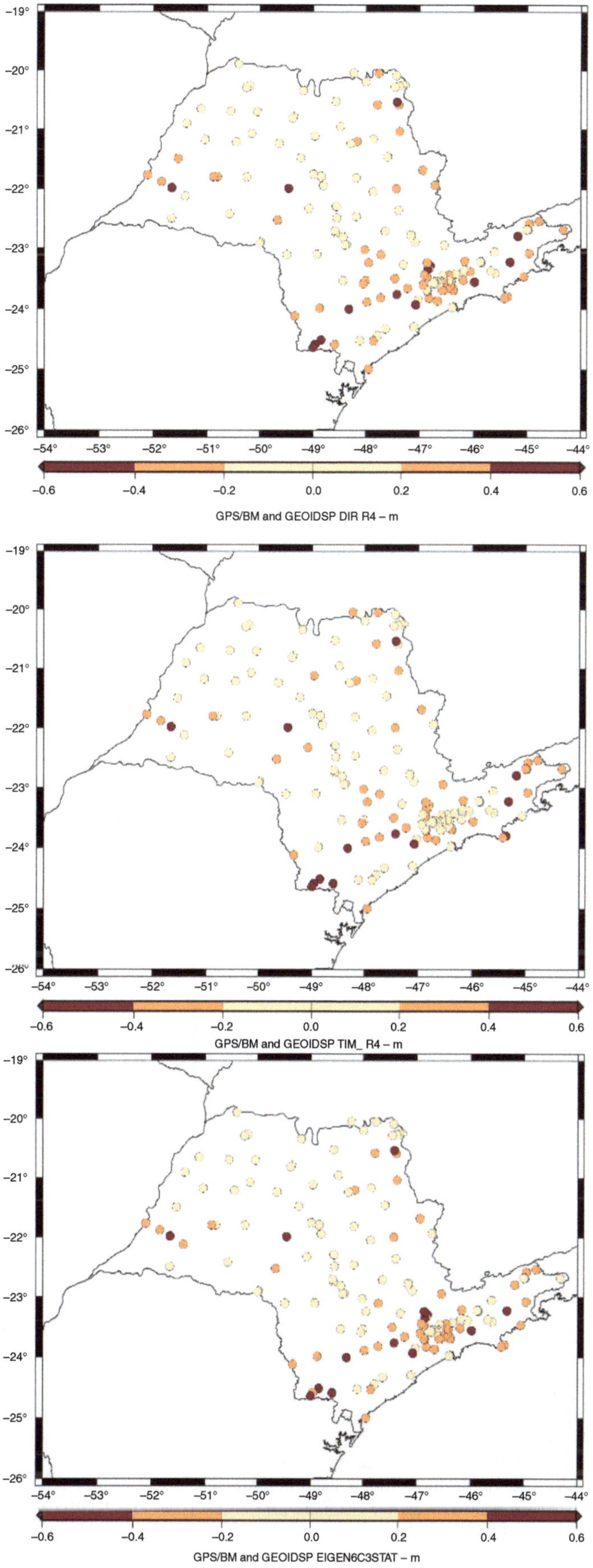

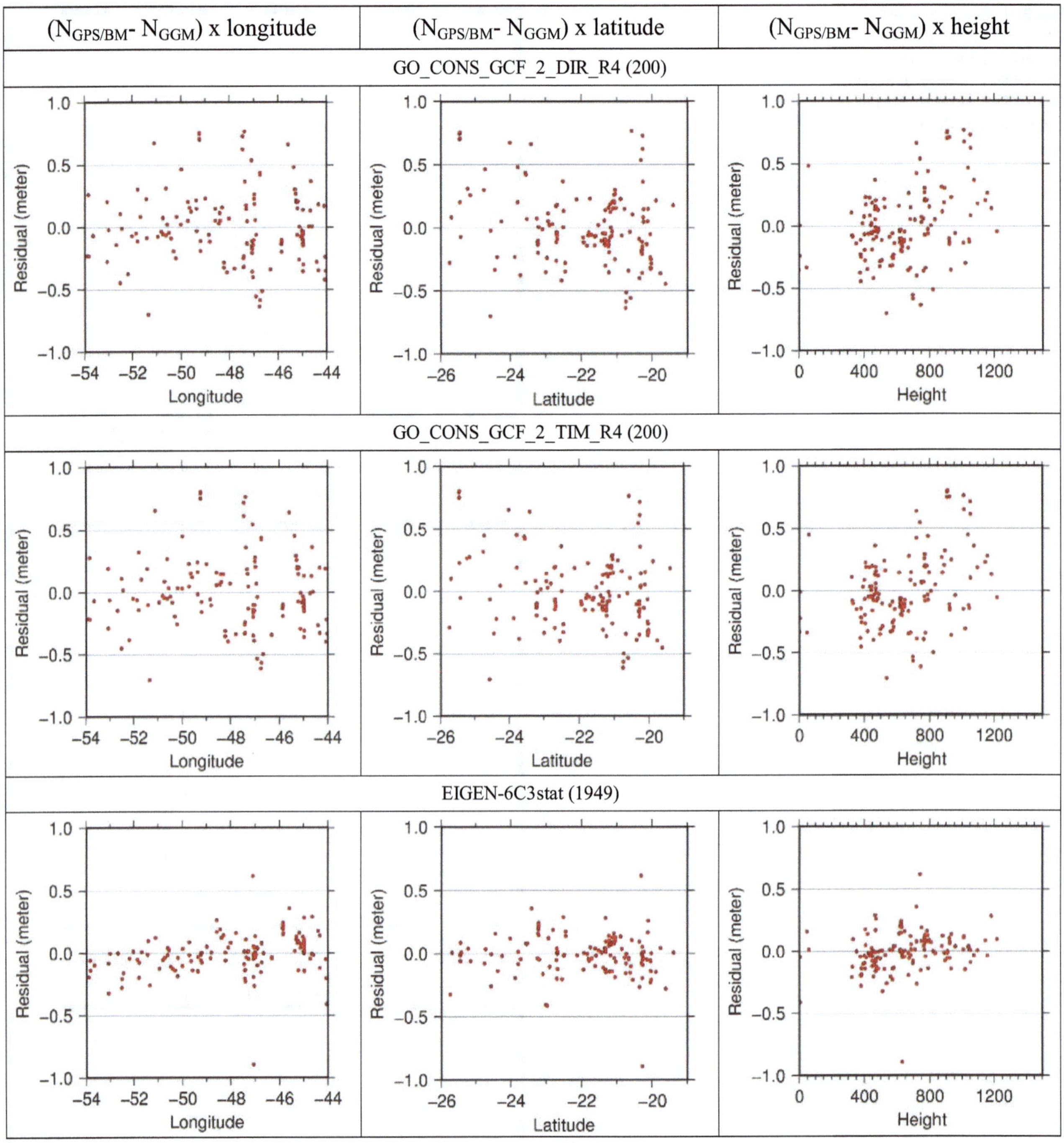


**Fig. 4** Comparisons between GPS/RN geoid undulation and GGM height anomaly in terms of latitude, longitude and height

It is worth mentioning that it is important to cover the lacks of gravity data in the State of São Paulo and its surrounding area. In this way, the geoid model would be improved.

**Acknowledgements** The first author acknowledges CAPES for the grant that allowed him to participate in the IGFS meeting. The state of São Paulo benefited from a sharp improvement in the distribution of gravity data through the thematic project sponsored by FAPESP (Process: 2006/04008-2).

## References

Andersen OB (2010) The DTU10 gravity field and mean sea surface. Second international symposium of the gravity field service – IGFS2 20–22 September 2010 Fairbanks, Alaska. http://www.space.dtu.dk/english/Research/Scientific_data_and_models/Global_Mean_sea_surface

Blitzkow D, Matos ACOC, Campos IO, Ellmann A, Vaníček P, Santos MC (2008) An attempt for an Amazon geoid model using Helmert gravity anomaly. In: Sideris MG (Org) Observing our changing earth, 1 edn, vol 133. Springer, Berlin, pp 187–194

Blitzkow D, Matos ACOC, Guimarães GN, Lobianco MCB (2010) Recent progress of the geoid in South America. 42° Reunión del Consejo Directivo del Instituto Panamericano de Geografía e Historia (IPGH), Lima, Peru. http://www.sirgas.org/fileadmin/docs/Boletines/Bol15/42_Blitzkow_et_al_Geoid_in_South_America.pdf

Bruinsma SL, Foerste C, Abrikosov O, Marty JC, Rio MH, Mulet S, Bonvalot S (2013) The new ESA satellite-only gravity field model via the direct approach. Geophys Res Lett 40:3607–3612

Ellmann A, Vaníček P (2007) UNB applications of Stokes-Helmert's approach to geoid computation. J Geodyn 43:200–213

Farr TG, Rosen PA, Caro E, Crippen R, Duren R, Hensley S, Kobrick M, Paller M, Rodriguez E, Roth L, Seal D, Shaffer S, Shimada J, Umland J, Werner M, Oskin M, Burbank D, Alsdorf D. (2007) The shuttle radar topography mission. Rev Geophys 45(2). doi:10.1029/2005RG000183/pdf

Featherstone WE (2003) Software for computing five existing types of deterministically modified integration kernel for gravimetric geoid determination. Comput Geosci 29:183–193, http://www.iamg.org/CGEditor/index.html

Förste C, Flechtner F, Schmidt R, Meyer U, Stubenvoll R, Barthelmes F, König R, Neumayer K, Rothacher M, Reigber C, Biancale R, Bruinsma S, Lemoine J, Raimondo J (2006) Mean global gravity field model from the combination of satellite mission and altimetry/gravimetry surface data – EIGEN-GL04C. Geophys Res Abst 8

Förste C, Bruinsma S, Flechtner F, Marty J-C, Lemoine J-M, Dahle C, Abrykosov O, Neumayer K-H, Biancale R, Barthelmes F, Balmino G (2012) A preliminary update of the direct approach GOCE processing and a new release of EIGEN-6C, AGU 2012 Fall Meeting San Francisco, USA. http://fallmeeting.agu.org/2012/eposter/g31b-0923/

Gruber T, Visser PNAM, Ackermann C, Hosse M (2011) Validation of GOCE gravity field models by means orbit residuals and geoid comparisons. J Geod 85:845–860

Lemoine FG, Pavlis NK, Kenyon SC, Rapp RH, Pavlis EC, Chao BF (1998a) New high-resolution model developed for earth' gravitational field. EOS Trans AGU 79(9):117–118. doi:10.1029/98EO00076/pdf

Lemoine FG, Kenyon SC, Factor JK, Trimmer RG, Pavlis NK, Chinn DS, Cox CM, Klosko SM, Luthcke SB, Torrence MH, Wang YM, Williamson RG, Pavlis EC, Rapp RH, Olson TR (1998b) The development of the joint NASA GSFC and the national imagery and mapping agency (NIMA) geopotential model EGM96, NASA/TP-1998-206861. National Aeronautics and Space Administration, Maryland, USA. http://bowie.gsfc.nasa.gov/697/staff/lemoine/EGM96_NASA-TP-1998-206861.pdf

Martinec Z (1998) Boundary-value problems for gravimetric determination of a precise geoid. Lecture Notes Earth Sci 73

Martinec Z, Vaníček P (1994) Direct topographical effect of Helmert's condensation for a spherical approximation of the geoid. Manuscr Geod 19:257–268

Matos ACOC, Blitzkow D (2008) Modelagem Digital de Terrenos (MDT) de 3′′ para a América do Sul Pos-Doc Report – Escola Politécnica, Universidade de São Paulo, São Paulo. http://www.ptr.poli.usp.br/ltg/proj/proj26.htm

Pail R, Bruinsma S, Migliaccio F, Förste C, Goiginger H, Schuh W-D, Höck E, Reguzzoni M, Brockmann JM, Abrikosov O, Veicherts M, Fecher T, Mayrhofer R, Krasbutter I, Sanso F, Tscherning CC (2011) First GOCE gravity field models derived by three different approaches. J Geod 85:819–843

Toustou D (1991) Chaîne de validation interactive de données gravimétriques: DIVA. Note technique No. 10, Bureau Gravimétrique International (in French)

# Accurate Approximation of Vertical Gravity Gradient Within the Earth's External Gravity Field

Dongming Zhao, Shanshan Li, Huan Bao, and Qingbin Wang

**Abstract**

Vertical gravity gradient plays an important role in the research of the Earth's gravity field. However, the measurement of the vertical gravity gradient is a hard work. With the fast development of the Earth's gravity field modeling technique, it is possible to accurately approximate the vertical gravity gradient with the aid of the gravity field model as well as increasing gravity anomalies and rich terrain data. In the paper, a theoretical analysis was made on the computation of the vertical gravity gradient firstly, and then three methods, the gravity potential model method, the remove-restore method, and the point mass method, were used to accurately approximate the anomaly of the vertical gravity gradient. Tests of the three methods were made using some actual measurements of vertical gravity gradient over some area in China, and analyses were also made. Comparisons among the three methods show that the point mass method has the highest accuracy in approximation. At the end of the paper, some issues on the vertical gravity gradient to be further investigated were proposed.

**Keywords**

Point mass model • Remove-restore method • The Earth's gravity field model • Vertical gravity gradient

## 1 Introduction

Vertical gravity gradient is the rate at which the Earth's gravity varies along the plumb line. As another important parameter in gravimetry and explorational geophysics, the variation and distribution of vertical gravity gradient is an effective indicator of the Earth's figure, underground structure as well as its activities. In gravimetry, vertical gravity gradient and its corrections are more and more important, due to the fact that requirements on accuracy improvement have become higher and higher.

Literatures that discuss the vertical gravity gradient always concentrate on three aspects: observation method of vertical gravity gradient, inverse computation of vertical gravity gradient to determine the density, shape and depth, and the derivation of vertical gravity gradient using terrain and ground gravity anomaly. As far as the derivation of vertical gravity gradient is concerned, there are principally three kinds of methods.

The first kind is forward modeling methods which yield the gravitational gradients on the basis of direct applications of Newtonian gravitational potential theory. With a model for the mass density (as simple as a constant value) gradients are easily calculated using various approximations to Newton's density integrals (Jekeli and Zhu 2006). The density variations of the subsurface masses generally are difficult to

D. Zhao (✉) • S. Li • H. Bao • Q. Wang
Zhengzhou Surveying and Mapping Institute, 450052 Zhengzhou, Henan Province, China
e-mail: zhaodongming@cntv.cn

D. Zhao
State Key Laboratory of Geo-Information Engineering, Xi'an, China



S. Jin, R. Barzaghi (eds.), *IGFS 2014*, International Association of Geodesy Symposia 144,
DOI 10.1007/1345_2015_113

model and approximations imply corresponding errors in the forward model of the gradients.

The second type of forward model is the solution to the classical boundary-value problem for the potential. Stokes's integral is well known in geodesy as relating surface (say, geoid) values of the gravity anomaly to the disturbing potential. Clearly, by twice differentiating this integral we obtain a corresponding relationship between gravity anomalies and gravitational gradient disturbances. This alternative method to derive the gradient tensor from gravity data was already addressed by Mickus and Hinojosa (2001).

The third kind is the so-called combination method, say, the remove-restore method. Such methods or techniques always appear in gravimetric geodesy (Forsberg 1984; Tziavos et al. 1988; Sjoberg 1995; Tsoulis 2003; Xiong 2004; Zhang and Bian 2005).

In the paper, the determination of vertical gravity gradient is discussed from the perspective of the Earth's gravity field approximation, while the actual effect of nearby buildings on the anomaly of vertical gravity gradient is not the central topic here. In the research area three methods including gravity field model based method, Stokes integral method and the point mass method are used to approximate the vertical gradient of the disturbing gravity. In order to find a suitable method for the approximation of vertical gravity gradient, results of approximation of each method are analyzed.

## 2 Anomaly of Vertical Gravity Gradient

The vertical gravity gradient at any point on the Earth's surface consists of two parts theoretically. The first part is the vertical gradient of normal gravity, which is actually caused by the rotating ellipsoid with layered homogeneous density along the plumb line that the Earth is supposed to be; and the second part is caused by the difference of the mass distribution of the actual Earth and that of the ideal rotating ellipsoid, or called anomaly of vertical gravity gradient.

### 2.1 Vertical Gradient of Normal Gravity

In case that the Earth is viewed as a rotating ellipsoid with homogeneous mass distribution, the vertical gradient of normal gravity at a point with height $h$ above the ellipsoid surface is (Torge 1989; Wang 2003)

$$\frac{\partial \gamma}{\partial h} = -3086\,(1 + 0.0007\cos 2\varphi) + 1.44 \times 10^{-3} h \quad (1)$$

where $\gamma$ is normal gravity; $\varphi$ is the geocentric latitude; the unit of $h$ is meter, and the unit of normal gravity gradient is E(Eötvös,$10^{-9}\,s^{-2}$).

### 2.2 Anomaly of Vertical Gravity Gradient

The anomaly of vertical gravity gradient is induced by the non-homogeneous and not-layered mass distribution of the actual Earth. The factors such as terrain, ground objects, non-homogeneity of mass density around the station have a direct effect on the vertical gravity gradient. Therefore, the anomaly of vertical gravity gradient on the station point can be divided into three parts: the effect of terrain undulation, the effect of residual mass of non-homogeneous underground density, and the effect of buildings around the station. If $A$ stands for the effect of terrain undulation; $B$ for the effect of underground density anomaly and $C$ for the effect of ground buildings, then the vertical gravity gradient on the station point can be expressed as (Wang 2003)

$$\frac{\partial g}{\partial h} = \frac{\partial \gamma}{\partial h} + A + B + C \quad (2)$$

The above formula accounts for the fact that, if mass distribution of the actual Earth and the non-homogeneity of the mass above and below the ground surface are considered, vertical gradient of normal gravity cannot be used to replace the actual vertical gravity gradient on the station. On the other hand, from the above formula the following expression can be derived

$$A + B + C = \frac{\partial g}{\partial h} - \frac{\partial \gamma}{\partial h} = \frac{\partial (g - \gamma)}{\partial h} = \frac{\partial \delta g}{\partial h} \quad (3)$$

where $\delta g$ is the vertical component (or radial component) of the disturbing gravity on the station. In the following discussions, spherical approximation is considered, and $\partial/\partial n = \partial/\partial h$ holds, which means that the normal line coincides with the height direction.

Expression (3) shows that the anomaly of vertical gravity gradient can be approximated as a whole. In (Wang 2003), the calculations of the three parts, $A$, $B$ and $C$ are discussed respectively in a direct way by modeling underground substance and nearby buildings. However, here the sum of the three parts is calculated by approximating the Earth's external gravity field, which is the point of the paper.

### 2.3 Manipulation of the Anomaly of Vertical Gravity Gradient

At common points, the anomaly of vertical gravity gradient leads to the difference between the vertical gradient of actual gravity and the vertical gradient of normal gravity, and what's more important, due to the non-homogeneity of underground mass density as well as ground buildings, accurate formula for the calculation and derivation of vertical gradient of the actual gravity cannot be established. Especially, as far as absolute gravimetry and relative gravimetry with high accuracy are concerned, the vertical gradient of normal gravity cannot be used to replace the vertical gradient of actual gravity in the presence of ground objects and buildings, and a measurement of vertical gravity gradient becomes necessary. However, expression (3) indicates that, if the vertical gradient of the disturbing gravity at a point can be calculated with certain accuracy, then the numerical approximation of the anomaly of vertical gravity gradient can be obtained with the same uncertainty. The reason is that the theory and method have been constructed for the accurate approximation of the disturbing gravity in recent years. Here we list two existing methods.

The first kind of method is based on the Earth's gravity field model. According to the series expansion of the Earth's disturbing gravity potential by (Heiskanen and Moritz 1967), the second derivative of the disturbing gravity $T$ with respect to the radial direction can be directly formulated

$$\frac{\partial \delta g_{\rho}}{\partial \rho} = \frac{\partial^2 T}{\partial \rho^2} = \frac{fM}{\rho^3}\left[\sum_{n=2}^{n_{\max}}\sum_{m=0}^{n}(n+1)(n+2)\left(\frac{a}{\rho}\right)^n\left(\overline{C}^*_{nm}\cos m\lambda + \overline{S}_{nm}\sin m\lambda\right)\overline{P}_{nm}(\sin\varphi)\right] \tag{4}$$

where $(\varphi, \lambda, \rho)$ are geocentric coordinates; $f$ is the gravity constant; $\overline{C}^*_{nm}$ is the difference between the fully normalized geo-potential coefficient and the coefficient of the potential generated by a reference ellipsoid; $\overline{S}_{nm}$ is fully normalized spherical harmonic coefficients; $a$ is the equatorial radius of the Earth, and $\overline{P}_{nm}(\sin\varphi)$ is the fully normalized associated Legendre function.

Expression (4) indicates that the anomaly of the vertical gravity gradient can be directly calculated using a known gravity potential model, say, EGM2008.

The second kind of method is based on the twice differentiating of the Stokes's Integral (Heiskanen and Moritz 1967). Here we present the final formula directly

$$\frac{\partial \delta g_{\rho}}{\partial \rho} = T_{\rho\rho} = \frac{R^2}{4\pi}\iint_{\sigma}\frac{\partial^2 S(\rho,\psi)}{\partial \rho^2}\Delta g d\sigma \tag{5}$$

where the kernel is the second derivative of the generalized Stokes kernel with respect to the radial direction. For the sake of saving room, the detailed expressions of the above formulae are not listed.

For the computation of expression (5), global gravity anomalies are needed, which in fact cannot be satisfied. To calculate $T$ as well as its partial derivatives the remove-and-restore principle is always applied. The relative descriptions are also omitted here.

## 3 Approximation of the Anomaly of Vertical Gravity Gradient Using Point Mass Method

Recent years have seen the growing importance of the Earth's external disturbing gravity field in more and more fields, which causes more attention to the research of external disturbing gravity field. As the anomaly of vertical gravity gradient is the first order derivative of the disturbing gravity with respect to the radial direction, we can obtain approximation methods of the anomaly of vertical gravity gradient directly from existing approximation methods of the disturbing gravity. Here the point mass method is given more attention due to the fact that it is based on Bjerhammar boundary value problem.

From Keldysh-Lavtentiev theorem, a canonical harmonic function $T^*$ outside a sphere that lies within the Earth can approximate the disturbing potential $T$ outside the Earth. Here $T^*$ is supposed to be the gravity potential generated by a monolayer covering the inside sphere.

As Fig. 1 shows, the Earth's surface is $\Sigma$; $\sigma$ is a sphere within $\Sigma$ with radius $R_B$ whose center coincides with that of the Earth, also called Bjerhammar sphere. On the sphere is a monolayer with density $\mu$. From the potential theory (Heiskanen and Moritz 1967), the monolayer generates a gravity potential $T^*$, and at point $P$ the gravity potential can be expressed as

$$T^*_P = f\iint_{\sigma}\frac{\mu}{r}d\sigma \tag{6}$$

In which $f$ is the gravity constant, and $r$ is the distance from point P to $d\sigma$ with $r^2 = \rho^2 + R_B^2 - 2\rho R_B\cos\psi$. As $T^*$ is canonical outside $\sigma$, density $\mu$ should be determined in some way in order that $T^*$ be able to approximate the actual disturbing potential $T$.

To make $T^* = T$ outside $\Sigma$, $T^*$ must satisfy the following boundary condition (Moritz 1980)

$$\left.-\frac{\partial T^*}{\partial \rho} - \frac{2T^*}{\rho}\right|_{\Sigma} = \Delta g \tag{7}$$

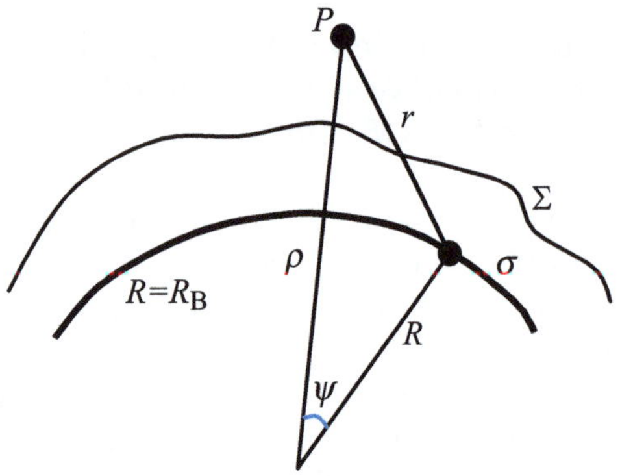


**Fig. 1** Bjerhammar sphere

By substituting (6) into (7), the following expression can be derived

$$f\iint_{\sigma}\left[\frac{1}{2\rho_Q}\left(\frac{\rho_Q^2-R_B^2}{r_Q^3}-\frac{3}{r_Q}\right)\right]\mu d\sigma=\Delta g_Q \qquad Q\in\Sigma \tag{8}$$

where $r_Q$ is calculated via the geocentric radius $\rho_Q$ of point $Q$.

For the integral expression (8), if the unknown density $\mu$ can be solved for, then the disturbing potential outside $\Sigma$ can be determined. Since the monolayer potential $T^*$ and the disturbing potential $T$ satisfy the boundary value condition on boundary $\Sigma$ at the same time, $T^*$ must equal $T$ outside the boundary $\Sigma$ by the uniqueness of the solution of the third boundary value problem, that is $T_P \equiv T_P^*$ with $P$ outside $\Sigma$.

As the known gravity anomalies on ground $\Sigma$ are discrete, numerical method should be used to solve the integral equation. Now dividing the sphere layer $\sigma$ into $K$ areas $\Delta\sigma_j$ ($j=1,2,\ldots,K$), and assuming the density $\mu$ of each area to be constant, then the disturbing potential is

$$T_P^*=f\sum_{j=1}^{K}\frac{1}{r_{Pj}}\mu_j\Delta\sigma_j=f\sum_{j=1}^{K}\frac{m'_j}{r_{Pj}}=\sum_{j=1}^{K}\frac{m_j}{r_{Pj}} \tag{9}$$

where $r_{Pj}$ is the distance from point P to the middle point of $\Delta\sigma_j$; $m'_j=\mu_j\Delta\sigma_j$ is the mass of area $\Delta\sigma_j$. Then the above expression represents the gravity potential generated by mass points with mass $m'_j$ or $m_j=fm'_j$, and the points are located in the middle of each area.

The discrete form of the integral expression (8) is

$$\frac{1}{2\rho_i}\sum_{j=1}^{K}\left(\frac{r_i^2-R_B^2}{r_{ij}^3}-\frac{3}{r_{ij}}\right)m_j=\Delta g_i \tag{10}$$

where $\Delta g_i$ and $\rho_i$ are gravity anomaly and geocentric radius of the i$^{th}$ ground point respectively; $r_{ij}$ is the distance between the i$^{th}$ ground gravity anomaly point and the j$^{th}$ mass point. When there are $N$ known gravity anomalies, we have the following equations

$$AM=\Delta G \tag{11}$$

where $A$ is a matrix containing $N\times K$ coefficients; $\Delta G$ is a vector containing $N$ gravity anomalies.

When $K$ unknown point masses $m_j(j=1,2,\ldots K)$ are solved for from equations (11), the disturbing gravity potential $T_P \equiv T_P^*$ can then be determined using (9). Then by taking the second derivative of the disturbing potential with respect to the radial direction, the anomalous vertical gravity gradient can be derived

$$\frac{\partial\delta g_\rho}{\partial\rho}=T_{\rho\rho}=-\sum_{j=1}^{K}\left(\frac{r_j^2-3(\rho-R_j\cos\psi_j)^2}{r_j^5}\right)m_j \tag{12}$$

### 3.1 Comparisons with Actual Observations

Located near Zhengzhou City, Henan Province of China, the test area is close to a $6'\times 6'$ or $10\times 10$ km rectangle in shape. The longitude and latitude of the top left (or north west) vertex of the test area are 113°38′ and 34°46′ respectively. In the test area, 16 points with equal space interval, say 2 km, were selected, and vertical gravity gradients of the 16 points were obtained through measurement with two CG-5 relative gravimeters, and then the vertical gradients of disturbing gravity were determined. The observed anamalous VGGs were obtained by subtracting vertical gradient of normal gravity from the observed vertical gravity gradients.

Then, the three methods proposed in the paper, which are gravity potential model (GPM) method, remove-and-restore based Stokes integral (SI) method, and point mass (PM) method were used to approximate the vertical gradient of the disturbing gravity. For each method, comparison with the actual measurements was also made, with the difference listed in the following table. To make results more obvious, the results are also shown by Fig. 2.

In Table 1, data in the fourth column are the difference of the vertical gradient of disturbing gravity calculated using a high degree gravity potential model and the actual observations; the sixth column are the difference of the vertical gradient of disturbing gravity calculated using Stokes integral and the observations; the eighth column are the difference of the vertical gradient of disturbing gravity

**Fig. 2** Comparisons of results

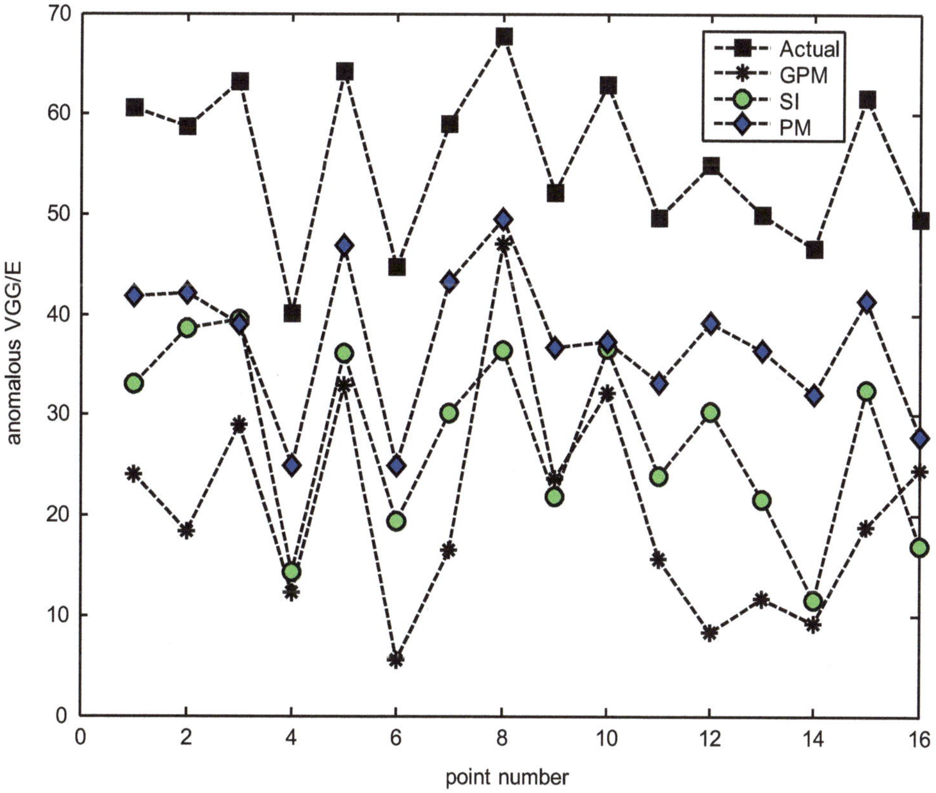


**Table 1** Comparisons of results from three methods with actual observations (unit: E)

| Point No. | Observations (observed anomalous VGG) | GPM results | Differences between GPM and observations | SI results | Difference between SI and observations | PM results | Difference between PM and observations |
|---|---|---|---|---|---|---|---|
| 1 | 60.6 | 24.1 | −36.5 | 33.1 | −27.5 | 41.8 | −18.8 |
| 2 | 58.7 | 18.4 | −40.3 | 38.6 | −20.1 | 42.1 | −16.6 |
| 3 | 63.2 | 29 | −34.2 | 39.5 | −23.7 | 39.1 | −24.1 |
| 4 | 40.1 | 12.3 | −27.8 | 14.4 | −25.7 | 25 | −15.1 |
| 5 | 64.2 | 32.9 | −31.3 | 36.2 | −28 | 46.9 | −17.3 |
| 6 | 44.8 | 5.6 | −39.2 | 19.4 | −25.4 | 24.9 | −19.9 |
| 7 | 59 | 16.5 | −42.5 | 30.2 | −28.8 | 43.3 | −15.7 |
| 8 | 67.7 | 47.1 | −20.6 | 36.4 | −31.3 | 49.6 | −18.1 |
| 9 | 52.1 | 23.7 | −28.4 | 21.9 | −30.2 | 36.8 | −15.3 |
| 10 | 62.9 | 32.2 | −30.7 | 36.6 | −26.3 | 37.3 | −25.6 |
| 11 | 49.7 | 15.6 | −34.1 | 23.9 | −25.8 | 33.3 | −16.4 |
| 12 | 55 | 8.4 | −46.6 | 30.4 | −24.6 | 39.2 | −15.8 |
| 13 | 50 | 11.7 | −38.3 | 21.6 | −28.4 | 36.5 | −13.5 |
| 14 | 46.7 | 9.2 | −37.5 | 11.5 | −35.2 | 32.1 | −14.6 |
| 15 | 61.7 | 18.9 | −42.8 | 32.5 | −29.2 | 41.4 | −20.3 |
| 16 | 49.6 | 24.5 | −25.1 | 16.9 | −32.7 | 27.8 | −21.8 |
| Mean | | | −35.35 | | −27.75 | | −16.95 |
| Std | | | 7.07 | | 3.67 | | 3.48 |

calculated using a multi-layer point mass model and the observations. Through statistics analysis, the mean of the difference resulting from a high degree gravity potential model is the largest, and the mean of the difference resulting from a multi-layer point mass model is the smallest. In the calculation, the Earth's gravity potential model, EGM2008 was used, which is equivalent to using mean gravity anomalies of $5'$ resolution, however, the gravity potential model cannot take surrounding terrain into account, which led to the largest difference. While in the multi-layer point mass model method, not only ground gravity anomalies with multi resolutions, but also terrain data with high resolution were used, which led to the smallest difference.

## 4 Conclusion

With the actual measurements of the vertical gravity gradient in test area, three methods that approximate the vertical gradient of disturbing gravity were proposed and tested. The three methods were gravity potential model method, Stokes integral method and multi-layer point mass method. Through calculations and comparisons with actual measurements, among the three proposed methods the multi-layer point mass method gives the best approximation results. Tests also showed that the terrain around the station point contributes a lot to the vertical disturbing gravity gradient. The effect of terrain along with the buildings around station on vertical gravity gradient is our future work.

**Acknowledgements** The work in the paper is financially supported by the open fund project No. SKLGIE2013-Z-1-1 from State Key Laboratory of Geo-Information Engineering, and the Scientific Research Fund project No.2014601102 from Zhengzhou Surveying and Mapping Institute.

## References

Forsberg R (1984) A study of terrain reductions, density anomalies and geophysical inversion methods in gravity field modeling: report 355. Department of Geodetic Science and Surveying, The Ohio State University, Columbus

Heiskanen W, Moritz H (1967) Physical geodesy. W. H. Freeman and Co., San Francisco

Jekeli C, Zhu L (2006) Comparison of methods to model the gravitational gradients, from topographic data bases. Geophys J Int 166:999–1014

Mickus KL, Hinojosa JH (2001) The complete gravity gradient tensor derived from the vertical component of gravity: a Fourier transform technique. J Appl Geophys 46:159–174

Moritz H (1980) Advanced physical geodesy. Abacus, Tunbridge Wells Kent

Sjoberg L (1995) On the quasigeoid to geoid separation. Manuscr Geod 20:182–192

Torge W (1989) Gravimetry. Walter de Gruyter, Berlin

Tsoulis D (2003) Terrain modeling in forward gravimetric problems: a case study on local terrain effects. J Appl Geophys 54:145–160

Tziavos IN, Sideris MG, Forsberg R, Schwarz KP (1988) The effect of the terrain on airborne gravity and gradiometry. J Geophys Res 93:9173–9186

Wang Q (2003) Gravitology. Seismological Press, Beijing

Xiong L (2004) Gravity gradiometry in geophysical prospecting. In: Zhu Y, Sun H (eds) Progress in geodesy and geophysics. Hubei Science & Technology, Wuhan, pp 160–172

Zhang C, Bian S (2005) Determination of disturbing gravity vertical gradient in earth's surface. Prog Geophys 12:969–973

# New Geoid of Greenland: A Case Study of Terrain and Ice Effects, GOCE and Use of Local Sea Level Data

Rene Forsberg and Tim Jensen

**Abstract**

Making an accurate geoid model of Greenland has always been a challenge due to the ice sheet and glaciers, and the rough topography and deep fjords in the ice free parts. Terrestrial gravity coverage has for the same reasons been relatively sparse, with an older airborne survey of the interior being the only gravity field data over the interior, and terrain and ice thickness models being insufficient both in terms of resolution and accuracy. This data situation has in the later years changed substantially, first of all due to GOCE, but also new airborne gravity and ice thickness data from the NASA IceBridge mission, and new terrain models from ASTER, SPOT-5 and digital photogrammetry.

In the paper we use all available data to make a new geoid of Greenland and surrounding ocean regions, using remove-restore techniques for ice and topography, spherical FFT techniques and downward continuation by least squares collocation. The impact of GOCE and the new terrestrial data yielded a much improved geoid, as evidenced by comparison to GPS measurements along fjords, which serves as a proxy for GPS leveling data, and comparisons to new GPS leveling data in Iceland. The comparisons show significant improvements over EGM08 and older geoid models, and also highlight the problems of global sea level models, especially in sea ice covered regions, and the definition of a new consistent vertical datum of Greenland.

**Keywords**

Geoid • Gravity • Greenland

## 1 Introduction

Greenland presents a major challenge for geoid determination, due to sparse gravity data, rugged topography, deep fjords, ice sheets and glaciers, and the absence of GPS leveling. Current gravimetric geoids date back to year 2000 (for the "official" GPS-fitted version), to intermediate North Atlantic Geoid models (Forsberg et al. 2004), and to EGM2008 (Pavlis et al. 2012), which incorporated all available gravity field information up to that point in time. All geoid models have up to now, however, had many local errors, especially due to lack of good digital elevation model data in the ice free part, due to lack of data over deep fjords, and due to lack of ice sheet thickness in marginal regions.

Several new data sets have aided in improving this situation in recent years, especially new high-resolution digital DEM data, gravity and radar and ice thickness data from NASA IceBridge airborne campaigns, and GOCE satellite data. Combined with the need for a new vertical datum for the towns and ice free regions of Greenland, for precise GPS surveying in support of infrastructure such as hydropower and mining development, the computation of a new geoid

R. Forsberg (✉) • T. Jensen
National Space Institute, Technical University of Denmark, Elektrovej 327, DK2800 Lyngby, Denmark
e-mail: rf@space.dtu.dk


S. Jin, R. Barzaghi (eds.), *IGFS 2014*, International Association of Geodesy Symposia 144,
DOI 10.1007/1345_2015_50

model is therefore timely. This also accentuates the need for a new vertical datum, since all towns and settlements have local height systems based on historical tide gauge data of varying quality, with no leveling connections between towns. Combined with large climate-induced land uplift values in many regions, this make the definition of a consistent height system between towns and settlements quite a challenge.

In this paper we will therefore outline a new geoid determination, incorporating all available new gravity, DEM, ice thickness and satellite data, and compare to existing GPS and tide gauge height information in towns, and – in the absence of leveling – compare the new geoid to fjord GPS-tide gauge profiles and Iceland leveling, demonstrating major improvements.

## 2 Geoid Determination Method

The methodology for the new Greenland gravimetric geoid is based on remove-restore techniques. The anomalous gravity potential $T$ is split into three parts

$$T = T_{EGM} + T_{RTM} + T_{res} \tag{1}$$

where $T_{EGM}$ is the contribution of a combined GOCE/EGM2008 spherical harmonic expansion, $T_{RTM}$ the terrain effect, and $T_{res}$ the residual field

$$T_{EGM} = \frac{GM}{r}\sum_{n=2}^{N}\left(\frac{R}{r}\right)^{n} \times \sum_{m=0}^{n}\left(C'_{nm}\cos m\lambda + S_{nm}\sin m\lambda\right) P_{nm}(\sin\varphi) \tag{2}$$

where GM is the mass of the earth times the gravitational constant, and R the earth radius. The spherical harmonic coefficients of the anomalous gravity field $C'_{nm}$ and $S_{nm}$ are taken from EGM2008 up to a maximal degree 360 after subtraction of the normal field coefficients, but with coefficients of the ESA GOCE Release 4 "direct" model (Pail et al. 2011) substituted in the harmonic degree band 80–200 (with a gradual linear blending of EGM08 and GOCE R4 in the bands 80–90 and 180–200). This approximate approach has been used to improve the mid-wavelength bands of EGM2008 with GOCE data. The EGM2008 model has not been used to the full degree, since better high-resolution data is now available in Greenland than was available for EGM2008.

The reference field and the geoid determination will be done rigorously as a quasigeoid computation, i.e. referring to the geoid to the surface of the terrain by

$$\zeta = \frac{T(\varphi, \lambda, H)}{\gamma_0} \tag{3}$$

where $\zeta$ is the height anomaly (quasigeoid) and $\gamma_0$ normal gravity. For the same reason the reference field $T_{EGM}$ is computed as a spatial "sandwich grid" at 3 separate elevations of 0, 3 and 8 km, and then interpolated to the actual height of a gravity observation point or a quasi-geoid computation point (some IceBridge transit flight legs were at a flight elevation of 10 km).

Terrain reductions are done by the RTM method using space-domain prism integration (Forsberg 1984), based on a reference DEM formed from the detailed DEM by a low-pass filtering at a corresponding resolution to the used reference field. To handle the terrain and ice together in the RTM method, equivalent rock topography was used over the ice covered regions, i.e.

$$h^{EQ} = h^{DEM} - \frac{2.67}{0.92} t^{ice} \tag{4}$$

where $t^{ice}$ is the thickness of the ice, $h^{DEM}$ the surface elevation, and $h^{EQ}$ the equivalent rock topography, assuming a rock density of 2.67 g/cm$^3$ (2,670 kg/m$^3$) and ice 0.92 kg/m$^3$. The use of rock-equivalent topography for terrain reductions, rather than the full 3-D mass field, is justified by the flat nature of the interior of the ice sheet (i.e., ice effects can with good approximation be treated as a Bouguer plate effect; the terrain effect from a Bouguer plate is independent of actual heights, only total mass per area).

The DEM used for Greenland was based on IceSat data, augmented in the ice-free and marginal ice zone parts by ASTER and SPOT5 stereo photogrammetry (Howat et al. 2014), edited and converted to an orthometric height DEM, and supplemented with DEM data from national mapping projects in northern Greenland, Iceland and Arctic Canada regions. Ice thickness data in Greenland were taken from the compilation of radio echo sounding by Bamber et al. (2013), and rock-equivalent heights subsequently computed within regions classified as ice by national mapping masks. Due to lack of radar echo sounding data in some margin areas and outlying ice caps, significant errors in the terrain reductions are expected there. For Iceland radar ice thickness data were provided by the University of Iceland for the three main ice caps. The composite rock-equivalent DEM was computed at 15″ resolution, and shown in Fig. 1.

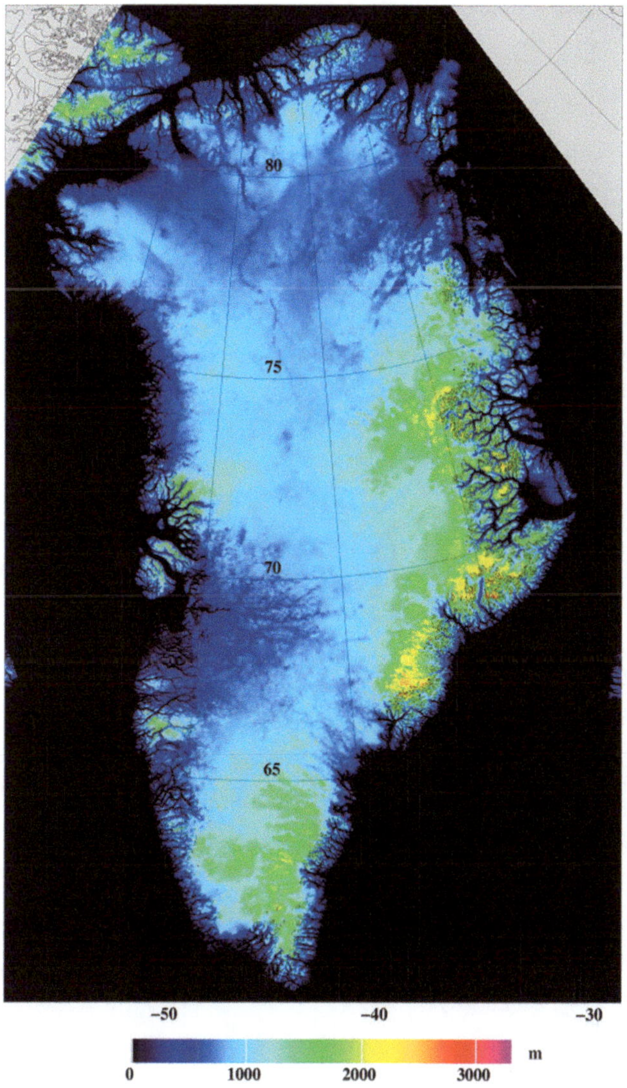


**Fig. 1** Rock-equivalent topography DEM of Greenland, highlighting coastal mountains. (the actual height of the ice sheet is more than 3,000 m in the center of Greenland)

After reduction of data for reference model and RTM, airborne and surface gravimetry data is downward continued and gridded at the surface of the rock-equivalent terrain by least-squares collocation using the planar logarithmic covariance function (Forsberg 1987), used in a block scheme of 1° resolution blocks with 50% overlap to neighboring blocks. The conversion from gravity to geoid is subsequently done using bandwise spherical Fourier techniques, for details of the methods see Forsberg and Sideris (1993). In the method the Stokes integral is evaluated by a series of convolutions, each accurate around a certain reference latitude $\varphi$

$$\begin{aligned}\zeta_{res} &= S_{ref}(\Delta\varphi, \Delta\lambda) * [\Delta g_{res}(\varphi, \lambda)\sin\varphi] \\ &= F^{-1}[F(S_{ref})\,F(\Delta g_{res}\sin\varphi)]\end{aligned} \quad (5)$$

where F[·] is the two-dimensional Fourier transform, $S_{ref}$ the Stokes function at the reference latitude and $\Delta g_{res}$ the residual gravity anomaly.

To prevent the inherently highly accurate GRACE/GOCE gravity field information in the EGM model to be "overruled" by the influence from terrestrial gravity data, *modified* Stokes functions are used. The modified Wong-Gore formulation used here is of form

$$S_{mod}(\psi) = S(\psi) - \sum_{n=2}^{N_2} \alpha(n)\frac{2n+1}{n-1}P_n\cos(\psi) \quad (6)$$

where the "GRACE-transition" coefficient $\alpha(n)$ increase linearly from 0 to 1 between degrees $N_1$ and $N_2$

$$\alpha(n) = \begin{cases} 1 & for\ \ 2 \le n \le N_1 \\ \frac{N_2-n}{N_2-N_1} & for\ N_1 \le n \le N_2 \\ 0 & for\ \ \ \ N_2 \le n \end{cases},\ n = 2,\ldots,N \quad (7)$$

The estimation of $N_1$ and $N_2$ can only be done empirically, but values in the range 180–200 would be expected, based on the error characteristics of GRACE and GOCE; these values have therefore been used here.

After the spherical Fourier transformation from gravity to quasigeoid, and restoring of the terrain and EGM-effects on the geoid, the classical geoid N is obtained by

$$\zeta - N \approx -\frac{\Delta g_B}{\gamma_0}H \quad (8)$$

where $\Delta g_B$ is the Bouguer anomaly, readily computed from the reduced anomalies by restoring of the reference field and the Bouguer effect of the smooth reference DEM. It should be noted that the formula (8) is *exact* if Helmert-type normal density orthometric heights are used; for details see Heiskanen and Moritz (1967).

## 3 Gravity Data and Gravimetric Geoid

The new Greenland gravimetric geoid was computed in the geographic region 58–85°N, 78–7°W, at a $1.2' \times 3'$ grid (about 2 km resolution). The gravity data used originate from a large variety of national and international land, marine and airborne surveys, including the Greenland-wide first airborne survey of the US Naval Research Laboratory 1991–1992 (Brozena 1991), DTU Space airborne surveys of coastal and Arctic Ocean regions 1998–2009, and NASA IceBridge gravity data 2010–2012 (Cochran and Bell 2010).

All data in the database were quality checked and transformed to a common, geoid-based gravity reference system, and subsequently reduced for EGM and RTM terrain effects.

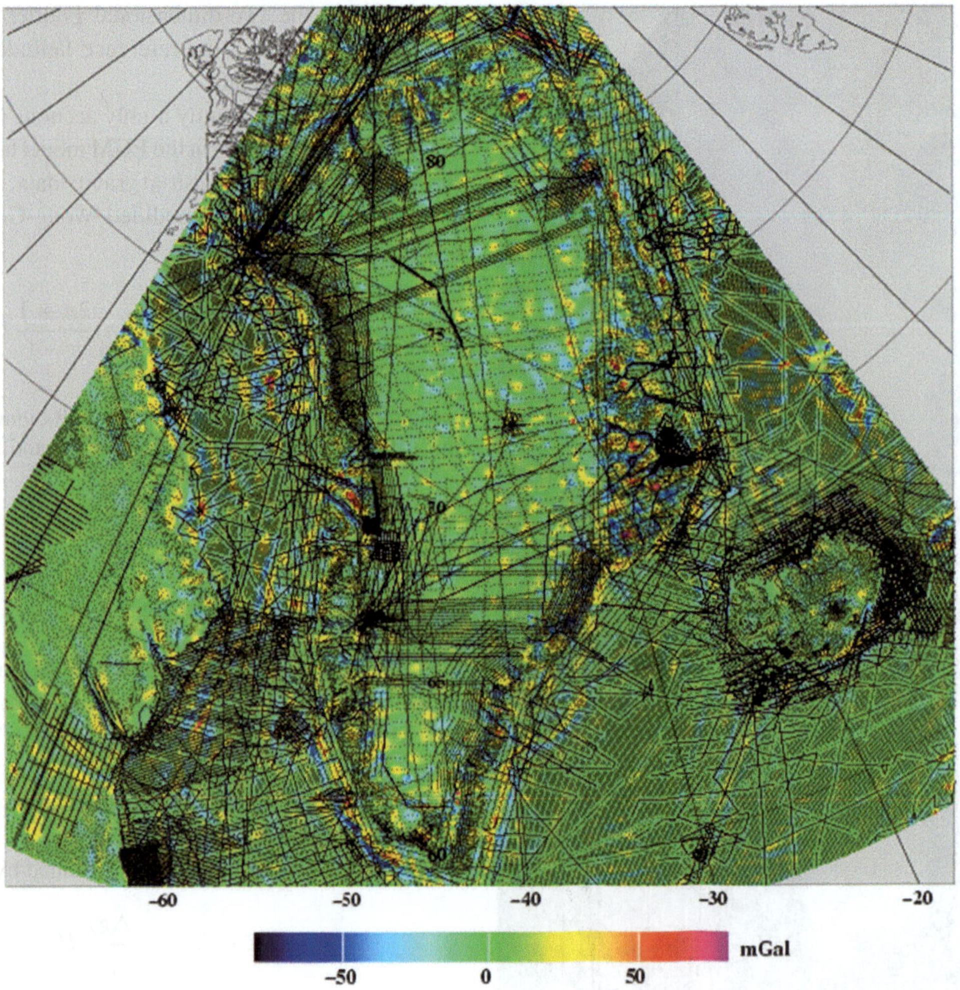


**Fig. 2** Gravity data coverage overlaid on EGM-and RTM-reduced gravity grid

**Table 1** Statistics of gravity data reductions for selected, thinned data sets (mGal)

| | | Original data | | Reduced data | |
|---|---|---|---|---|---|
| Data | No. of points | Mean | Std.dev. | Mean | Std.dev. |
| Land gravity | 8,503 | −16.8 | 44.8 | −4.6 | 16.0 |
| IceBridge airborne | 56,921 | 10.5 | 44.3 | −1.7 | 12.9 |
| NRL airborne | 32,010 | 16.4 | 38.0 | 0.1 | 11.5 |
| DTU airborne | 64,265 | 9.8 | 38.1 | 0.3 | 16.1 |

Figure 2 shows a plot of all used gravity data, overlaid on the gridded, reduced and downward continued data set, taking into account the varying standard deviations of the various data. For marine areas with no marine gravity data, DTU10 satellite altimetry gravity was used as fill-in. Especially the IceBridge data, with low-level flights along the ice margins, major outlet glaciers and fjords, have made a major difference relative to the data used in EGM2008. Table 1 shows the statistics for the data reductions of some main data sets.

It is seen from Table 1 that all airborne data sets, after reductions, give a consistent near-zero bias, and small standard deviation. For the land data, the bias likely express the uneven distribution of the land gravity points, mostly done by helicopter on mountains and nunataks, and fjord/local ice cap effects. The r.m.s. statistics of data have been used as a guide to the determination of the covariance function used for the downward continuation.

After the Fourier transformation, the restoral of terrain and EGM effects, and the conversion from quasigeoid to geoid, a final gravimetric geoid is obtained (Fig. 3).

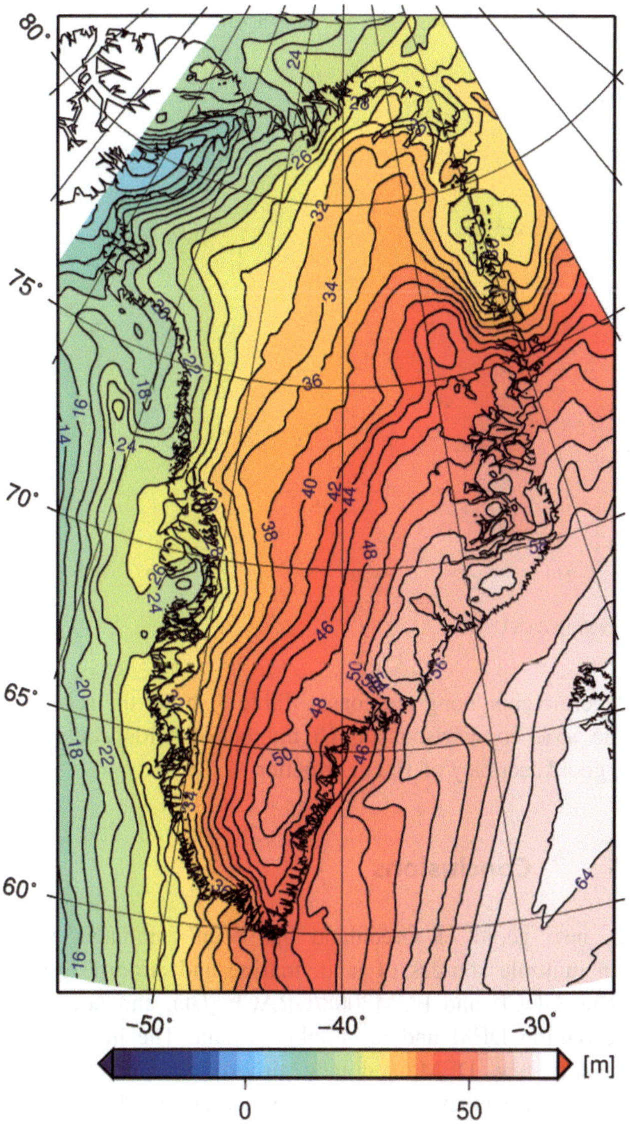


**Fig. 3** New gravimetric geoid of Greenland from surface, airborne and satellite gravity

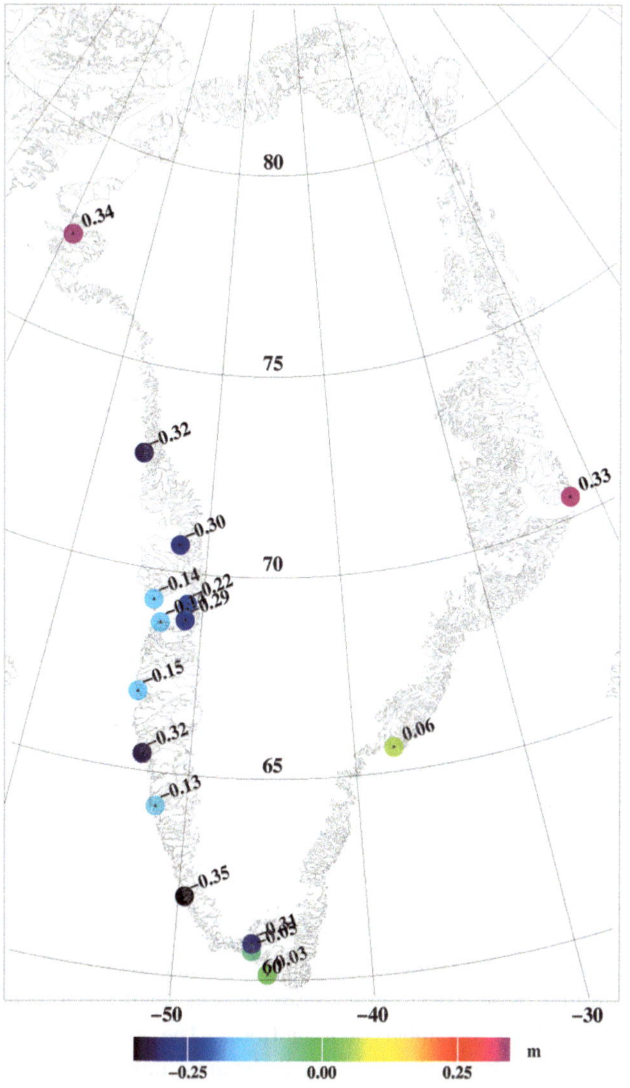


**Fig. 4** Differences (m) between GPS/leveling (Asiaq) and the new geoid is the Greenland towns

## 4 Evaluation of the New Geoid

Validating the new geoid is difficult, as no GPS leveling exists between settlements. The Greenland height system is based on local tide gauge measurements in the towns, of varying duration from few days to years, and all dating back to the 1960s or earlier. Figure 4 shows the offsets of the new geoid in the main towns, based on precise first order GPS measurements (REFGR), and the local tide gauge heights from the Greenland Survey (ASIAQ). With land uplift values up to several mm/year, and spatially changing mean sea level, is it quite natural that the geoid would not fit; basically the present vertical datum of Greenland is a historical mean sea level system, and not geoid-based.

Since the new geoid of Greenland also covers Iceland, and Iceland recently has done a releveling of whole country, the fit of the geoid to Iceland GPS leveling can give an independent check of accuracy. Iceland shares with Greenland the mountainous terrain and several large ice caps, but has a relatively good gravity coverage, and a coherent leveling network, cf. Fig. 5. Table 2 shows the fit of the geoid to Iceland GPS, showing a major improvement over EGM2008.

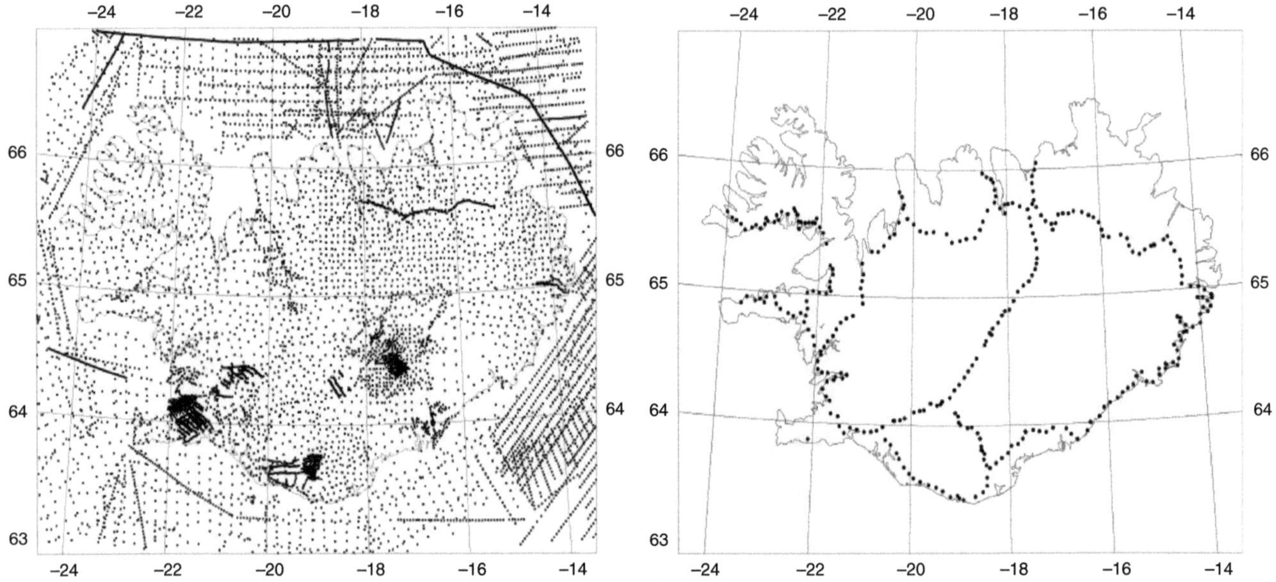


**Fig. 5** Gravity coverage of Iceland (*left*), and recently observed GPS and leveling data (*right*)

**Table 2** Comparison between GPS-leveling data and geoids in Iceland (unit: m)

| Comparison (311 GPS points) | Mean | Std.dev. |
|---|---|---|
| Original GPS-leveling data | 65.54 | 0.794 |
| Difference GPS minus new geoid | −0.10 | 0.037 |
| Difference GPS/lev minus EGM08 | −0.12 | 0.113 |

To get an estimate of geoid accuracy in Greenland, profiles with GPS observations of sea level have been done along a number of long fjords. The long fjords in Greenland have relatively small dynamic topography, and detailed oceanographic observations of the Godthåbsfjord region has confirmed changes in mean dynamic topography at the level of 5 cm and less (Mortensen, Greenland Climate Research Centre, pers.comm.). Therefore the fjords, except for the tides, can be treated as an approximation to an equipotential surface. Because wind conditions along Greenland fjords are often calm, it is relatively simple to determine instantaneous heights of the fjord relative to a GPS antenna to an accuracy level of few cm by local leveling.

By making *relative* tidal measurements, with temporary tide gauges positioned temporarily along the fjords, and combined with GPS observations, a "quasi" GPS/leveling data set can be observed in a matter of few days. This has been done along several fjords in Greenland, at typical profile lengths of 100–150 km, either in winter where fjords are frozen (and kinematic GPS "tide gauges" can be put directly on the sea ice, cf. Fig. 6 for an example), or in summer with slightly less accurate local GPS observations of apparent sea level heights along the coasts. Table 3 shows the results of the geoid comparisons along a number of fjords, confirming a geoid accuracy of 10 cm or better.

## 5 Conclusions

A new geoid of Greenland has been computed, based on multiple sources of land, marine and airborne gravity data, GOCE and EGM2008/GRACE data, and new high-resolution DEM and ice thickness data. The inclusion of GOCE, new DEM, and new airborne gravity data from IceBridge, has given a much more accurate geoid than earlier geoid models, confirmed also by new hydropower surveys (Asiaq, pers.comm.). With the use of GOCE/GRACE reference model data, and modified Stokes kernels preventing local data from changing the long-wavelength satellite information, the geoid accuracy at long wavelength should be similar to GOCE, i.e. at the cm-level for wavelengths longer than some 400–500 km. Local accuracy depends on the gravity data coverage and the errors coming from unknown fjord depths or local ice cap thickness. Results from Iceland and fjord profiles point to errors in the 5–10 cm range.

Given the uncertainty in the existing vertical datum, it is clear that a new geoid-based height system of Greenland would be superior for GPS use, with higher intrinsic accuracy, similar to the recently adopted new vertical reference system of Canada. Discussions are therefore currently ongoing with Greenland and Danish authorities for the possible adoption of a new height system.

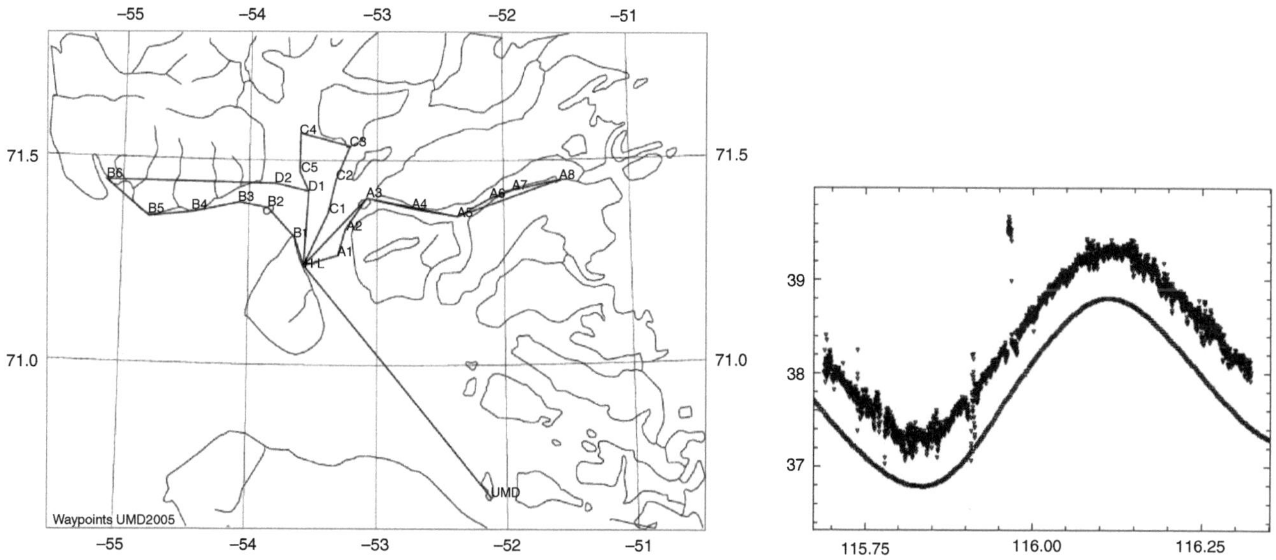


**Fig. 6** Example of a 160 km long winter fjord profile in central W Greenland (Uummannaq area). *Left*: GPS points on ice, *right*: example of a GPS (*upper curve*) and local tide gauge observation (*lower curve*) of sea level at the reference point (x-axis: time in Julian days, y-axis height in m)

**Table 3** Comparison of the Greenland geoid along selected fjord profiles (unit: m)

| Comparison fjord | Mean | Std.dev. | EGM08 st.dev. |
|---|---|---|---|
| Svartenhuk-Kangerlussuaq (NW Greenland) | −0.24 | 0.073 | 0.302 |
| Ameralik (Nuuk; SW Greenland) | −0.08 | 0.117 | 0.158 |
| Tasermiut (South Greenland) | 0.05 | 0.127 | 0.137 |
| Ammassalik Fjord (Tasiilaq, East Grenland) | −0.14 | 0.089 | 0.139 |

**Acknowledgements** Gravity data for the new Greenland geoid have been contributed by NASA IceBridge, Iceland Geodetic Survey, Geodetic Survey Division, Canada, and numerous government and commercial surveys in Greenland. GPS and leveling data in Iceland were provided by Gudmundur Valsson, Iceland Geodetic Survey.

## References

Bamber J, Griggs JA, Hurkmans RTWL, Dowdeswell JA, Gogineni SP, Howat I, Mouginot J, Paden J, Palmer S, Rignot E, Steinhage D (2013) A new bed elevation dataset for Greenland. Cryosphere 7:499–510. doi:10.5194/tc-7-499-2013

Brozena J (1991) The greenland aerogeophysics experiment: airborne gravity, topographic and magnetic mapping of an entire continent. In: Colombo (ed) From Mars to Greenland: charting gravity with space and airborne instruments. IAG Symposium Series 110. Springer Verlag, pp 203–214

Cochran JR, Bell RE (2010) IceBridge Sander AIRGrav L1B geolocated free air gravity anomalies 2010-12. Boulder. National Snow and Ice Data Center, Colorado USA

Forsberg R (1984) A study of terrain reductions, density anomalies and geophysical inversion methods in gravity field modelling. Reports of the Department of Geodetic Science and Surveying, No. 355, The Ohio State University, Columbus, Ohio

Forsberg R (1987) A new covariance model for inertial gravimetry and gradiometry. J Geophys Res 92(B2):1305–1310

Forsberg R, Sideris MG (1993) Geoid computations by the multi-band spherical FFT approach. Manuscript Geod 18:82–90

Forsberg R, Olesen A, Vest A, Solheim D, Hipkin R, Omang O, Knudsen P (2004) Gravity field improvements in the north atlantic region. Proceedings GOCE Workshop, ESA-ESRIN

Heiskanen W, Moritz H (1967) Physical geodesy. W.H. Freeman and Co, San Francisco

Howat IM, Negrete A, Smith BE (2014) The Greenland Ice mapping project (GIMP) land classification and surface elevation datasets. Cryosphere 8:1509–1518. doi:10.5194/tc-8-1509-2014

Pail R, Bruinsma SL, Migliaccio F, Förste C, Goiginger H, Schuh WD, Höck E, Reguzzoni M, Brockmann JM, Abrikosov O, Veicherts M, Fecher T, Mayrhofer R, Krasbutter I, Sanso F, Tscherning CC (2011) First GOCE gravity field models derived by three different approaches. J Geod 85(11):819–843. doi:10.1007/s00190-011-0467-x

Pavlis NK, Holmes SA, Kenyon S, Factor JK (2012) The development and evaluation of the earth gravitational model 2008 (EGM2008). J Geophys Res. doi:10.1029/2011JB008916

# Egyptian Geoid Using Best Estimated Response of the Earth's Crust due to Topographic Loads

Hussein A. Abd-Elmotaal

**Abstract**

In this paper, the Egyptian gravimetric geoid is computed using the best estimated response of the earth's crust due to the topographic loads. It has been proved that both the inverse isostasy approach and the direct isostasy approach (with Kelvin function kei $x$) give practically the same response of the earth's crust due to topographic loads. The window remove-restore technique (Abd-Elmotaal and Kühtreiber, J Geod 77(1–2):77–85, 2003) has been used to avoid the double consideration of some of the topographic-isostatic masses in the neighbourhood of the computational point. The tailored geopotential model EGTGM2014 (Abd-Elmotaal, Egyptian geoid using ultra high-degree tailored geopotential model. Proceedings of the 25th international federation of surveyors FIG congress, 2014) has been used for the long wavelength contributions of the earth's gravity field. The gravimetric geoid is computed for Egypt using Stokes' integral in the frequency domain by 1-D FFT technique. For the sake of comparison, another geoid for Egypt using EGM2008 and Airy floating hypothesis has been computed. The computed geoids are scaled/fitted to the GPS-levelling derived geoid. The internal precision of the computed geoids is almost the same and it is at the level of 3 cm. The external accuracy of the geoid computed by the best estimated response of the earth's crust is better by 4 dm.

**Keywords**

Egypt • Geoid • Isostasy • Kelvin function • Tailored geopotential model

## 1 Introduction

In order to compute a precise geoid, one needs to model the behaviour of the earth's crust due to the topographic loads. The plate loading theory with the exact solution of the bent plat, represented by the Kelvin function kei $x$ (Abramowitz and Stegun 1965), proved to be the best *direct* model of the behaviour of the earth's crust due to the topographic loads (cf. Abd-Elmotaal 1993).

H.A. Abd-Elmotaal (✉)
Civil Engineering Department, Faculty of Engineering, Minia University, Minia, Egypt
e-mail: abdelmotaal@lycos.com

Dorman and Lewis (1970) proposed that the density anomaly is linearly related to the topography by a convolution of the topography and an isotropic kernel function. Accordingly, it has been proved that the attraction of the compensating masses is also a convolution of the topography and an isotropic isostatic response function. Such an isostatic response function can be determined by deconvolution (Abd-Elmotaal 2004). Applying the principle of inverse isostasy (Moritz 1990), it has been proved that such an isostatic response function, representing the *inverse* model of the behaviour of the earth's crust due to topographic loads, behaves as the Kelvin function kei $x$ (Abd-Elmotaal 2013). Hence both direct and inverse models, mentioned above, represent the best model describing the response of the earth's crust due to topographic loads. Related numerical



S. Jin, R. Barzaghi (eds.), *IGFS 2014*, International Association of Geodesy Symposia 144,
DOI 10.1007/1345_2015_194

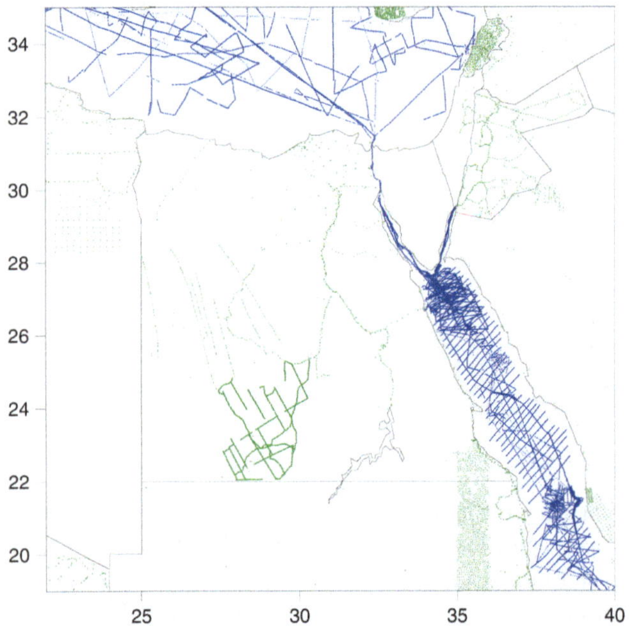


**Fig. 1** Distribution of the available free-air gravity anomalies for Egypt

results may be found, e.g., in Lewis and Dorman (1970), Bechtel et al. (1987), Hein et al. (1989).

The main aim of this paper is to compute a geoid model for Egypt employing such a best model describing the behaviour of the earth's crust due to topographic loads.

The used gravity, Digital Height Models (DHMs) and GPS data sets are described. The window remove-restore technique (Abd-Elmotaal and Kühtreiber 2003) has been outlined. The plate loading theory with the exact solution of the bent plat is described. The Moho depths for Egypt according to the plate loading theory have been computed. The tailored geopotential reference model EGTGM2014 (Abd-Elmotaal 2014), complete to degree and order 2160, has been used. The EGM2008 geopotential model (Pavlis et al. 2012), complete to degree and order 2160, has also been used to compute a gravimetric geoid for Egypt for comparison purposes. The computed geoids are fitted to the GPS-levelling derived geoid. A wide comparison between the computed geoids has been carried out.

## 2 Used Data

### 2.1 Gravity

Figure 1 shows the distribution of the free-air gravity anomalies for Egypt available for the current investigation. The distribution of the free-air gravity anomaly stations on land is very poor. Many areas are empty. The distribution of the data points on sea is significantly better. The distribution of the data points at the Red sea is better than that at the Mediterranean sea.

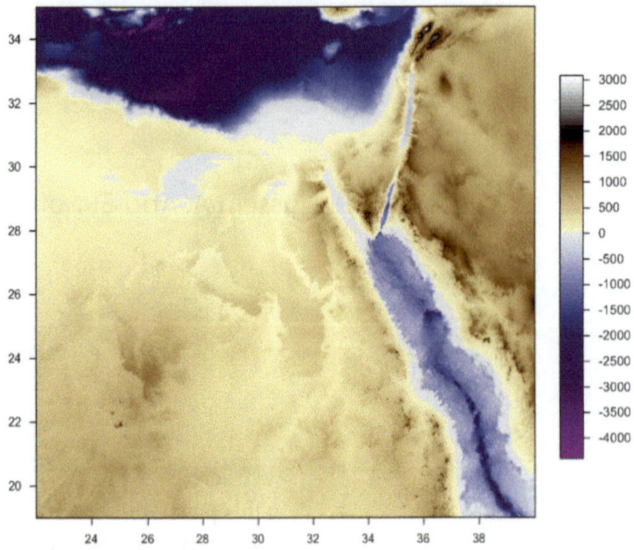


**Fig. 2** The 3″ × 3″ EGH13S03 Digital Height Model. Units in m

The total number of the available gravity anomaly values is 102,418. The free-air gravity anomalies range between −210.60 and 314.99 mgal with an average of −27.58 mgal and a standard deviation of about 50.65 mgal. Highest values are in sea regions.

### 2.2 Digital Height Models

For the terrain reduction computation, a set of fine and coarse Digital Height Models DHMs is needed. The fine EGH13S03 3″ × 3″ and the coarse EGH13S30 30″ × 30″ DHM's (Abd-Elmotaal et al. 2013) are used for the current investigation. They cover the window $19° \le \phi \le 35°$, $22° \le \lambda \le 40°$. Figure 2 illustrates the EGH13S03 fine DHM. These heights range between −4392 and 3084 m with an average of −18.4 m and a standard deviation of about 992.4 m.

### 2.3 GPS Stations

The distribution of the most reliable GPS stations with known orthometric height available for the current investigation is shown in Fig. 3. The distribution is fairly good, but the total number of the GPS stations is too small with respect to Egypt's surface area.

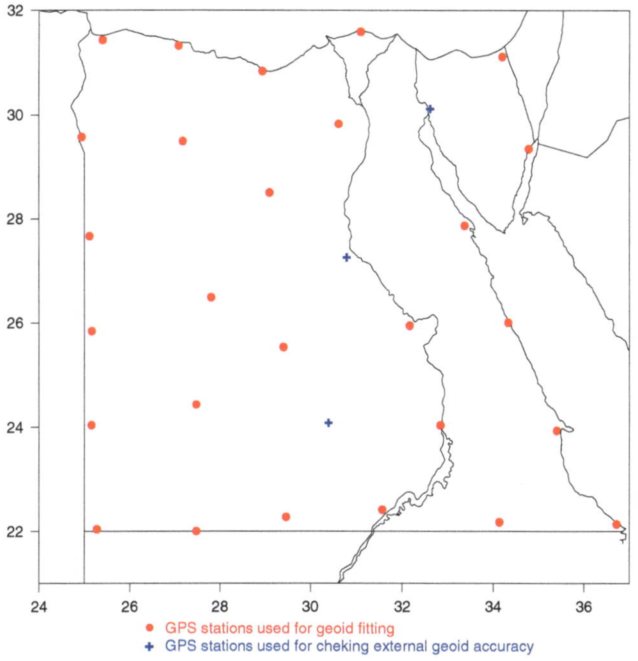


**Fig. 3** Distribution of the GPS stations with known orthometric height in Egypt

## 3 The Window Technique

Within the well-known remove-restore technique (Forsberg 1984), the effect of the topographic-isostatic masses is removed from the gravity anomalies and then restored to the resulting geoidal heights. Hence, the reduced gravity anomalies in the framework of the remove-restore technique is computed by

$$\Delta g_{red} = \Delta g_F - \Delta g_{TI} - \Delta g_{GM} , \tag{1}$$

where $\Delta g_F$ stands for the free-air anomalies, $\Delta g_{TI}$ is the effect of the topographic-isostatic masses, and $\Delta g_{GM}$ is the effect of the reference geopotential model on the gravity anomalies. Thus the final computed geoid $N$ within the remove-restore technique can be expressed by:

$$N = N_{GM} + N_{\Delta g} + N_{TI} , \tag{2}$$

where $N_{GM}$ gives the contribution of the reference field, $N_{\Delta g}$ gives the contribution of the reduced gravity anomalies, and $N_{TI}$ gives the contribution of the topography and its compensation (the indirect effect).

The traditional way of removing the effect of the topographic-isostatic masses faces a theoretical problem. A part of the influence of the topographic-isostatic masses is removed twice as it is already included in the global

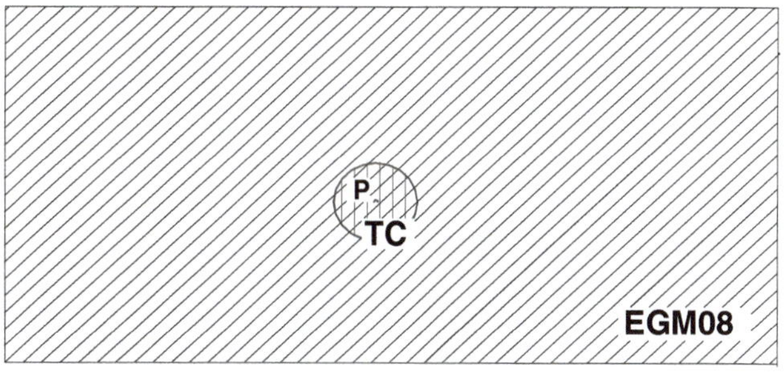


**Fig. 4** The traditional remove-restore technique

reference field. This leads to some double consideration of that part of the topographic-isostatic masses. Figure 4 shows schematically the traditional gravity reduction for the effect of the topographic-isostatic masses. The short-wavelength part depending on the topographic-isostatic masses is computed for a point $P$ for the masses inside the circle (say till 167 km around the computational point $P$; denoted by $TC$ in Fig. 4). Removing the effect of the long-wavelength part by a global earth's gravitational reference field normally implies removing the influence of the global topographic-isostatic masses (shown as the full rectangle in Fig. 4). The double consideration of the topographic-isostatic masses inside the circle (double hatched) is then seen.

A possible way to overcome this difficulty in computing geoidal heights from topographic-isostatic reduced gravity anomalies is to adapt the used reference field due to the effect of the topographic-isostatic masses for a fixed local data window (denoted by the small rectangle in Fig. 5) (Abd-Elmotaal and Kühtreiber 2003). Figure 5 shows the advantage of the window remove-restore technique. Let us concentrate on the influence of the topographic-isostatic masses in the reduction process. Consider a measurement at point $P$, the short-wavelength part depending on the topographic-isostatic masses is now computed by using the masses of the whole data area (the small rectangle in Fig. 5). The adapted reference field is created by subtracting the effect of the topographic-isostatic masses of the data window, in terms of potential coefficients (cf. Abd-Elmotaal and Kühtreiber 1999), from the reference field coefficients (adapted reference field is then represented by the area of the big rectangle, representing the globe, minus the area of the small rectangle, representing the local data area; cf. Fig. 5). Thus, removing the long-wavelength part by using this adapted reference field does not lead to a double consideration of a part of the topographic-isostatic masses (there is no double hatched area in Fig. 5).

The remove step of the window remove-restore technique can then mathematically be written as

$$\Delta g_{red} = \Delta g_F - \Delta g_{TI\,win} - \Delta g_{GM\,Adapt} , \tag{3}$$

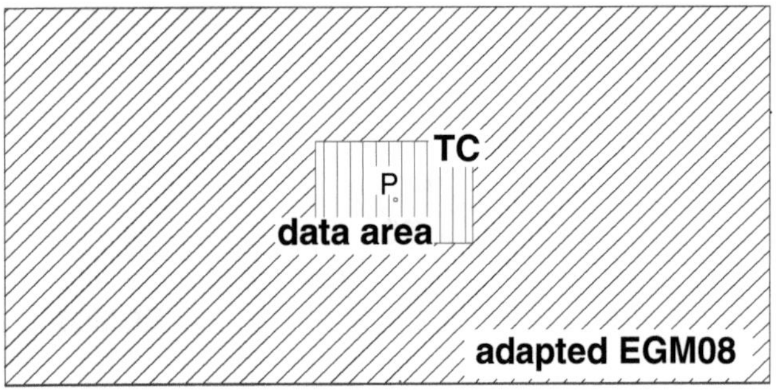


**Fig. 5** The window remove-restore technique

where $\Delta g_{GM\,Adapt}$ is the contribution of the adapted reference field and $\Delta g_{TI\,win}$ stands for the effect of the topographic-isostatic masses for a fixed data window. The restore step of the window remove-restore technique can be written as

$$N = N_{GM\,Adapt} + N_{\Delta g} + N_{TI\,win}\,, \tag{4}$$

where $N_{GM\,Adapt}$ gives the contribution of the adapted reference field and $N_{TI\,win}$ gives the contribution of the topography and its compensation (the indirect effect) for the same fixed data window as used for the remove step.

## 4 The Plate Loading Theory

Postulate that the earth's crust of density $\rho_\circ$ is represented by an infinite thin plate floating on a denser mantle of density $\rho_1$ bent by the topographic load, Fig. 6. To get a horizontal upper surface, the dotted hollow area appearing in Fig. 6 should be filled. This represents the so-called indirect effect. The equation of equilibrium for that floating bent plate, taking the indirect effect into account, is then given by Abd-Elmotaal (1993) and Turcotte and Schubert (1982)

$$D\Delta^2 z = P - g(\rho_1 - \rho_\circ)z\,, \tag{5}$$

where $D$ is the cylindrical rigidity of the plate, $z$ is the downward displacement, $g$ is the gravity and $\Delta$ denotes the two dimensional Laplace operator defined as

$$\Delta \equiv \frac{\partial^2}{\partial x^2} + \frac{\partial^2}{\partial y^2}\,. \tag{6}$$

Let us consider a point unit mass load concentrated at the origin $O$. Then, outside the origin $O$, $P = 0$ and (5) is reduced to

$$D\Delta^2 z = -g(\rho_1 - \rho_\circ)\,z\,, \tag{7}$$

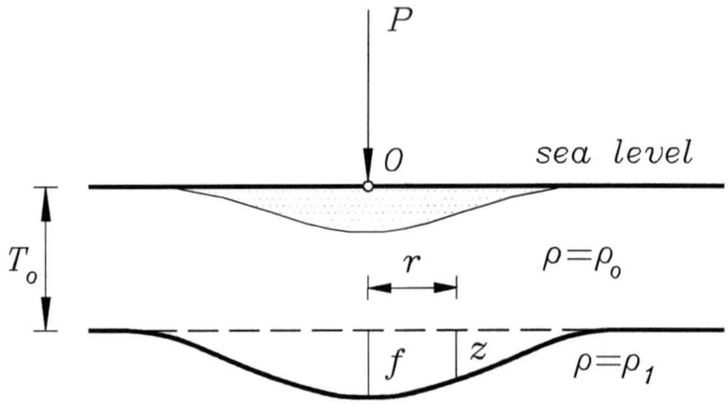


**Fig. 6** *Bending curve* of the earth's crust due to topographic loads

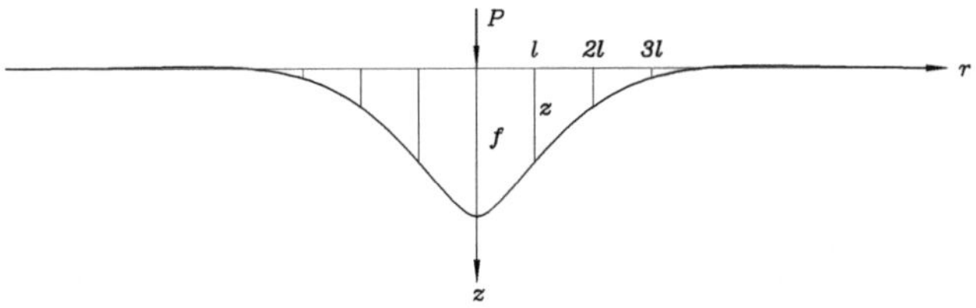


**Fig. 7** Exact solution of the thin plate bent by the topographic loads as given by Kelvin function kei $x$

or

$$\Delta^2 z + l^{-4} z = 0 \tag{8}$$

with

$$l = \sqrt[4]{\frac{D}{g(\rho_1 - \rho_\circ)}}\,. \tag{9}$$

Equation (9) gives the same expression for the so-called degree of regionality $l$ as given by Vening Meinesz (1940). It can be proved that the Kelvin Function kei $x$ gives an exact solution of (8) (Abd-Elmotaal 1993)

$$z = -\frac{4}{3} f \,\mathrm{kei}\, r\,, \tag{10}$$

where the factor $-4/3$ normalizes the function $z/f$ at the origin $O$, $z$ denotes the vertical displacement, $r$ is the distance from the origin $O$ and $f$ is the maximum bending at the origin given by Vening Meinesz (1940)

$$f = \frac{1}{8(\rho_1 - \rho_\circ)l^2}\,. \tag{11}$$

The exact solution of the thin plate bent by the topographic loads, computed by (10), is given in Fig. 7. More details can be found in Abd-Elmotaal (1993), Moritz (1990), and Abd-Elmotaal (1991).

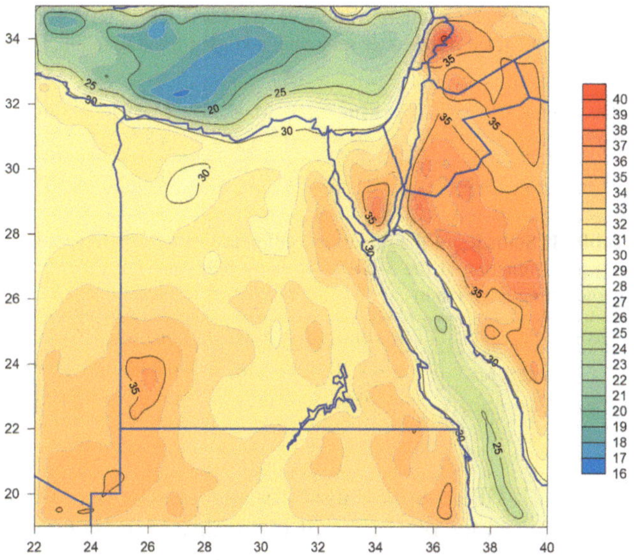

**Fig. 8** Fine 30″ × 30″ Digital Moho Model for Egypt employing the plate loading theory. Units in km

It should be noted that an alternative solution for the spherical earth can be found in Brotchie and Silvester (1969).

## 5 Moho Depths

The depth of the Mohorovičić discontinuity (shortly Moho depths) $H_{Moho}$ can be determined within the plate loading theory assuming that the earth's crust is in its elastic stage by (Abd-Elmotaal 1993)

$$H_{Moho} = T_{\circ} + \iint_{\sigma} z d\sigma \,, \tag{12}$$

where $T_{\circ}$ is the normal crustal thickness.

Fine 30″ × 30″ and coarse 3′ × 3′ Digital Moho Models (DMMs) for Egypt employing the plate loading theory have been created using (12). The integration appearing in (12) has been numerically evaluated and terminated to the available data window. Values of $l = 20$ km and $T_{\circ} = 30$ km have been used. Figure 8 illustrates the fine 30″ × 30″ DMM for Egypt employing the plate loading theory. These Moho depths range between 16.95 and 40.16 km with an average of 30.77 km and a standard deviation of about 4.48 km.

## 6 Gravity Reduction

The geoid computation in this investigation is carried out using the window remove-restore technique. The reduction step has been performed using (3). The contribution of the geopotential models to the gravity anomalies have been computed using the fast algorithm described in Abd-Elmotaal (1998). Alternative techniques are existing in, e.g., Rapp (1982), Tscherning et al. (1994).

**Table 1** Statistics of the gravity anomalies in Egypt (102,418 gravity stations). Units in mgal

| Gravity anomalies | Min | Max | Mean | Std |
|---|---|---|---|---|
| Free-air | −210.60 | 314.99 | −27.58 | 50.65 |
| Airy window | −99.15 | 122.49 | 2.31 | 20.46 |
| Moho window | −138.36 | 120.35 | 3.60 | 17.37 |

For the terrain reduction $\Delta g_{TI\,win}$, the topography and its compensation for the whole fixed data window (19° N $\leq \phi \leq$ 35° N; 22° E $\leq \lambda \leq$ 40° E) have been considered. A constant value for the density of the topography of 2.67 g/cm$^3$ is used. A value of 0.4 g/cm$^3$, which practically proved to give good results (cf. Abd-Elmotaal 2014), is used for the density contrast between the crust and mantle. The computation of the terrain reduction has been carried out using TC program, written originally by Forsberg (1984) after great improvements by Abd-Elmotaal and Kühtreiber (2003) and Abd-Elmotaal (2003).

Table 1 illustrates the statistics of the free-air and both Airy window (using EGM2008 geopotential model) and Moho window (using EGTGM2014 tailored geopotential model) isostatic anomalies. Here Moho window refers to using the plate loading theory to compute the Moho depths, while Airy window refers to using the well-known Airy floating hypothesis. Table 1 shows that using the plate loading theory to compute the Moho depths gives the best reduced anomalies. The standard deviation has dropped by more than 17% compared to that when using the Airy window technique with EGM2008 geopotential model.

## 7 Geoid Determination

The geoid for Egypt has been computed by 1-D FFT technique (Haagmans et al. 1993) using the FFTGEOID program by Sideris and Li (1993). The restore step (4) of the window remove-restore technique has been implemented.

Figure 9 shows the difference between the Airy window geoid computed by using the EGM2008 geopotential model with the Airy-Heiskanen isostatic model $N_{Airy\ win\ EGM2008}$ and the geoid derived from the combination of the GPS and levelling $N_{GPS}$. The difference, before removing the trend, ranges from −10.65 to 3.78 m with an average of −1.20 m and a standard deviation of 3.00 m. The structure of the differences shows a non-linear long-wavelength behaviour with a high east-west gradient.

Figure 10 shows the difference between the Moho window geoid computed by using the EGTGM2014 tailored geopotential model employing the plate loading

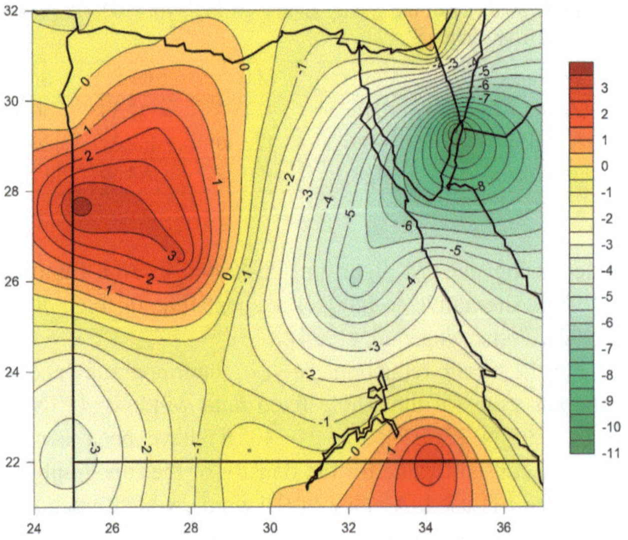


**Fig. 9** Geoid difference between $N_{Airy\ win\ EGM2008}$ and $N_{GPS}$ for Egypt. Contour interval: 0.5 m

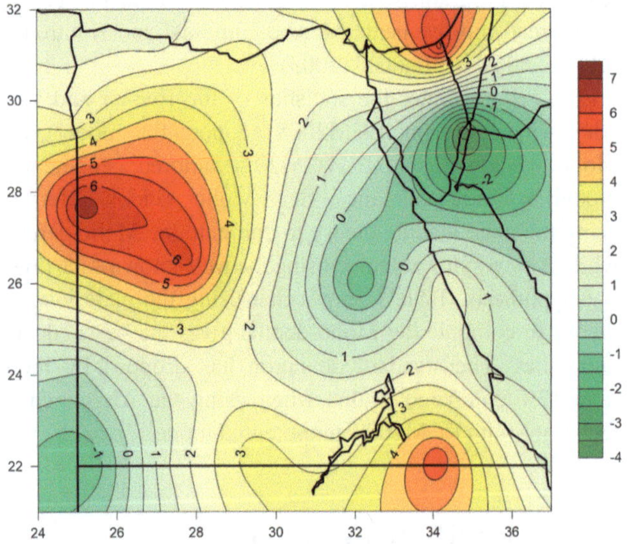


**Fig. 10** Geoid difference between $N_{Moho\ win\ EGTGM2014}$ and $N_{GPS}$ for Egypt. Contour interval: 0.5 m

theory $N_{Moho\ win\ EGTGM2014}$ and $N_{GPS}$. The difference, before removing the trend, ranges from −3.81 to 6.89 m with an average of 2.18 m and a standard deviation of 2.42 m. Figure 10 shows also a non-linear long-wavelength behaviour of the differences but with much less range and gradient. The range and standard deviation have dropped by about 35% and 24%, respectively, compared to those for the Airy window geoid using EGM2008 geopotential model.

The computed geoids have been fitted to the GPS-levelling derived geoid by removing a trend surface. A kriging trend function has been computed using only 27 GPS stations among the available 30 GPS stations in Egypt (cf. Fig. 3). Table 2 shows the statistics of the remaining differences after removing a kriging trend function

**Table 2** Statistics of the remaining differences at the 27 GPS stations used for the geoid fitting after removing a kriging trend function. Units in cm

| Geoid type | Min | Max | Mean | Std |
|---|---|---|---|---|
| $N_{Airy\ win\ EGM2008}$ | −9.2 | 9.8 | 0.2 | 3.41 |
| $N_{Moho\ win\ EGTGM2014}$ | −8.2 | 10.2 | 0.2 | 3.46 |

**Table 3** Statistics of the remaining differences after removing a kriging trend function at the 3 GPS stations which were not used for the geoid fitting. Units in m

| Geoid type | Min | Max | Mean | RMS |
|---|---|---|---|---|
| $N_{Airy\ win\ EGM2008}$ | −1.40 | 2.46 | 0.29 | 1.31 |
| $N_{Moho\ win\ EGTGM2014}$ | −0.86 | 2.11 | 0.31 | 0.92 |

at the 27 GPS stations used for the geoid fitting. This represents an internal check of the quality of the computed geoid. Table 2 shows that both geoids have practically the same internal precision.

Table 3 shows that statistics of the remaining differences after removing a kriging trend function at the 3 GPS stations which were not used for the geoid fitting. This represents an external check of the quality of the computed geoid. Table 3 shows that using the plate loading theory with EGTGM2014 tailored geopotential model improves the external geoid accuracy by about 23%, and the range of the remaining differences has dropped by about 30%.

It should be noted that the external geoid accuracy can hardly be estimated by only 3 GPS stations. The values here are only for comparison purpose. It is, however, believed that the actual accuracy of the computed geoid is far better, but this needs more GPS stations to fairly judge the geoid external accuracy.

Figure 11 shows the Moho window gravimetric geoid for Egypt using the plate loading theory and the tailored EGTGM2014 geopotential model fitted to 27 GPS stations using the kriging trend function. The values of the geoid range between 5.38 and 23.51 m with an average of 14.40 m and standard deviation of 3.79 m.

## 8 Conclusion

The plate loading theory, being proved to give the best estimated response of the earth's crust due to topographic loads, has been employed to generate the Moho depths for Egypt for the first time. The generated Moho depths for Egypt by the plate loading theory show a smooth structure. These Moho depths have then been used within the window remove-restore technique to compute a gravimetric geoid for Egypt using the 1-D FFT technique. Using the Moho depths computed by the plate loading theory gives better residual gravity anomalies (unbiased with a smaller standard deviation). The standard deviation has dropped by about 17%.

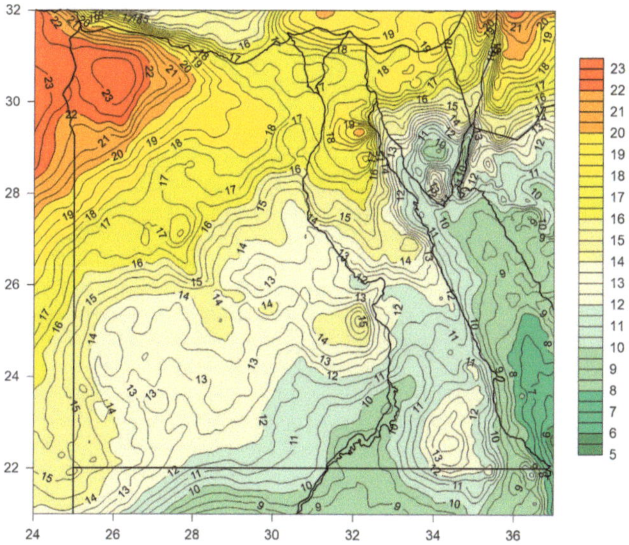


**Fig. 11** Gravimetric geoid for Egypt using the plate loading theory and the tailored EGTGM2014 geopotential model with window technique fitted to 27 GPS stations using the kriging trend function. Contour interval: 0.5 m

The geoid computed using Moho depths generated by applying the plate loading theory has been compared with a geoid computed by applying the well-known Airy-Heiskanen isostatic model. The computed geoids have been fitted to the GPS-levelling derived geoid by removing a trend surface. A kriging trend function has been computed using only 27 GPS stations among the available 30 GPS stations in Egypt (cf. Fig. 3). The internal precision of the fitted geoids is very good (about 3 cm) and it is rather equal for both geoids. Using the Moho depths generated by applying the plate loading theory and the EGTGM2014 tailored geopotential model improves the external geoid accuracy by about 23%, and the range of the remaining differences has dropped by about 30%.

Finally, it should be noted that the distribution of the GPS stations in Egypt is fairly good, however, the number of the GPS stations should significantly be increased for a better geoid fitting and assessment.

**Acknowledgements** This project was supported financially by the Science and Technology Development Fund (STDF), Egypt, Grant No. 366.

## References

Abd-Elmotaal HA (1991) Gravity anomalies based on the Vening Meinesz isostatic model and their statistical behaviour. Mitteilungen der geodätischen Institute der Technischen Universtät Graz 72

Abd-Elmotaal HA (1993) Vening Meinesz Moho depths: traditional, exact and approximated. Manuscr Geodaet 18(4):171–181

Abd-Elmotaal HA (1998) An alternative capable technique for the evaluation of geopotential from spherical harmonic expansions. Boll Geod Sci Affini 57(1):25–38

Abd-Elmotaal HA (2003) Implementing seismic Moho depths in geoid computation. Surv Rev 37(289):235–245

Abd-Elmotaal HA (2004) Isostatic response of the earth's crust derived by inverse isostasy. J Geodyn 37(2):139–153. doi:10.1016/j.jog.2004.01.002

Abd-Elmotaal HA (2013) Behaviour of earth's crust due to topographic loads derived by inverse and direct isostasy. NRIAG J Astron Geophys 2:196–202. doi:10.1016/j.nrjag.2013.12.005

Abd-Elmotaal HA (2014) Egyptian geoid using ultra high-degree tailored geopotential model. In: Proceedings of the 25th international federation of surveyors FIG congress, Kuala Lumpur, 16–21 June 2014. www.fig.net/pub/fig2014/papers/ts02a/TS02A_abd-elmotaal_6856.pdf

Abd-Elmotaal HA, Abd-Elbakhy M, Ashry M (2013) 30 Meters digital height model for Egypt. In: VIII Hotine-Marussi symposium, Rome, 17–22 June 2013

Abd-Elmotaal HA, Kühtreiber N (1999) Improving the geoid accuracy by adapting the reference field. Phys Chem Earth Pt A 24(1):53–59

Abd-Elmotaal HA, Kühtreiber N (2003) Geoid determination using adapted reference field, seismic Moho depths and variable density contrast. J Geod 77(1–2):77–85

Abramowitz M, Stegun IA (1965) Handbook of mathematical functions, with formulas, graphs, and mathematical tables. Dover Publications, New York

Bechtel TD, Forsyth DW, Swain CJ (1987) Mechanisms of isostatic compensation in the vicinity of the east African rift, Kenya. Geophys J Roy Astron Soc 90:445–465

Brotchie JF, Silvester R (1969) On crustal flexure. J Geophys Res 74:5240–5252

Dorman LM, Lewis BTR (1970) Experimental isostasy: 1. Theory of the determination of the earth's isostatic response to a concentrated load. J Geophys Res 75:3357–3365

Forsberg R (1984) A study of terrain reductions, density anomalies and geophysical inversion methods in gravity field modelling. Ohio State University, Department of Geodetic Science and Surveying, Rep 355

Haagmans R, de Min E, van Gelderen M (1993) Fast evaluation of convolution integrals on the sphere using 1D FFT, and a comparison with existing methods for Stokes' integral. Manuscr Geodaet 18(4):227–241

Hein GW, Eissfeller B, Ertel M, Hehl K, Jacoby W, Czerwek D (1989) On gravity prediction using density and seismic data. Institute of Astronomical and Physical Geodesy, University FAF Munich

Lewis BTR, Dorman LM (1970) Experimental isostasy: 2. An isostatic model for the USA derived from gravity and topographic data. J Geophys Res 75:3367–3386

Moritz H (1990) The figure of the Earth: theoretical geodesy and the Earth's interior. Wichmann, Karlsruhe

Pavlis N, Holmes S, Kenyon S, Factor J (2012) The development and evaluation of the earth gravitational model 2008 (EGM2008). J Geophys Res 117(B04406). doi:10.1029/2011JB008916

Rapp RH (1982) A Fortran program for the computation of gravimetric quantities from high degree spherical harmonic expansions. Ohio State University, Department of Geodetic Science, Rep 334

Sideris MG, Li YC (1993) Gravity field convolutions without windowing and edge effects. Bull Geod 67(2):107–118. doi:10.1007/BF01371374

Tscherning CC, Knudsen P, Forsberg R (1994) Description of the GRAVSOFT package. Geophysical Institute, University of Copenhagen, Technical Report

Turcotte DL, Schubert G (1982) Geodynamics: applications of continuum physics to geological problems. Wiley, New York

Vening Meinesz FA (1940) Fundamental tables for regional isostatic reduction of gravity values. Publ Netherlands Acad Sci, sec 1 Dl. 17(3):1–44

# Part IV

# Mass Movements in the Earth System

# An Investigation on the Closure of the Water Budget Methods Over Volta Basin Using Multi-Satellite Data

Vagner G. Ferreira and Zibrila Asiah

**Abstract**

The terrestrial water budget over a particular hydrological domain (e.g., a drainage basin) plays an important role in understanding the interactions among the energy and water cycles. In this work, we assess the closure of the water budget methods [i.e., terrestrial water budget (TWB) and the coupled atmospheric-terrestrial water budget (ATWB)] as well as the control of total water storage gain over Volta Basin of West Africa. To achieve this, we explore the available satellite and data products: GRACE-derived terrestrial water storage changes (TWSC), satellite altimetry, TRMM-measured rainfall, MODIS-estimated evaporation, atmospheric moisture storage and divergence (ERA-Interim reanalysis data), and in situ discharge. The closure of water budgets are assessed by comparison with GRACE-derived TWSC. The results indicate that the closure of water budgets over the Basin, considering the agreement with GRACE-derived TWSC, present an RMSE of 33.72 and 48.22 mm/month for TWB and ATWB, respectively. This implies that the net precipitation (precipitation minus evaporation) computed from ERA-Interim reanalysis data presents high uncertainties over the Basin. Further, the significant accumulated water gain of 162.84 mm for the period of January 2003–December 2012, is 48% controlled by Lake Volta.

**Keywords**

Discharge • GRACE • Hydro-geodesy • Mass transport • Water budget methods

## 1 Introduction

Volta Basin of West Africa has an area of 417,382 km$^2$ and a population of 20 million people, where Lake Volta is one of the most important physiographic features with a submerged area of 8,500 km$^2$ (it shrinks and swells during the dry and wet seasons, respectively) due to the water impoundment of Aksombo Dam, Ghana. The Basin has experienced extreme climatic conditions and is highly vulnerable to droughts and floods (cf., Andam-Akorful et al. 2014). However, estimating the impacts of droughts and floods on the Volta Basin's water availability from observations is difficult because limited in situ measurements (especially for groundwater and soil moisture) are available. Nevertheless, with many Earth observation technologies available for monitoring the components of the water cycle and their evolution, it seems to be possible to close the terrestrial water budget over a desired river basin as shown by, for example, Sheffield et al. (2009).

The terrestrial water budget equation is based on the principle of conservation of mass applied to some part of the water cycle as (e.g., Yirdaw et al. 2008):

$$\frac{dS}{dt} = P - E - \nabla \cdot \mathbf{R}, \quad (1)$$

V.G. Ferreira (✉) • Z. Asiah
School of Earth Sciences and Engineering, Hohai University, Nanjing, China
e-mail: vagnergf@hhu.edu.cn


S. Jin, R. Barzaghi (eds.), *IGFS 2014*, International Association of Geodesy Symposia 144,
DOI 10.1007/1345_2015_137

where $S$ is the water volume stored per unit area; $P$ is the areal mean rate of precipitation; $E$ is the areal mean rate of evaporation (describing all process of vaporization); and $\nabla \cdot \mathbf{R}$ is the lateral transport of water. If the area is a natural river basin, bounded by natural divides, the outflow terms are generally larger than inflow terms thus, $\nabla \cdot \mathbf{R}$ represents the discharge $R$. The quantities in the Eq. (1) can be obtained from a multitude of products from satellite missions (see, e.g., Sheffield et al. 2009; Sneeuw et al. 2014).

Although $S$ provides a key measure of the continental water cycle and available water resources in a given region, only with the advent of the Gravity Recovery and Climate Experiment (GRACE) it became available (Jin et al. 2012). The terrestrial water storage (TWS) and its change ($dS/dt$, henceforth abbreviated as TWSC) derived from GRACE measurements has been used in several applications by considering the Eq. (1). For instance, GRACE-derived TWSC, in conjunction with precipitation and discharge observations, has been used for estimating evaporation as a residual (see, e.g., Rodell et al. 2004; Ramillien et al. 2006; Long et al. 2014; Zeng et al. 2014; Andam-Akorful et al. 2014). Furthermore, if precipitation and evaporation (or precipitation minus evaporation $P - E$, named net precipitation) are available in conjunction with GRACE-derived TWSC, it is possible to compute the total discharge for a particular drainage basin as shown by many authors (see, e.g., Syed et al. 2005; Ferreira et al. 2013; Hassan and Jin 2014a; Lorenz et al. 2014; Sneeuw et al. 2014).

Previous studies using GRACE-derived TWS mentioning Volta Basin, have reported an increase in TWS (Ferreira et al. 2012; Ahmed et al. 2014) and the high annual amplitudes of TWS in comparison with other regions in Africa (Ferreira et al. 2012; Hassan and Jin 2014b). Nevertheless, only Ahmed et al. (2014) and Moore and Williams (2014) have drawn the attention to the water impoundment of Akosombo Dam and its impact on the GRACE-derived TWS. Additionally, Ferreira et al. (2014) assessed the closure of terrestrial water budget over Volta Basin considering sink terms ($E + R$) employing GRACE, Tropical Rainfall Measuring Mission (TRMM), and Global Land Data Assimilation System (GLDAS) while Andam-Akorful et al. (2014) assessed the closure in terms of evaporation.

To the best of our knowledge no one appears to have quantified the control of terrestrial water storage within the Volta Basin. Thus, the aim of the present work is twofold: (1) to find what is the control of terrestrial water storage gain over the Basin, and (2) to assess the closure of the water budget methods over the Basin. To achieve this, we explored the available satellite products such as GRACE-derived TWSC, rainfall (TRMM), water level (Satellite Altimetry), evaporation (MODerate resolution Imaging Spectroradiometer – MODIS driven by Surface Energy Balance Algorithm for Land – SEBAL), and atmospheric moisture storage and divergence (ERA-Interim reanalysis data).

## 2 Material and Methods

To estimate the water storage and to assess the closure of water budget methods over Volta Basin, several data sets have been used. The time span of January 2003 to December 2012 was adopted because the data set records overlap for this period. Since all data listed below (Sects. 2.1–2.5) present different spatial resolutions, we have made them compatibly with GRACE by converting to spherical harmonics coefficients (SHCs) up to degree and order (d/o) 90 and applied the same filter scheme (i.e., DDK3). In practical applications, it would not be necessary apply the same filter scheme, however, since we are interested in comparison of the different methods using different data sets, it seems to be reasonable bring them all to the same spectral resolution (i.e., d/o 90, which is equivalent to a grid cell of $\sim$220 km at the equator).

### 2.1 GRACE Level 2 Products

The Release 05a (RL05a) Level 2 products (L2), i.e., SHCs from GeoForschungsZentrum (GFZ), were used for this study up to d/o 90. The choice of GFZ L2 products is due to the fact that the calibrated uncertainties of the SHCs are available. The time span of the SHCs covered the period from January 2003 to January 2013, with the SHCs for June, 2003, January and June of 2011, and May and October of 2012 missing. GRACE satellites do not recover the degree-1 coefficients, so we replaced them from the results in Swenson et al. (2008). Additionally, GRACE is insensitive to the zonal degree two coefficient of the gravity field ($C_{2,0}$) due to the orbit configuration. Thus, $C_{2,0}$ has been replaced from the results provided by Cheng and Tapley (2004). For each monthly solution, the long-term mean of January 2003 to January 2013 was removed from the monthly SHCs. Additionally, GRACE gravity fields at high order coefficients exhibit a high level of noise which is known as "stripes" in spatial domain. In order to obtain coherent results, the convolution filter coefficients DDK3 (Kusche 2007), was applied to the residual SHCs derived from GRACE observations. An exhaustive comparison of the suitability of the filter methods available can be found in Werth et al. (2009).

Estimates of monthly terrestrial water storage anomalies ($\delta S$) were obtained from the residual coefficients using an integration approach described in Wahr et al. (1998). Missing GRACE derived terrestrial water storage anomalies were estimated using the previous and the next months (Ramillien et al. 2006). A regional average of the terrestrial water storage anomaly ($\delta S$) can then computed by:

$$\delta S(t) = \frac{1}{A}\sum_{i=1}^{n}\delta S(\varphi_i,\lambda_i,t)A_i, \quad (2)$$

where $n$ is the number of cells within a particular basin, $A_i$ is the area of the cell $i$ and $A$ is the total area of the basin. Given that the difference between $S$ and $\delta S$ is a constant value (mean of the study period), the following equation can be derived from numerical differentiation using the center difference (two-sided difference):

$$\frac{dS(t)}{dt} \approx \frac{\delta S(t+\Delta t) - \delta S(t-\Delta t)}{2\Delta t}, \quad (3)$$

where $dS/dt$ is the approximation of the water mass variations of month $t$ as in Eq. (1), and $\Delta t0$ is the time interval (1 month).

## 2.2 Tropical Rainfall Measuring Mission (TRMM)

TRMM is a joint mission between the United States (National Aeronautics and space Administration – NASA) and Japan (Japan Aerospace Exploration Agency – JAXA) designed to monitor tropical rainfall in the latitude range ±50°. In this work, we used monthly averaged 3B43 V7 rainfall rate products with a spatial resolution of 0.25°, which are inferred from not only the TRMM observations, but also employs data from a number of other satellites and ground-based rain gauge data (Huffman et al. 2007). The data was obtained from NASA's Goddard Earth Sciences and Data and Information Service Center (GES DISC). Over Volta Basin, Thiemig et al. (2012) found that interpolated rainfall derived from ground observations agrees well with TRMM, exhibiting only a slight underestimation of 11%.

## 2.3 MODIS Global Evapotranspiration Project (MOD16)

The MOD16 global evaporation data, are provided by the Earth Observing System of the NASA (NASA/EOS) as part of global project. The estimates are derived from MODIS-based vapor pressure deficit, solar radiation and air temperature, as well as a network of eddy towers and global meteorological data (Mu et al. 2011). The MOD16 data is available at 8-day, monthly, and annual intervals. Analysis for this study however is based on monthly products with a spatial resolution of 0.5°. Andam-Akorful et al. (2014) reported that evaporation estimations inferred from MODIS represent relatively lower uncertainties of 4 mm/month over Volta Basin.

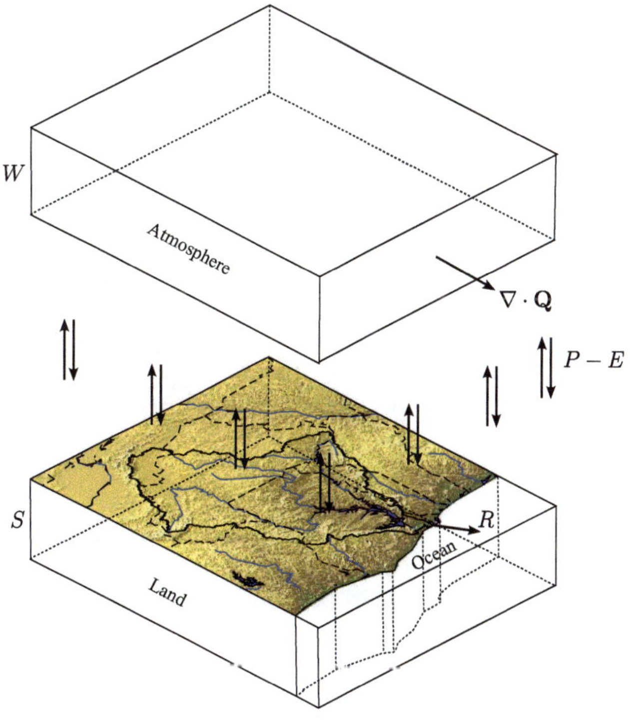


**Fig. 1** Schematic diagram of the terrestrial and atmosphere water budgets. Modified from Yirdaw et al. (2008)

## 2.4 Precipitable Water and Vapor Flux Divergence

The specific humidity ($q$), the eastern ($u$), and the northern ($v$) direction winds from ERA-Interim (Dee et al. 2011), the latest global atmospheric reanalysis produced by the European Centre for Medium-Range Weather Forecasts (ECMWF), were used to calculate the vapor flux divergence $\nabla \cdot \mathbf{Q}$ and precipitable water $W$. These data are necessary to solve the atmospheric water budget in terms of precipitation minus evaporation ($P - E$) as:

$$-(P-E) = \frac{\partial W}{\partial t} + \nabla \cdot \mathbf{Q}. \quad (4)$$

The terms $W$ and $\nabla \cdot \mathbf{Q}$ in the right hand side of (4) were computed respectively by using the Eqs. (5) and (2) of Yirdaw et al. (2008). Figure 1 shows a diagram of the various components of the terrestrial and atmosphere water budgets where they interchange the $P - E$ values on the land surface.

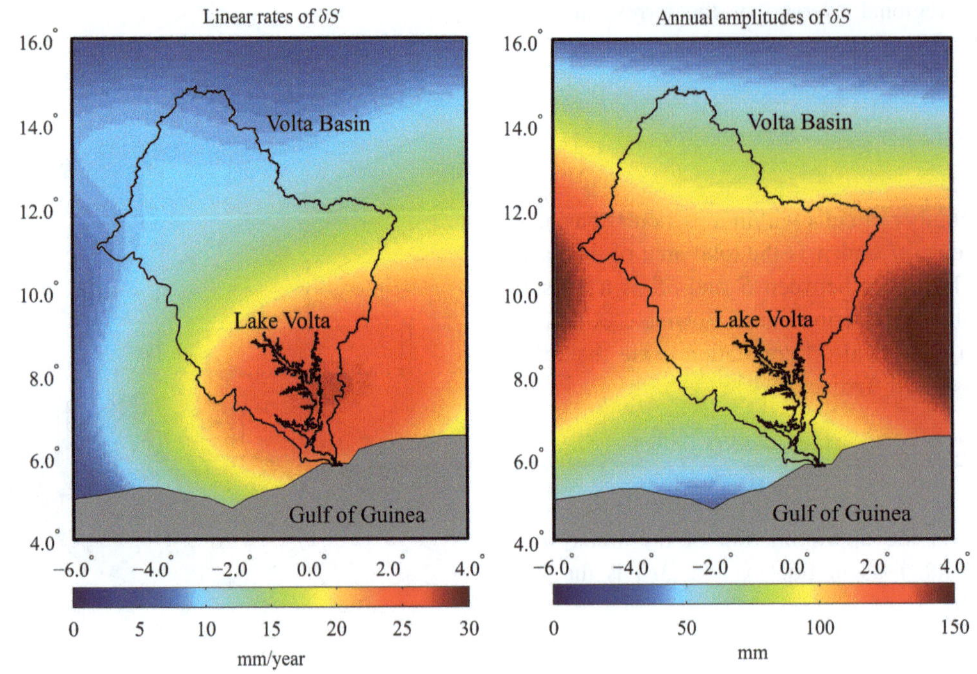


**Fig. 2** Inter-annual changes in $\delta S$ (*right*) in mm/year and annual amplitudes of $\delta S$ (*left*) in mm for the period from January 2003 to December 2012

## 2.5 Lake Volta Levels from Satellite Altimetry

From the available satellite altimetry missions (e.g., Topex/Poseidon, Jason-1 and Jason-2) since 1992, the U.S. Department of Agriculture's Foreign Agricultural Service (USDA-FAS), in co-operation with NASA and the University of Maryland, are routinely monitoring lake and reservoir height variations for many large lakes around the world (Birkett et al. 2011). The Lake Volta water level variations, obtained from USDA, were converted to mass changes expressed in terms of equivalent water height (EWH) by considering a global grid mask of constant EWH equal one within the lake and zero elsewhere (see, e.g., Forootan et al. 2014). This global grid mask was then expanded to SHCs up to d/o 90. These SHCs were then filtered by using DDK3 and subsequently used to synthesize the lake's mask. Then, each monthly lake level height was used to scale the synthesized mask providing the monthly mass changes due to the water impoundment at Lake Volta. The Table 4 of Ričko et al. (2012) presents a root mean square error (RMSE) of 0.54 m for USDA products over Lake Volta, which expressed as EWH is approximately 4 mm over the Volta Basin.

## 2.6 In Situ Discharge

In addition to the satellite-derived datasets, monthly discharge rates from Akosombo Dam in the terrestrial water budget and atmospheric-terrestrial water budget methods. The Akosombo Dam, through the dam's powerhouse and spillways, regulates the main outflow from Volta Basin to the Gulf of Guinea (Atlantic Ocean). The data were obtained from the Water Research Institute of Ghana covering the time span from January 2003 to December 2012 and the records are complete. These data from Akosombo Dam were not smoothed (SHCs and DDK3), as for example, for the rainfall data. Among all data used here, the uncertainty in discharge is often negligible w.r.t. the other hydrological quantities, thus a conservative value of 10% can be assumed.

# 3 Results and Discussions

## 3.1 GRACE-Derived Mass Changes

Following the computation of the terrestrial water storage anomalies ($\delta S$) in a grid format, maps showing the trends and annual signals were generated by simultaneously fitting a time series at each grid point by using a mean, linear trend, annual sine and cosine, and semi-annual sine and cosine. Figure 2 shows the inter-annual changes in $\delta S$ (left panel) and the annual amplitudes of $\delta S$ (right panel) as well. The largest $\delta S$ rates were found at Lake Volta where it indicates a mass increase up to 26 mm/year. Further, the positive rate of GRACE-derived $\delta S$ is most likely caused by the water impoundment at the Lake Volta. Moreover, it seems that the northern parts experience more drier periods

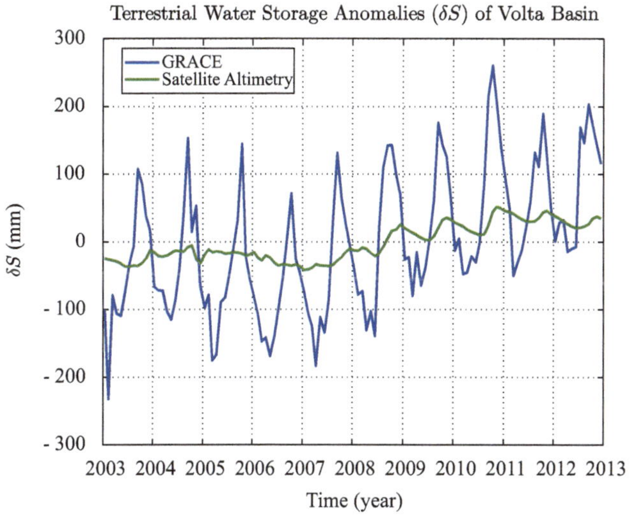


**Fig. 3** Comparison of the contribution of Lake Volta level changes derived from satellite altimetry and expressed in terms of $\delta S$ (in *green*), averaged over the entire Basin, with the averaged $\delta S$ variations derived from GRACE (in *blue*) for the entire Basin

relative to the southern parts. It makes sense if we take into account the annual rainfall varies greatly across the basin, from 1,500 mm in the south to 400 mm in the north. The spatial distribution of the annual amplitude is dominated in the tropical transition zone by large amplitudes in the $\delta S$. This tropical transition zone is characterized with two rainfall seasons close to each other.

In order to assess the water mass changes due to the water impoundment at Lake Volta, the lake level heights from satellite altimetry were converted in $\delta S$ (cf., Sect. 2.5) and compared with GRACE-derived $\delta S$, both averaged over the Volta River basin using Eq. (2). Figure 3 shows the comparison between GRACE-derived ($\delta S_{\text{GRACE}}$) and Satellite Altimetry (SA)-derived ($\delta S_{\text{SA}}$) signals, it is possible to note that both show the increasing in water storage over the basin. For the time window of January 2003 to December 2012, Volta Basin shows an increase in $\delta S$ of 16.28 mm/year as observed by GRACE. However, water storage due to Lake Volta measured from Satellite Altimetry is 7.89 mm/year, which contributes with approximately 48% to the total water mass gain within the Basin.

While Lake Volta covers approximately 2.0% of the Volta Basin's area, it accounts for 6.5% of seasonal $\delta S$ variations on the extracted GRACE response with a phase lag of 77.8 days. This phase lag is mainly explained by considering that the hydrological processes from upstream of the Basin take a while to reach the lake (mass transport from upstream to downstream). Another import aspect related to $\delta S$ shown in Figs. 2 and 3 is that in estimating groundwater trends over West Africa for example, one should carefully considering the contribution of Lake Volta (Moore and Williams 2014). Additionally, Forootan et al. (2014) pointed out that for forecasting $\delta S$ over West Africa it is necessary to remove Lake Volta's contribution, otherwise the amplitudes of forecasted $\delta S$ over the lake would be overestimated.

## 3.2 Closure of the Water Budgets in Volta Basin

Figure 4 (top panel) shows the time series for precipitation (TRMM), evaporation (MOD16) and net precipitation (Era-Interim) – after filtered using SHCs up to d/o 90 and DDK3 – and in situ discharge. The bottom panel shows the derivative of terrestrial water storage anomalies ($\delta S$) with respect to time ($t$) which is also expressed in terms of fluxes as the other quantities (i.e., mm/month). It is possible to note the same seasonal patterns among of the hydrological quantities, for example, precipitation has annual amplitude of 97.62 mm/month and it peaks at day 206.97 (around July 24). The amplitude ratio between terrestrial water storage changes ($dS/dt$) and precipitation is 0.52 with a phase lag of $-17.14$ days (precipitation lags terrestrial water storage changes). The amplitude ratio and phase lag for evaporation is, respectively, 0.37 and 17.38 days, and for net precipitation ($P - E$) it is 0.52 and $-17.68$ days.

To evaluate the performance of the different approaches, i.e., terrestrial water budget (TWB) versus coupled atmospheric-terrestrial water budget (ATWB) methods, we computed the imbalance in the water budget methods as:

$$\delta_{\text{TWB}} = \frac{dS}{dt} - (P - E - R_{\text{obs}}) \tag{5}$$

and

$$\delta_{\text{ATWB}} = \frac{dS}{dt} - \left[ - \left( \frac{\partial W}{\partial t} + \nabla \cdot \mathbf{Q} \right) - R_{\text{obs}} \right]. \tag{6}$$

The variables $\delta_{\text{TWB}}$ and $\delta_{\text{ATWB}}$ are, respectively, the water budget imbalances of the terrestrial water budget and coupled atmospheric-terrestrial water budget. We use the mean, RMSE, the percentage bias (PBIAS) and the Nash-Sutclife Efficiency (NSE) coefficient (see, e.g., Sneeuw et al. 2014, and references therein for details) to assess the closure of water budget methods. The PBIAS value provides information about the long term imbalance of the water budget while the NSE provides information on the monthly time scales (cf., Lorenz et al. 2014). Table 1 shows these metrics for assessing the closure of the water balance in Volta Basin.

Overall, the results (cf., Table 1) show that the terrestrial water budget approach seems to underestimate (positive PBIAS) while the coupled atmospheric-terrestrial water budget approach overestimate (negative PBIAS) the mean

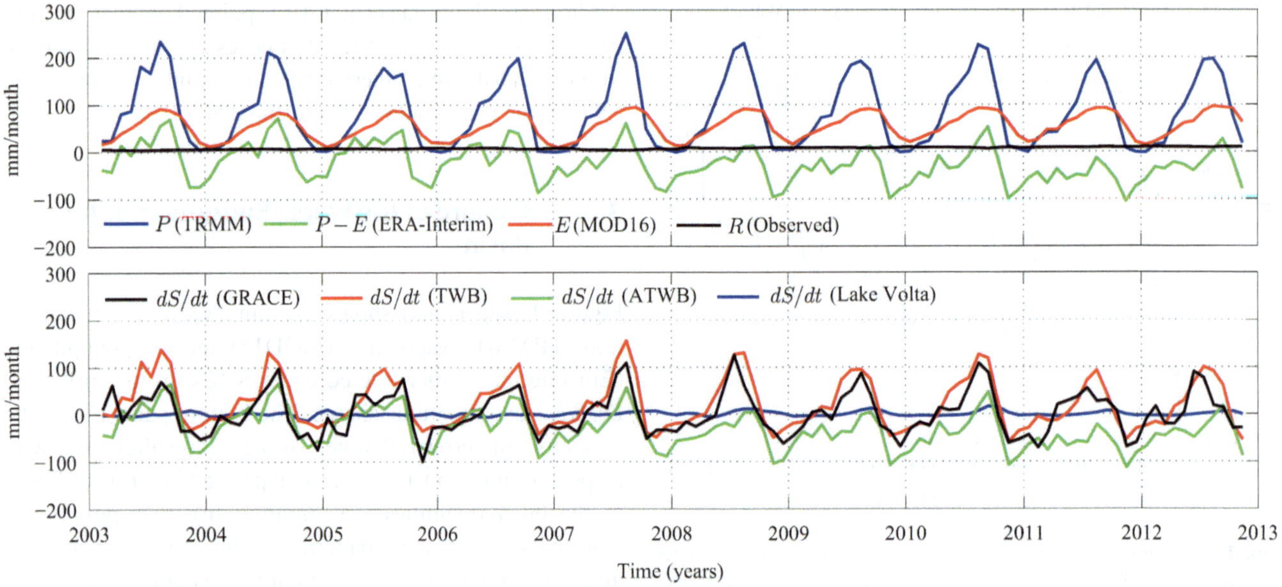


**Fig. 4** The *top panel* shows the precipitation ($P$), evaporation ($E$), net precipitation ($P - E$), and observed discharge ($R$); the *bottom panel* shows the GRACE-derived $dS/dt$, SA-derived $dS/dt$, TWB-estimated $dS/dt$, and ATWB-estimated $dS/dt$

**Table 1** Statistical comparison between the time series of GRACE-derived $dS/dt$ (reference) and estimated $dS/dt$ as shown in Eqs. (5) and (6) in terms of imbalance in water budget methods

| Statistics | $\delta_{TWB}$ | $\delta_{ATWB}$ |
|---|---|---|
| Mean (mm/month) | −20.63 | 33.04 |
| RMSE (mm/month) | 33.72 | 48.22 |
| PBIAS (%) | 67.56 | −108.21 |
| NSE | 0.45 | −0.12 |

$dS/dt$ – not a steady state over the time window. Despite the huge imbalances in terms of PBIAS (optimal value of PBIAS is 0.0), it seems that terrestrial water budget approach performs better than the coupled atmospheric-terrestrial water budget. From Fig. 4 it is also possible to note that TWB-estimated $dS/dt$ time series present the same patterns as GRACE-derived $dS/dt$, which are also expressed in terms of NSE shown in Table 1 (optimal value of NSE is 1.0). This implies that the closure of the terrestrial water budget at monthly time-scales is possible considering the data and methodology applied in this study. As we can see from Fig. 2, Volta Basin is under significant water storage increase of which water impoundment of Lake Volta plays a partial role. This implies in a low and nearly constant discharge throughout the study time span (cf., the black curve in Fig. 4, top panel). Thus, considering the effect of Lake Volta storage on the closure of the TWB, it was found a slight improvement of 3.4% in terms of NSE and 2.5% in terms of PBIAS.

Furthermore, it should be mentioned that systematic errors between each technique, such as GRACE (see, e.g., Jin et al. 2012; Jin and Feng 2013), Satellite Altimetry (see, e.g., Ričko et al. 2012; Jin et al. 2014), MODIS, TRMM, global atmospheric reanalysis fields (ERA-Interim), and observed discharge impact the closure of the water budget methods. In this context, the standard deviation (STD) of GRACE-derived TWSC ($dS/dt$) time series is approximately 45.79 mm/month and the GRACE estimated error (including measurement error and leakage) is 38.85 mm/month over Volta Basin, which provides a signal-to-noise ratio (SNR) of 1.18. However, the relative error of the TWB-estimated $dS/dt$ is approximately 11.92 mm/month considering the uncertainties in $P$ (TRMM), $E$ (MODIS), and $R$ (observed) with an STD of 53.18 mm/month, which provides a SNR of 4.46.

## 4 Conclusions

We have used the space-borne data such as evaporation (MODIS), rainfall (TRMM), terrestrial water storage (GRACE), satellite altimetry (USDA), and ancillary data such as precipitable water and vapor flux divergence (ERA-Interim) as well as observed discharge to evaluate the closure of the water budget methods over Volta Basin and to quantify the water mass gain due to the contribution of Lake Volta. The amplitude of subsurface water storage, due to the water impoundment of Lake Volta, represents approximately 93% of the amplitude of TWS anomalies (at annual time scale) over Volta Basin. Further, a significant wetting trend within the Volta Basin is apparent, with an accumulated gain of 162.84 mm over the time window of January 2003 to December 2012, where 48% is mostly controlled by Lake Volta. Finally, it is possible to close the terrestrial water budget over the Basin, while the converse was observed for the use of coupled atmospheric-terrestrial budget method, probably due to the low quality atmospheric data (i.e., precipitable water and moisture flux divergence) since GRACE data and the in situ discharge are the same for both methods.

**Acknowledgements** V.G. Ferreira acknowledges the financial support provided by National Natural Science Foundation of China (Grant No. 41204016). Prof. Dr.-Ing. Jürgen Kusche is appreciated for providing the DDK filter coefficients. We would like to thank the Editor-in-Chief and the three anonymous reviewers for their comments, which improved this manuscript. We are grateful for the data used in this study as described in Section 2.

## References

Ahmed M, Sultan M, Wahr J, Yan E (2014) The use of GRACE data to monitor natural and anthropogenic induced variations in water availability across Africa. Earth Sci Rev 136(0):289–300. doi:10.1016/j.earscirev.2014.05.009

Andam-Akorful SA, Ferreira VG, Awange JL, Forootan E, He XF (2014) Multi-model and multi-sensor estimations of evapotranspiration over the Volta basin, West Africa. Int J Climatol. doi:10.1002/joc.4198

Birkett C, Reynolds C, Beckley B, Doorn B (2011) From research to operations: the USDA global reservoir and lake monitor. In: Vignudelli S, Kostianoy AG, Cipollini P, Benveniste J (eds) Coastal altimetry. Springer, Berlin/Heidelberg, pp 19–50. doi:10.1007/978-3-642-12796-0_2

Cheng M, Tapley BD (2004) Variations in the earth's oblateness during the past 28 years. J Geophys Res Solid Earth 109(B9):B09402. doi:10.1029/2004JB003028

Dee DP, Uppala SM, Simmons AJ, Berrisford P, Poli P, Kobayashi S, Andrae U, Balmaseda MA, Balsamo G, Bauer P, Bechtold P, Beljaars ACM, van de Berg L, Bidlot J, Bormann N, Delsol C, Dragani R, Fuentes M, Geer AJ, Haimberger L, Healy SB, Hersbach H, Hólm EV, Isaksen L, Kållberg P, Köhler M, Matricardi M, McNally AP, Monge-Sanz BM, Morcrette JJ, Park BK, Peubey C, de Rosnay P, Tavolato C, Thépaut JN, Vitart F (2011) The ERA-interim reanalysis: configuration and performance of the data assimilation system. Q J R Meteorol Soc 137(656):553–597. doi:10.1002/qj.828

Ferreira VG, Gong Z, Andam-Akorful SA (2012) Monitoring mass changes in the Volta river basin using GRACE satellite gravity and TRMM precipitation. Boletim de Ciencias Geodesicas 18(4):549–563. doi:10.1590/S1982-21702012000400003

Ferreira VG, Gong Z, He X, Zhang Y, Andam-Akorful SA (2013) Estimating total discharge in the Yangtze river basin using satellite-based observations. Remote Sens 5(7):3415–3430. doi:10.3390/rs5073415

Ferreira VG, Andam-Akorful SA, He X, Xiao R (2014) Estimating water storage changes and sink terms in Volta Basin from satellite missions. Water Sci Eng 7(1):5–16. doi:10.3882/j.issn.1674-2370.2014.01.002

Forootan E, Kusche J, Loth I, Schuh WD, Eicker A, Awange J, Longuevergne L, Diekkrüger B, Schmidt M, Shum C (2014) Multivariate prediction of total water storage changes over West Africa from multi-satellite data. Surv Geophys 35(4):913–940. doi:10.1007/s10712-014-9292-0

Hassan AA, Jin S (2014a) Lake level change and total water discharge in East Africa Rift Valley from satellite-based observations. Global Planet Chang 117(0):79–90. doi:10.1016/j.gloplacha.2014.03.005

Hassan AA, Jin S (2014b) Water cycle and climate signals in Africa observed by satellite gravimetry. IOP Conf Ser Earth Environ Sci 17(1):012149. doi:10.1088/1755-1315/17/1/012149

Huffman GJ, Bolvin DT, Nelkin EJ, Wolff DB, Adler RF, Gu G, Hong Y, Bowman KP, Stocker EF (2007) The TRMM multisatellite precipitation analysis (TMPA): quasi-global, multiyear, combined-sensor precipitation estimates at fine scales. J Hydrometeorol 8(1):38–55. doi:10.1175/JHM560.1

Jin S, Feng G (2013) Large-scale variations of global groundwater from satellite gravimetry and hydrological models, 2002–2012. Global Planet Change 106(0):20–30. doi:10.1016/j.gloplacha.2013.02.008

Jin S, Hassan A, Feng G (2012) Assessment of terrestrial water contributions to polar motion from GRACE and hydrological models. J Geodyn 62(0):40–48. doi:10.1016/j.jog.2012.01.009 [earth Rotation]

Jin S, Feng G, Andersen O (2014) Errors of mean dynamic topography and geostrophic current estimates in China's marginal seas from GOCE and satellite altimetry. J Atmos Ocean Technol 31(11):2544–2555. doi:10.1175/JTECH-D-13-00243.1

Kusche J (2007) Approximate decorrelation and non-isotropic smoothing of time-variable GRACE-type gravity field models. J Geodesy 81(11):733–749. doi:10.1007/s00190-007-0143-3

Long D, Longuevergne L, Scanlon BR (2014) Uncertainty in evapotranspiration from land surface modeling, remote sensing, and GRACE satellites. Water Resour Res 50(2):1131–1151. doi:10.1002/2013WR014581

Lorenz C, Kunstmann H, Devaraju B, Tourian MJ, Sneeuw N, Riegger J (2014) Large-scale runoff from landmasses: a global assessment of the closure of the hydrological and atmospheric water balances. J Hydrometeorol 15(6):2111–2139. doi:10.1175/JHM-D-13-0157.1

Moore P, Williams SDP (2014) Integration of altimetric lake levels and GRACE gravimetry over Africa: inferences for terrestrial water storage change 2003–2011. Water Resour Res 50(12):9696–9720. doi:10.1002/2014WR015506

Mu Q, Zhao M, Running SW (2011) Improvements to a MODIS global terrestrial evapotranspiration algorithm. Remote Sens Environ 115(8):1781–1800. doi:10.1016/j.rse.2011.02.019

Ramillien G, Frappart F, Güntner A, Ngo-Duc T, Cazenave A, Laval K (2006) Time variations of the regional evapotranspiration rate from gravity recovery and climate experiment (GRACE) satellite gravimetry. Water Resour Res 42(10):W10403. doi:10.1029/2005WR004331

Ričko M, Birkett CM, Carton JA, Crétaux JF (2012) Intercomparison and validation of continental water level products derived from satellite radar altimetry. J Appl Remote Sens 6(1):061710–061710. doi:10.1117/1.JRS.6.061710

Rodell M, Famiglietti JS, Chen J, Seneviratne SI, Viterbo P, Holl S, Wilson CR (2004) Basin scale estimates of evapotranspiration using GRACE and other observations. Geophys Res Lett 31(20):L20504. doi:10.1029/2004GL020873

Sheffield J, Ferguson CR, Troy TJ, Wood EF, McCabe MF (2009) Closing the terrestrial water budget from satellite remote sensing. Geophys Res Lett 36(7):L07403. doi:10.1029/2009GL037338

Sneeuw N, Lorenz C, Devaraju B, Tourian M, Riegger J, Kunstmann H, Bárdossy A (2014) Estimating runoff using hydro-geodetic approaches. Surv Geophys 35(6):1333–1359. doi:10.1007/s10712-014-9300-4

Swenson S, Chambers D, Wahr J (2008) Estimating geocenter variations from a combination of GRACE and ocean model output. J Geophys Res Solid Earth 113(B8):B08410. doi:10.1029/2007JB005338

Syed TH, Famiglietti JS, Chen J, Rodell M, Seneviratne SI, Viterbo P, Wilson CR (2005) Total basin discharge for the Amazon and Mississippi river basins from GRACE and a land-atmosphere water balance. Geophys Res Lett 32(24):L24404. doi:10.1029/2005GL024851

Thiemig V, Rojas R, Zambrano-Bigiarini M, Levizzani V, De Roo A (2012) Validation of satellite-based precipitation products over sparsely gauged African river basins. J Hydrometeorol 13(6):1760–1783. doi:10.1175/JHM-D-12-032.1

Wahr J, Molenaar M, Bryan F (1998) Time variability of the Earth's gravity field: hydrological and oceanic effects and their possible detection using GRACE. J Geophys Res 103(B12):30205–30229. doi:10.1029/98JB02844

Werth S, Güntner A, Schmidt R, Kusche J (2009) Evaluation of GRACE filter tools from a hydrological perspective. Geophys J Int 179(3):1499–1515. doi:10.1111/j.1365-246X.2009.04355.x

Yirdaw S, Snelgrove K, Agboma C (2008) GRACE satellite observations of terrestrial moisture changes for drought characterization in the Canadian Prairie. J Hydrol 356(1–2):84–92. doi:10.1016/j.jhydrol.2008.04.004

Zeng Z, Wang T, Zhou F, Ciais P, Mao J, Shi X, Piao S (2014) A worldwide analysis of spatiotemporal changes in water balance-based evapotranspiration from 1982 to 2009. J Geophys Res Atmos 119(3):1186–1202. doi:10.1002/2013JD020941

# Application of Independent Component Analysis in GRACE-Derived Water Storage Changes Interpretation: A Case Study of the Tibetan Plateau and Its Surrounding Areas

Hanjiang Wen, Zhenwei Huang, Youlei Wang, Huanling Liu, and Guangbin Zhu

**Abstract**

Independent component analysis (ICA) is applied to decompose the water storage changes derived from 132 months (2003.01 to 2013.12) Gravity Recovery and Climate Experiment (GRACE) measurements over Tibetan Plateau. The results are then compared with those from NOAH and WaterGAP Global Hydrology Model (WGHM) hydrological models. Our assessments indicate that the decomposed components from the water storage changes and hydrological models agree well, indicating the ICA's relatively robust performance in separating independent pattern from water storage observations with few a priori information.

**Keywords**

GRACE • Hydrological Models • Independent Component Analysis • Water Storage Changes

## 1 Introduction

Extracting information about the large-scale water storage change is very useful to improve the efficiency of water resources development and to understand the water cycle. Gravity Recovery and Climate Experiment (GRACE) gravity mission provides gravity field products, which can be used to monitor the worldwide terrestrial water storage changes (Tapley et al. 2004). However, GRACE measures total gravity changes, which are mainly caused by storage changes in various storage compartments such as the terrestrial water storage changes, surface water, and groundwater compartments. Therefore, signal separation techniques are required to enhance a better understanding the source of variability derived from GRACE products (Forootan et al. 2014).

Several methods exist that can be used to extract specific patterns from multi-dimensional data are currently used, most of which try to represent a large portion of the variance in the data to reduce the dimension. Among these methods, Empirical Orthogonal Function (EOF, also known as Principal Component Analysis) and its extension methods such as rotated EOF (REOF) and MSSA have made a wide range of applications (Jeffers 1967; Niu et al. 2002; Price et al. 2006; Rangelova et al. 2007; Schrama et al. 2007; Zotov and Shum 2010).

In the decomposition process, the methods mentioned above only use the empirical covariance matrix or correlation matrix of the data sets, that contains only up to second order statistical information, while ignoring the higher-order statistical information. In general, when the probability distribution function of the observations follows a Gaussian distribution, nearly no information is available in higher-order statistical moments (Hyvarinen 1999). When the distribution function follows a non-Gaussian distribution, using

H. Wen • Z. Huang (✉) • Y. Wang
Chinese Academy of Surveying and Mapping, Beijing 100830, China
e-mail: huangzw@casm.ac.cn

H. Liu
Chinese Academy of Surveying and Mapping, Beijing 100830, China

School of Geodesy and Geomatics, Wuhan University, Wuhan 430079, China

G. Zhu
Satellite Surveying and Mapping Application Center, NASG, Beijing 101300, China


S. Jin, R. Barzaghi (eds.), *IGFS 2014*, International Association of Geodesy Symposia 144,
DOI 10.1007/1345_2015_187

the covariance only might not be enough to measure the statistical relationships between different samples (e.g., storage observations). However, one can see that the probability distribution function of GRACE-derived water storage time series is often non-Gaussian (Forootan and Kusche 2012). Further, with uncorrelated and orthogonal hypothesis, the EOF analysis provides the signal components which might not be necessarily interpretable (Jolliffe 2003).

To incorporate more statistical information from the probability distribution function, Cardoso (1992) proposed to add higher-order statistical moments to the EOF method, which is the origin of the Independent Component Analysis (ICA) method. ICA is originally developed for solving blind source separation problems with no or little priori information about the source signals. It has been widely applied in the biomedical field, face recognition, mobile phone communications and other fields (Makeig et al. 1996; Hyvärinen et al. 2004; Bartlett 2001). In recent years, ICA has also been applied in the separation of seismic data, processing of GPS multipath error and GRACE satellite data (Lv et al. 2007; Luo et al. 2012; Guo et al. 2014; Frappart et al. 2010; Forootan and Kusche 2012, 2013; Boergens et al. 2014). In this paper, we follow the approach in Forootan and Kusche (2012).

In this paper, we will first illustrate the data that was used in this study. Then the ICA method that will be used to separate water cycle signals from GRACE-derived water storage changes for better interpretation of detected anomalies over the Tibetan Plateau and its surrounding areas (20° to 40° N, 70° to 100° E) was introduced. To verify the reliability of the results, we decompose and compare the components deriving from the GRACE-derived water storage changes and hydrological models in the same region respectively. Finally, conclusions are summarized based on the comparison results.

## 2 Data

### 2.1 GRACE Gravity Field Products

Launched in March 2002, the GRACE mission is the first gravity satellites that can be studied directly for the Earth surface mass changes. The mission determines the variations of the Earth's gravity field which can be used to acquire the land water storage changes after removing other effects, including glacial isostatic adjustment process, mass changes within the ocean and atmosphere etc. Many studies have been done in these aspects, such as the detection to the mass changes of the Antarctic and to study global water storage changes from GRACE monthly gravity field models (Chen et al. 2006; Zhu et al. 2008).

In this paper, we used 132 months spherical harmonic gravity field models from January 2003 to December 2013 provided by Center for Space Research (CSR). The spherical harmonic coefficients of the GRACE RL05 data were computed to degree and order 60, the effects of tidal, atmosphere and oceans have been deducted in the calculation process. The gravity field variations of GRACE primarily reflect the water storage changes in land areas. The term $C_{20}$ was replaced by the Satellite Laser Ranging observation. The GRACE-derived land water storage changes were obtained after accounting for the glacial isostatic adjustment, decorrelation and Gaussian filter (with 350 km radius) (Chen et al. 2005; Duan et al. 2009; Wen et al. 2011). The results are prepared in $1° \times 1°$ grid.

### 2.2 Hydrological Models

NASA Goddard Earth Sciences Data and Information Services Center and the National Centers for Environmental Prediction jointly established the Global Land Data Assimilation System (GLDAS) which uses the latest ground modeling and space observation systems to provide parameters of the topography model data (Rodell et al. 2004). Based on NOAH land surface model, GLDAS hydrological model utilized rainfall and solar radiation observations as input parameters to calculate terrestrial water storage changes. It can also be used for verification of GRACE-derived water storage changes information. In this paper, the spatial resolution of the NOAH is the same as GRACE-derived water storage changes, and data is from January 2003 to December 2013, unfortunately the groundwater compartment is not included in the GLDAS.

WaterGAP Global Hydrology Model (WGHM) uses a conceptual framework to simulate continental water cycle. The model was originally developed in the study of continental water resources availability by Döll et al. (2003). Because the model provides the estimation of water mass change, which makes it useful to global water reserves and its dynamic hydrological analysis (Werth and Güntner 2010), WGHM hydrological model has been repeatedly used to compare changes in land water storage and GRACE-derived water storage (Schmidt et al. 2008). The WGHM model in the article uses the same spatial resolution as GRACE-derived water storage changes, from January 2003 to April 2012 that is available to us (personal communication).

## 3 ICA Method

If the equivalent water height (EWH) at $m$ grid points are derived from $n$ monthly gravity field models, the grids are arranged from north to south with longitudes and latitudes varying as. This time series of observations can be written as a matrix $\mathbf{X}$, as

$$\mathbf{X} = [\mathbf{x}_1, \mathbf{x}_2, \cdots, \mathbf{x}_m] \tag{1}$$

where $\mathbf{x}_i$ is the $i$-th column, represents the equivalent water height of $i$-th grid points at different months, i.e. $\mathbf{x}_i = [x_{1i}, \cdots, x_{ni}]^T, i = 1, \cdots, m$. The EOF decomposition of $\mathbf{X}$ can be expressed as

$$\mathbf{X} = \mathbf{P}\mathbf{E}^T \tag{2}$$

where **P** contains the principal components, **E** contains the eigenvectors of **X** that are normalized to unit vectors in the processing. The important characteristics of time series are contained in the eigenvalues and eigenvectors. Then, we sorted the eigenvectors in order according to corresponding magnitude of eigenvalues.

Independent Component Analysis (ICA) is a method searching for its intrinsic factor or component from multidimensional statistics. The main purpose of ICA is to make the principal components or eigenvectors as independent as possible. The ICA decomposition can be written as follows

$$\mathbf{X} = (\mathbf{P}\mathbf{R})\left(\mathbf{R}^T\mathbf{E}^T\right) = \mathbf{S}\mathbf{A}^T \tag{3}$$

Where **A** is called the mixing matrix, **S** is the source signal matrix and **R** is an orthogonal rotation matrix ($\mathbf{R}\mathbf{R}^T = \mathbf{I}$).

Suppose observations are composed of some unknown non-Gaussian distribution signals **S**, then the ICA method can separate independent sources from these mixed observations. Rotation matrix **R** is calculated from Joint Approximate Diagonalization of Eigenmatrixes (JADE) method proposed by Cardoso and Souloumiac (1993).

If the mean of column vector $\mathbf{z}_i, \mathbf{z}_j, \mathbf{z}_k, \mathbf{z}_l$ $(1 \le i, j, k, l \le n)$ are zero, its fourth-order cumulant is defined as

$$\begin{aligned} Cum\left(\mathbf{z}_i, \mathbf{z}_j, \mathbf{z}_k, \mathbf{z}_l\right) &= E\left(\mathbf{z}_i\mathbf{z}_j\mathbf{z}_k\mathbf{z}_l\right) - E\left(\mathbf{z}_i\mathbf{z}_j\right)E\left(\mathbf{z}_k\mathbf{z}_l\right) \\ &\quad - E\left(\mathbf{z}_i\mathbf{z}_k\right)E\left(\mathbf{z}_j\mathbf{z}_l\right) - E\left(\mathbf{z}_i\mathbf{z}_l\right)E\left(\mathbf{z}_j\mathbf{z}_k\right) \end{aligned} \tag{4}$$

For any $n \times n$ order matrix **M**, we can define a $n \times n$ fourth-order cumulant matrix **Q(M)** whose ($i$, $j$) element is given as

$$[\mathbf{Q}(\mathbf{M})]_{ij} = \sum_{k=1}^{n}\sum_{l=1}^{n} Cum\left(\mathbf{z}_i, \mathbf{z}_j, \mathbf{z}_k, \mathbf{z}_l\right)\mathbf{m}_{kl} \tag{5}$$

where $m_{kl}$ is the ($k$, $l$) elements of **M**. Because **Z** is centralized, we can get

$$\begin{aligned} \mathbf{Q}(\mathbf{M}) &= E\left\{\left(\mathbf{Z}^T\mathbf{M}\mathbf{Z}\right)\left(\mathbf{Z}\mathbf{Z}^T\right)\right\} - \mathbf{C}_Z\,\mathbf{tr}\left(\mathbf{M}\mathbf{C}_Z\right) \\ &\quad - \mathbf{C}_Z\mathbf{M}\mathbf{C}_Z - \mathbf{C}_Z\mathbf{M}^T\mathbf{C}_Z \end{aligned} \tag{6}$$

where $\mathbf{C}_\mathbf{Z}$ is the covariance matrix of **Z**, **tr** is the trace of a matrix. Then we can find a normalized orthogonal matrix **R** that is joint diagonalization for all fourth-order cumulant matrixes to make **Q(M)** diagonalized as possible (Cardoso 1999).

After the EOF decomposition, we can estimate the cumulant matrix **Q(M)** of **P** or **E** from Eq. (6), written as

$$F(\mathbf{R}) = \sum_{m=1}^{n^2} f\left(\mathbf{R}^T\mathbf{Q}\left(\mathbf{M}_m\right)\mathbf{R}\right) \tag{7}$$

where function $f$ represents the quadratic sum of the non-diagonal elements of matrix. The cumulant matrix **Q(M)** is optimized through the plane of rotation under the constraints of **R**, when squares of non-diagonal elements in (7) is the least, **R** is the optimal solution. Finally, the rotated independent components or spatial patterns are calculated, the corresponding principal components or spatial patterns are acquired from data projector (Forootan and Kusche 2012).

In this study, the temporal ICA method is used. It needs two steps to decompose the water storage changes by using temporal ICA method. First, EOF is conducted to decompose the water storage changes to obtain the dominant orthogonal modes in descending order. Second, the principal components are rotated temporally as mutually independent as possible by the rotation matrix **R** that determined by the relevant criterion based on Eq. (5). Then the spatial patterns are determined by rotating the corresponding spatial patterns.

## 4 Results

### 4.1 GRACE-Derived Water Storage Changes

We first analyze GRACE-derived water storage changes in Tibetan Plateau and its surrounding areas. The kurtosis of the time series is calculated as $E\left(X^4\right)/E\left(X^2\right)^2 - 3$ (where $E$ is the expectation operator). Results show that 62.7% of the absolute kurtosis value is greater than 0.5, which means that it is a non-Gaussian distribution.

The auto-covariance matrix of the sample is calculated from the centralized data, and it is diagonalized by eigenvalue decomposition method. The first eigenvalue takes 75.9% of the total energy, which is much larger than the other eigenvalues. The second, third and fourth eigenvalue takes 10.0%, 7.8% and 2.1% of the total energy respectively. The corresponding components of the first four eigenvalues contain about 95.8% of the total energy, so the other components are assumed to be insignificant.

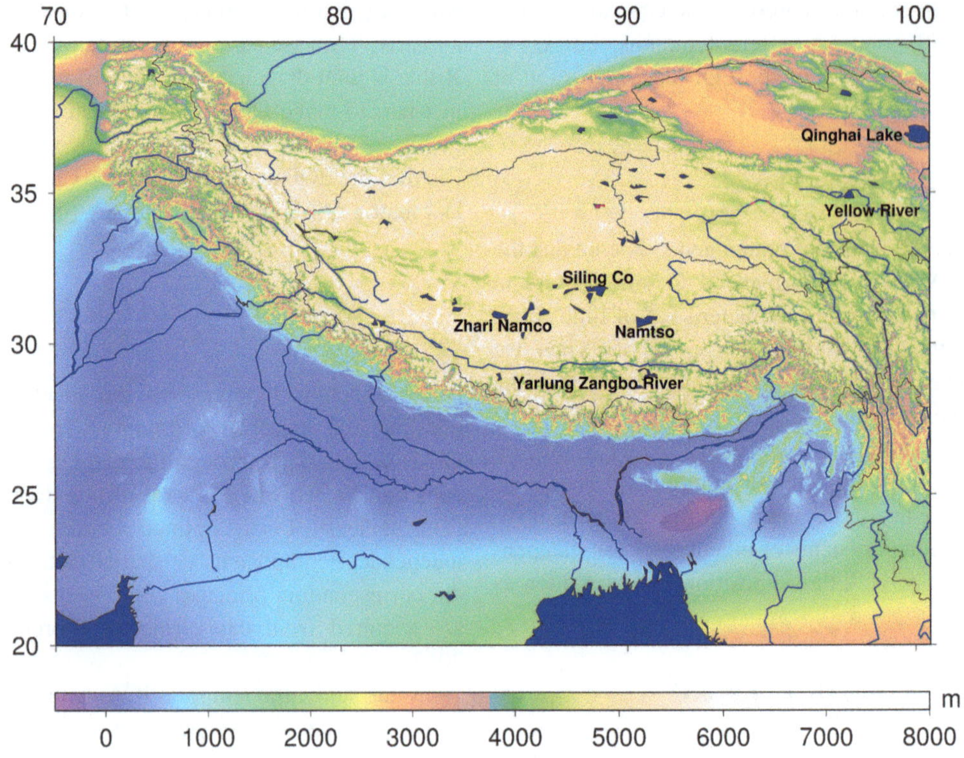


**Fig. 1** The major rivers and lakes (*Blue*) over Tibetan Plateau and its surrounding areas

### 4.2 The Results of ICA Method

Assuming that signals from the GRACE-derived water storage changes are as independent as possible, we uses temporal ICA method to analyze the GRACE-derived water storage changes in Tibetan Plateau and its surrounding areas. Figure 1 presents the map of Tibetan Plateau and its surrounding areas. The result of GRACE-derived water storage changes by ICA method in the Tibetan Plateau region was shown in Fig. 2. The temporal components were scaled by standard deviation, and the corresponding spatial patterns had also been multiplied by the same standard deviations to have a unit similar TWS fields. Then water storage changes can be reconstructed by multiplying the spatial components with the corresponding temporal components.

From Fig. 2, the first and second components display a strong annual period signal, the third component is a mixed signal of annual period and long-term trend, and the fourth component contains the annual and semi-annual period signal. IC1 (Independent Component 1, the same below) and IC2 that represents annual periodic signals are more obvious in the southern region, and periodic signals relatively weak in the northern region, which can be interpreted as the annual variation in precipitation. From the long-term trends of IC3, the water reserves of the Tibetan Plateau in western and southern regions reduced in January 2003 to December 2013 period, which is associated with the melting glaciers and shrinking snowpack. Meanwhile, the water reserves in central region of the Tibetan Plateau have increased that is associated with the increasing area of the plateau lakes. According to Zhang et al. (2013), the mean lake level increased 0.14 m/year for the 200 lakes over Tibetan Plateau and 0.20 m/year for the lakes in the central Tibetan Plateau with 4–7 years satellite altimetry observations, which might be associated with the increasing rainfall, melting snow, and the reducing evaporation of the region in recent years (Zhang et al. 2013).

## 5 Comparison with the Hydrological Models

Water storage changes from hydrological models represent significant non-Gaussian characteristics, because the existence of the complicated hydrological process, the hydrological data and model structure errors. The rate of absolute kurtosis value (greater than 0.5) calculated using NOAH or WGHM time series is about 68.7% and 71.4%, so showing that the outcomes of hydrological models are also of non-Gaussian distribution. Figure 3 shows that the RMSs of the GRACE-derived water storage changes, NOAH and WGHM hydrological models. The red curve is the 12-points weighted moving average results in order to remove the effects of seasonal variations. From Fig. 3, the water storage changes

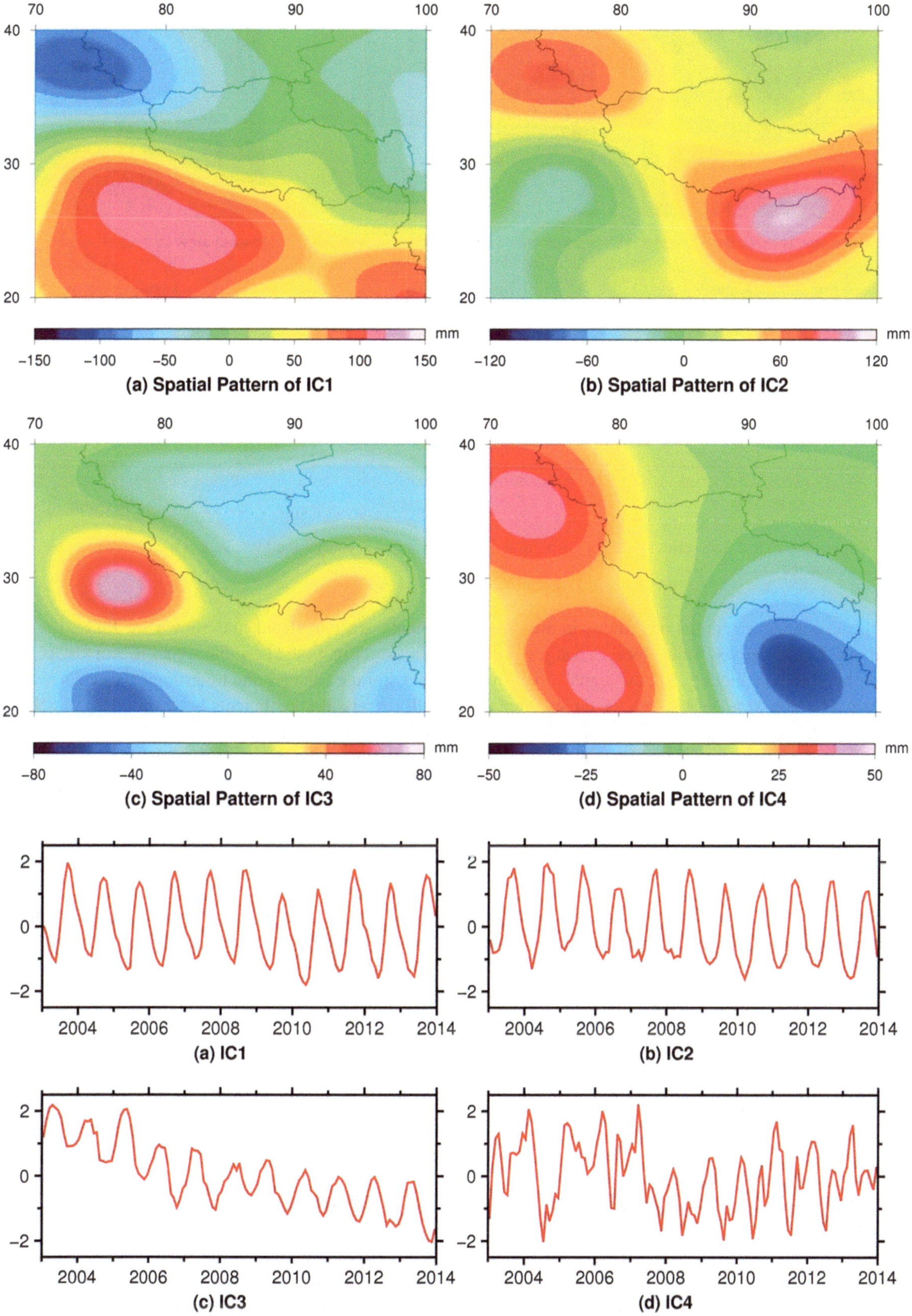


**Fig. 2** Spatial pattern (*up*) and time series (*down*) by ICA analysis using 132 months GRACE-derived water storage changes

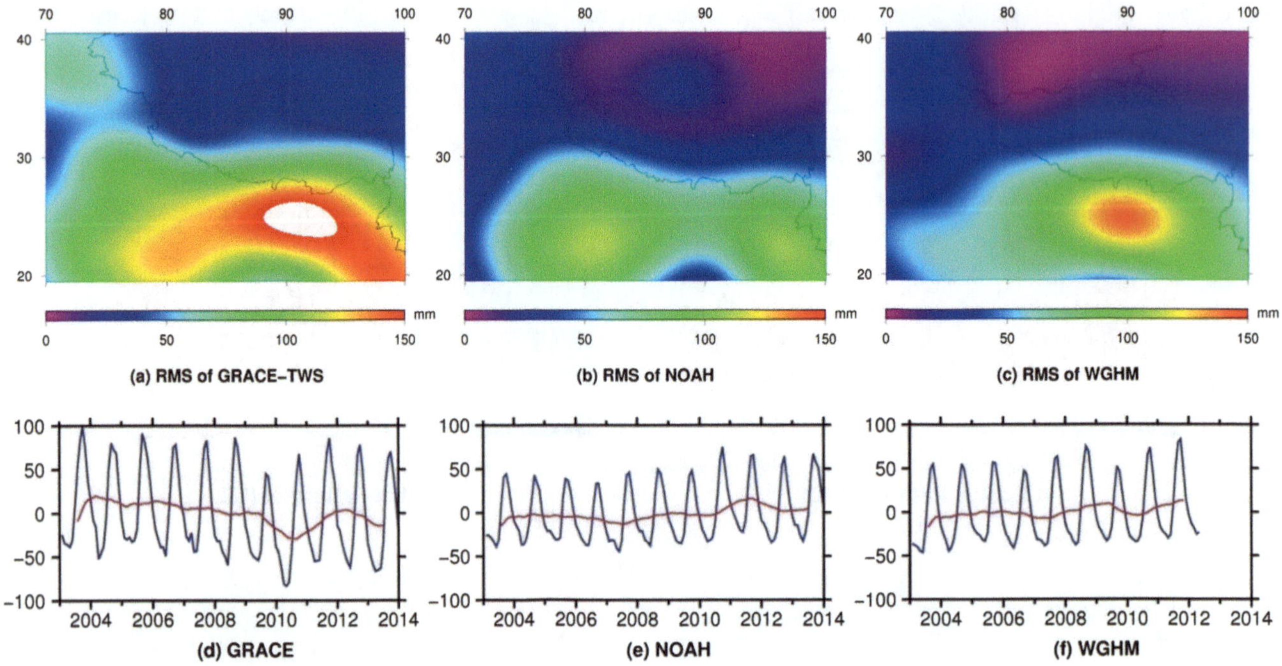


**Fig. 3** Signal strength (*up*) and time series (*down*) of GRACE-derived water storage changes and hydrological models, red curve is the moving average of the time series

show significant changes in annual periodicity in the region. The period of GRACE-derived water storage changes is very similar with hydrological models, but the signal strength of GRACE-derived water storage changes is larger than the hydrological models, which indicates that GRACE-derived water storage changes may include the groundwater changes and other factors.

## 5.1 Results of Hydrological Models

The independent patterns of NOAH and WGHM hydrological models extracted by ICA method are shown in Figs. 4 and 5. From Fig. 4, the first, third and the fourth components of NOAH hydrological model exhibit significant annual period, and the third component contains a long-term trend signal, the second component contains mainly semi-annual period signal. In Fig. 5, the four components of WGHM contain obvious annual periodic signal.

## 5.2 The Comparison Between Results from GRACE and the Hydrological Models

To verify the reliability of the results, time series of the independent components from both GRACE and hydrological models are compared and the results were shown in Fig. 6. The correlation coefficients between the corresponding components are calculated and shown in Table 1. As can be seen from Fig. 6, the periodic signal IC1 of GRACE-derived water storage changes and hydrological models results agree well, correlation coefficients between GRACE&NOAH and GRACE&WGHM are 0.884 and 0.877. IC1 is mainly annual periodic signal, but IC3 is mainly the long-term trend signal and some discrepancies exist between GRACE-derived water storage changes and hydrological models, this is likely due to the fact that: (1) the observations of hydrological models in Tibetan Plateau are less, but observations of the GRACE satellite can cover whole region; (2) hydrological models reflect major changes in surface water (including precipitation, snow melt, wetland changes, etc.), but GRACE-derived water storage changes also includes the groundwater changes, ignoring the groundwater changes in hydrology models may lead to some errors; (3) In addition, different models are calculated by different methods, the results will unavoidably introduce some errors which can be seen from the analysis results of NOAH and WGHM hydrological models.

# 6 Conclusions

In this paper, we studied the water storage changes of the Tibetan Plateau and its surrounding areas in large-scale by ICA method. The independent components results of GRACE-derived water storage changes and hydrological models are found being in good agreement. The correlation

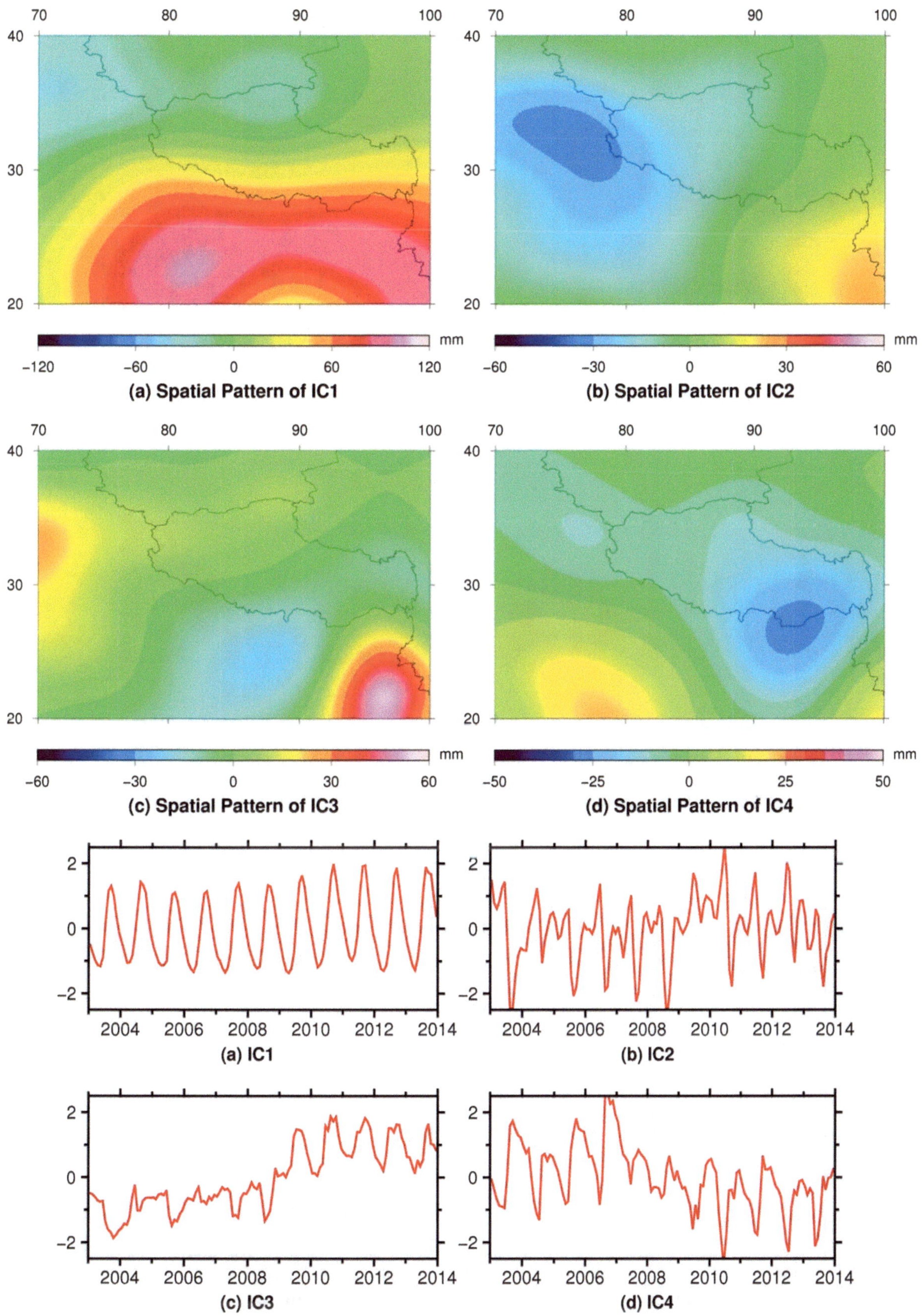


**Fig. 4** Spatial pattern (*up*) and time series (*down*) by ICA analysis using 132 months NOAH

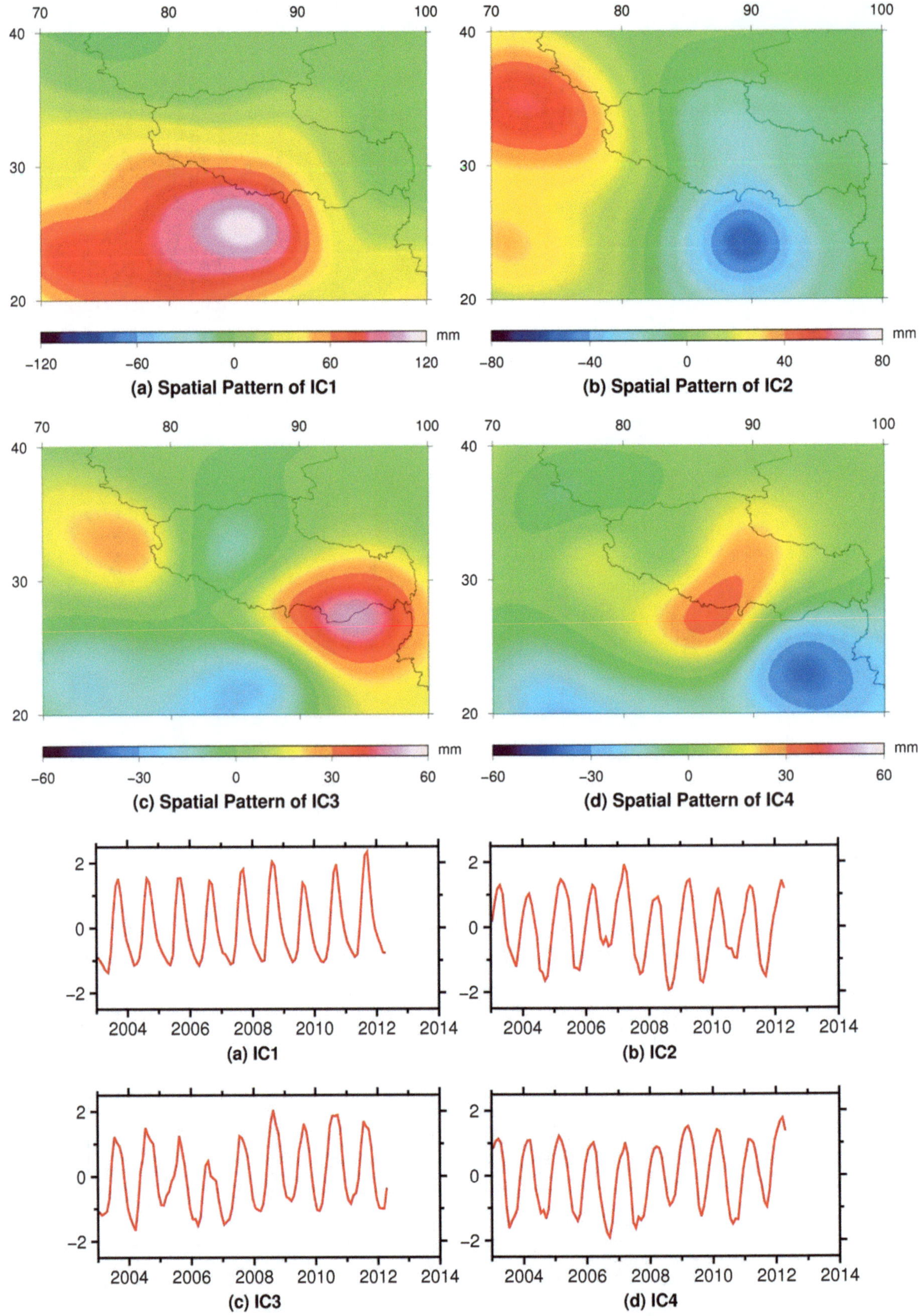


**Fig. 5** Spatial pattern (*up*) and time series (*down*) by ICA analysis using 112 months WGHM

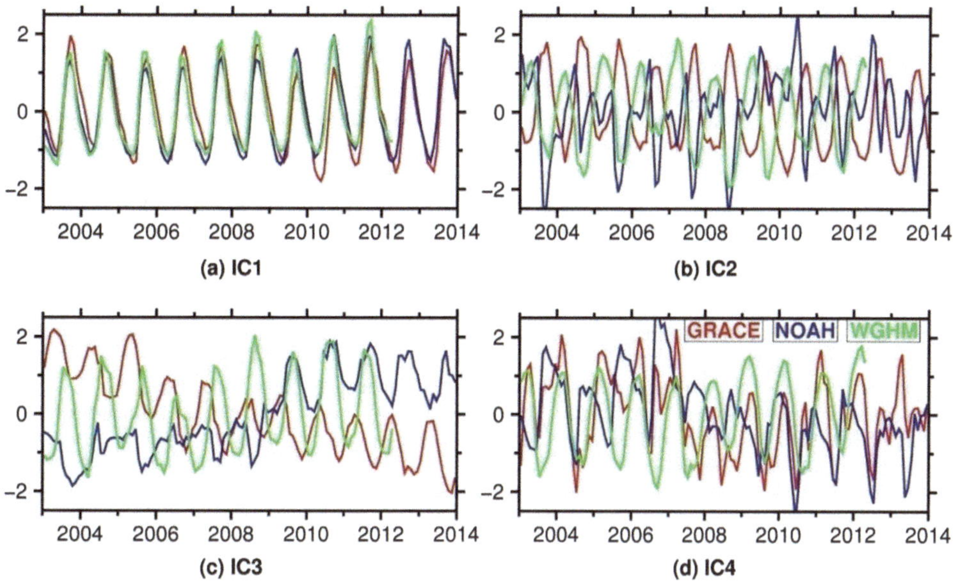


**Fig. 6** Comparison of time series results of GRACE and hydrological models by ICA analysis

**Table 1** Correlation coefficients of GRACE&NOAH and GRACE&WGHM

| Coefficients | IC1 | IC2 | IC3 | IC4 |
|---|---|---|---|---|
| GRACE&NOAH | 0.884 | −0.500 | −0.662 | 0.404 |
| GRACE&WGHM | 0.877 | −0.874 | 0.256 | 0.454 |

of IC1 from Table 1 is relevant strongly. From spatial patterns, the amplitudes of GRACE-derived water storage changes are larger than that of hydrological models, because the GRACE-derived water storage changes also may include groundwater changes. We can see that the water storage of the Tibetan Plateau in the southwest region has been reduced in recent years, which is likely due to the glaciers melting and snowpack shrinking caused by climate warming (Xu and Zhang 2013). With the analysis of temporal and spatial variation of water reserves, we could further study the impact of natural and anthropogenic factors on the regional climate.

ICA is able to separate independent signal components from data with a little priori information, which has a good prospect in information extraction of GRACE gravity observations and hydrological models. However, some problems still need to be solved in practical applications. How to utilize ICA method to analyze data more effectively, as well as the applicability of the method still need further study.

**Acknowledgements** The authors are grateful to Ehsan Forootan and two anonymous reviewers for their constructive comments which led to the improvement of the manuscript. The research is funded by National Key Basic Research Program (2013CB733302), National Natural Science Foundation of China(41274031, 41404014), Chinese Academy of Surveying and Mapping Fundamental Scientific Research Expenses (7771415).

## References

Bartlett MS (2001) Face image analysis by unsupervised learning. Kluwer, Boston. doi:10.1007/978-1-4615-1637-8

Boergens E, Rangelova E, Sideris MG, Kusche J (2014) Assessment of the capabilities of the temporal and spatiotemporal ICA method for geophysical signal separation in GRACE data. J Geophys Res Solid Earth 119(5):4429–4447

Cardoso JF (1992) Fourth-order cumulant structure forcing: application to blind array processing. In: Statistical signal and array processing, 1992 conference proceedings, IEEE Sixth SP Workshop on IEEE, pp 136–139. doi:10.1109/SSAP.1992.246830

Cardoso JF (1999) High-order contrasts for independent component analysis. Neural Comput 11(1):157–192. doi:10.1162/089976699300016863

Cardoso JF, Souloumiac A (1993) Blind beamforming for non-Gaussian signals. In: IEEE Proceedings F (Radar Signal Proc 140(6):362–370) IET Digital Library. doi:10.1049/ip-f-2.1993.0054

Chen JL, Rodell M, Wilson CR, Famiglietti JS (2005) Low degree spherical harmonic influences on Gravity Recovery and Climate Experiment (GRACE) water storage estimates. Geophys Res Lett 32(14). doi:10.1029/2005GL022964

Chen JL, Wilson CR, Blankenship DD, Tapley BD (2006) Antarctic mass rates from GRACE. Geophys Res Lett 33(11). doi:10.1029/2006GL026369

Döll, P, Kaspar, F, Lehner, B (2003) A global hydrological model for deriving water availability indicators: model tuning and validation. Journal of Hydrology, 270(1):105–134.

Duan XJ, Guo JY, Shum CK, Van Der Wal W (2009) On the post-processing removal of correlated errors in GRACE temporal gravity field solutions. J Geod 83(11):1095–1106

Forootan E, Kusche J (2012) Separation of global time-variable gravity signals into maximally independent components. J Geod 86(7):477–497. doi:10.1007/s00190-011-0532-5

Forootan E, Kusche J (2013) Separation of deterministic signals, using independent component analysis (ICA). Stud Geophys Geod 57:17–26. doi:10.1007/s11200-012-0718-1

Forootan E, Rietbroek R, Kusche J, Sharifi MA, Awange JL, Schmidt M, Famiglietti J (2014) Separation of large scale water storage patterns over Iran using GRACE, altimetry and hydrological data. Remote Sens Environ 140:580–595. doi:10.1016/j.rse.2013.09.025

Frappart F, Ramillien G, Maisongrande P, Bonnet MP (2010) Denoising satellite gravity signals by independent component analysis. Geosci Remote Sens Lett IEEE 7(3):421–425. doi:10.1109/LGRS.2009.2037837

Guo J, Mu D, Liu X, Yan H, Dai H (2014) Equivalent water height extracted from GRACE gravity field model with robust independent component analysis. Acta Geophys 62(4):953–972. doi:10.2478/s11600-014-0210-0

Hyvarinen A (1999) Survey on independent component analysis. Neur Comput Surv 2(4):94–128

Hyvärinen A, Karhunen J, Oja E (2004) Independent component analysis, vol 46. Wiley, New York

Jeffers JNR (1967) Two case studies in the application of principal component analysis. Appl Stat 225–236

Jolliffe IT (2003) A cautionary note on artificial examples of EOFs. J Clim 16(7):1084–1086. doi:10.1175/1520-0442(2003)016<1084:ACNOAE>2.0.CO;2

Luo F, Dai W, Tang C, Huang D, Wu X (2012) EMD-ICA with reference signal method and its application in GPS multipath. Acta Geod Cartogr Sin 41(3):366–371

Lv W, Chen Y, Zhang B (2007) Application of independent component analysis method in earthquake signal de-noise. Oil Geophys Prospect 42(2):132–136

Makeig S, Bell AJ, Jung TP, Sejnowski TJ (1996) Independent component analysis of electroencephalographic data. Adv Neur Inform Proc Syst 145–151

Niu T, Chen LX, Wang W (2002) REOF analysis of climatic characteristics of winter temperature and humidity on Xizang-Qinghai Plateau. J Appl Meteorol Sci 135:560–570

Price AL, Patterson NJ, Plenge RM, Weinblatt ME, Shadick NA, Reich D (2006) Principal components analysis corrects for stratification in genome-wide association studies. Nat Genet 38(8):904–909. doi:10.1038/ng1847

Rangelova E, van der Wal W, Braun A, Sideris MG, Wu P (2007) Analysis of gravity recovery and climateexperiment time-variable mass redistribution signals over North America by means of principal component analysis. J Geophys Res 112(F3). doi:10.1029/2006JF000615

Rodell M, Houser PR, Jambor UEA, Gottschalck J, Mitchell K, Meng CJ, Toll D (2004) The global land data assimilation system. Bull Am Meteorol Soc 85(3):381–394. doi:10.1175/BAMS-85-3-381

Schmidt R, Petrovic S, Güntner A, Barthelmes F, Wünsch J, Kusche J (2008) Periodic components of water storage changes from GRACE and global hydrology models. J Geophys Res Solid Earth 113(B8). doi:10.1029/2007JB005363

Schrama EJO, Wouters B, Lavalle DA (2007) Signal and noise in Gravity Recovery and Climate Experiment (GRACE) observed surface mass variations. J Geophys Res 112, B08407. doi:10.1029/2006JB004882

Tapley BD, Bettadpur S, Watkins M, Reigber C (2004) The gravity recovery and climate experiment: mission overview and early results. Geophys Res Lett 31(9). doi:10.1029/2004GL019920

Wen H, Zhu G, Cheng P, Chang X, Liu H (2011) The ice sheet height changes and mass variations in Antarctica by using ICESat and GRACE data. Int J Image Data Fusion 2(3):255–265. doi:10.1080/19479832.2010.491803

Werth S, Güntner A (2010) Calibration analysis for water storage variability of the global hydrological model WGHM. Hydrol Earth Syst Sci 14(1):59–78. doi:10.5194/hessd-6-4813-2009

Xu P, Zhang W (2013) Inversion of terrestrial water storage changes in recent years for Qinghai-Tiabetan plateau and Yarlung Zangbo River basin by GRACE. J Water Resour Water Eng 1:007

Zhang G, Yao T, Xie H, Kang S, Lei Y (2013) Increased mass over the Tibetan Plateau: from lakes or glaciers? Geophys Res Lett 40(10):2125–2130

Zhu G, Li J, Wen H, Wang J (2008) Study on variations of global continental water storage with grace gravity field models. J Geod Geodyn 28(5):39–44

Zotov L, Shum C (2010) Multichannel singular spectrum analysis of the gravity field data from GRACE satellites. In: Chakrabarti SK, Zhuk AI, Bisnovatyi-Kogan GS (eds) Astrophysics and cosmology after Gamow, vol 1206. Am Inst Phys, Odessa, Ukraine, pp 473–479. doi:10.1063/1.3292557

# Mass Variations in the Siberian Permafrost Region Based on New GRACE Results and Auxiliary Modeling

Akbar Shabanloui and Jürgen Müller

**Abstract**

GRACE (Gravity Recovery And Climate Experiment) determines the integral mass variations in the Earth system with a high spatial-temporal resolution. These mass variations should be adequately separated for better understanding of the individual signal contributions. In Siberia, the temporal mass variations are related to hydrological processes including thawing of permafrost layers. Permafrost layers with different thickness cover about 80% of Siberia. These frozen sheets play an important role for sea level rise and the global hydrological water cycle. In this study, the integral mass variations in Siberia are precisely estimated based on the new release of GRACE (RL05a) from GeoForschungsZentrum (GFZ) in Potsdam. In addition, various hydrological contributions (lake level variation, river run-off, etc.) can be estimated from different models and specific data. Here, mass variations in the Siberian permafrost region based on GRACE results and different hydrological models/data [i.e., GLDAS (Global Land Data Assimilation System) and GPCP (Global Precipitation Climatology Project)] are jointly investigated.

**Keywords**

GRACE • Mass variation • Hydrological model • GLDAS • GPCP • Permafrost thawing

## 1 Introduction

Since 2002, the GRACE satellite mission is in a near polar orbit for recovering the static and time-variable parts of the Earth's gravity field (Tapley et al. 2004). The time-variable part of the Earth's gravity field is related to mass re-distribution in the Earth system. GRACE monitors the sum of all mass variations occurring in the Earth's system. It cannot directly separate the different sources and effects. Therefore, an efficient separation technique and very precise background models have to be used for understanding mass variations in different regions over the globe. The provided time series of monthly gravity field solutions in the analysis centres (ACs) are already corrected for variations caused by atmosphere and ocean (Flechtner and Dobslaw 2014). Based on time-variable gravity field coefficients from GRACE, a number of studies to estimate secular trends and periodical changes for different regions have been successfully performed, such as mass balance in Greenland (Velicogna and Wahr 2013), hydrological mass variations in the Amazon basin (Werth et al. 2009), mass losses in Antarctica (Sasgen et al. 2013), mass variations in Siberia, Russia (Steffen et al. 2012; Vey et al. 2013; Chao et al. 2011), hydrological storage changes over Siberia (Frappart et al. 2006, 2011; Muskett and Romanovsky 2009), water depletion in India and China (Tiwari et al. 2014; Rodell et al. 2009; Zhong et al. 2009), terrestrial water budget in the Eurasian pan-Arctic (Landerer et al. 2010), hydrological mass variation in Australia (Seoane et al. 2013), global large-scale groundwater variations from satellite gravimetry and hydrological models (Jin and Feng 2013) and contribution of terrestrial water derived by GRACE in polar motion (Jin et al. 2012). A

A. Shabanloui (✉) • J. Müller
Institute of Geodesy, University of Hannover, Schneiderberg 50, 30167 Hannover, Germany
e-mail: shabanloui@ife.uni-hannover.de



S. Jin, R. Barzaghi (eds.), *IGFS 2014*, International Association of Geodesy Symposia 144,
DOI 10.1007/1345_2015_186

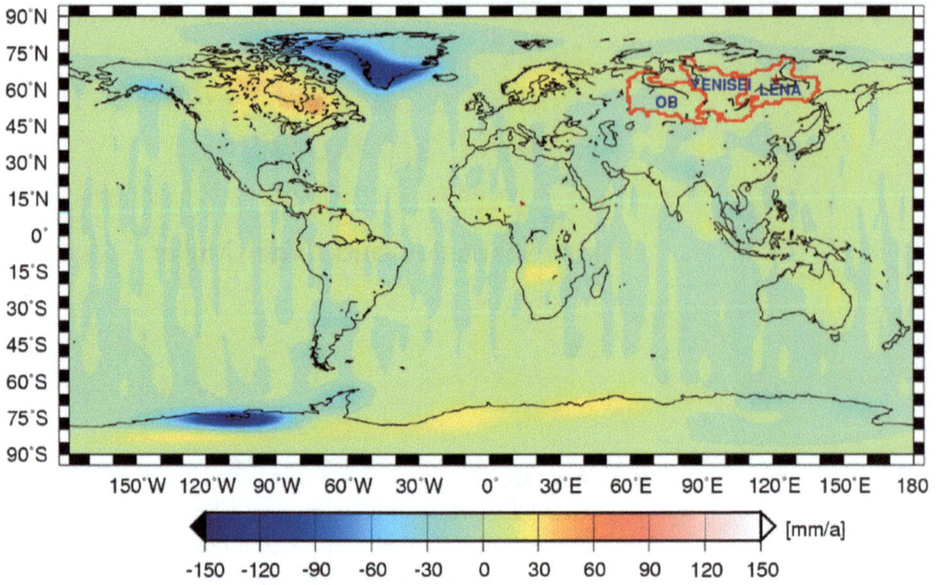


**Fig. 1** Secular trend estimation based on monthly gravity field solutions of GFZ-RL05a in the period of 2003–2013, zonal term $c_{2,0}$ replaced and a 1D Gaussian filter with radius 350 km applied. The study area and three river basins i.e. Lena, Yenisei and Ob in Siberia are indicated by *red* color

secular trend in terms of Equivalent Water Thickness (EWT) has been estimated for all those regions. Estimated periodical terms include semi-annual, annual and some further periods.

By increasing the time-span of GRACE observations, mass variation signals including secular and (long-) periodical contributions in different regions can be precisely estimated. One of the challenging regions, which is not sufficiently investigated so far, is the permafrost region in Siberia, Russia. During the past decades, the permafrost regime in the Siberian region has experienced significant changes in terms of permafrost thawing due to climate warming. As an example, Yang et al. (2013) found that the average fresh water discharge from the six largest Eurasian rivers to the Arctic ocean has increased by 7%. Lena, Yenisei and Ob are the three largest rivers in the Siberian permafrost region that flow into the Arctic ocean from south to north. Parts of the signal due to permafrost thawing can be detected based on the level (discharge) changes of those rivers. As the permafrost region covers almost 80% of Siberia with a maximum thickness of 1 km in Yakutia (central Siberia), it is expected that permafrost thawing due to climate change plays a key role in mass (water) balance in this region (Steffen et al. 2012). In addition, the permafrost has caught an extreme capacity of carbon-dioxide ($CO_2$) and nitrogen (N) that, if released, will play an important role for the Earth's ecosystem in the near future (Treat et al. 2014). Therefore, surface and sub-surface mass variations in Siberia based on the gravitational approach from GRACE monthly solutions are determined, using GRACE data covering almost 12 years. In addition, the GLDAS hydrological model (Rodell et al. 2004) has been used to obtain the soil-moisture contribution, and GPCP precipitation data for estimating precipitation trends in the target region.

## 2 GRACE Data and Analysis

### 2.1 Processing Procedure

The GRACE L1b data are officially processed by different analysis centres (ACs), e.g. GFZ, JPL and CSR, as well as by other institutes such as ITG (University of Bonn, Germany), AIUB (University of Bern, Switzerland) and TU Delft (Delft University of Technology, The Netherlands) to produce GRACE L2 data in terms of static and time-variable spherical harmonic coefficients. These processing centres apply different strategies and background models.

Therefore, also different mass variation signals are obtained from GRACE L2 products when used for further analysis in selected regions, e.g. basins with large signals such as in the Amazon. Processing of GRACE data from different ACs in the Siberian region with small mass variation signals delivers about the same results and no significant differences are detected (see Fig. 1). Thus, in this study, only the recently released GRACE (RL05a) products of GFZ are used to determine Total Water Storage Changes (TWSC) in the Siberian permafrost region. The following steps have been applied:

- **GSM data**: GRACE Level 2 monthly time-variable gravity field products are provided in terms of fully normalized geo-potential spherical harmonic coefficients. We used GFZ solutions from January 2003 to December 2013 (122 months).
- **Zonal term replacement**: The zonal term $c_{2,0}$ of the monthly gravity field coefficients is replaced by the solution from Satellite Laser Range observations (Cheng et al. 2013).

- **Filtering:** The monthly gravity field products are affected by correlated noise at higher frequencies. The correlated noise shows up in terms of striping effects in the mass variation results (e.g. gravity anomalies), when represented in a geographical plot. The striping errors are caused by the orbit design of GRACE-type configurations, incomplete reduction of non-tidal high frequencies of mass variations and limitation in the processing of the Earth gravity field. Therefore, an effective filter has to be applied to separate signal from noise. The normalized spherical harmonic coefficients can be smoothed in many different ways (Jekeli 1981; Swenson and Wahr 2006; Kusche 2007; Klees et al. 2008; Davis et al. 2008) of which we have tested several for Siberia.
- **Mass variation products:** For our purpose, the mass variations in the permafrost region are estimated in terms of EWT. Therefore, the filtered spherical harmonic coefficients from the previous step are used to determine the global TWSC based on an approach presented by Wahr et al. (1998).
- **Glacial Isostatic Adjustment (GIA) corrections:** Vey et al. (2013) showed that the ice history over Siberia does not reveal large-scale ice sheets during the last glacial maximum (LGM). In addition, by applying the two ice models RSES and ICE-5G with different lithospheric thicknesses, no significant GIA effects were found in central Siberia (Velichko et al. 2011; Vey et al. 2013).

GRACE monthly gravity field solutions contain the integral gravity variations that could be caused by changes in Atmosphere, Ocean, Hydrology, Ice and Solid Earth (AOHIS). High-frequency atmospheric and oceanic mass variations are removed during processing of GRACE L1 data (AOD1B products) (Flechtner and Dobslaw 2014). Thus, the mass variations that are contained in GRACE L2 data have ice, hydrology and solid Earth changes as potential sources. Depending on the location of the Region Of Interest (ROI), one or more Signals Of Interest (SOI) can be related to the mass changes. For example, mass variations in the Siberian permafrost region are related to hydrological (water) variations that are only partly caused by permafrost thawing but also by precipitation and run-off variations. To estimate secular and periodical contributions, the following equation is used (Ogawa 2010):

$$EWT(\lambda, \phi, t) = a+bt+\sum_{f=1}^{4}(c_f \cos(\omega_f t)+s_f \sin(\omega_f t))+\varepsilon. \quad (1)$$

$EWT(\lambda, \phi, t)$ represents the equivalent water thickness at the node position $(\lambda, \phi)$ at time $t$. The parameters $a$ and $b$ are bias and secular trend. The periodical parts are expressed in terms of cosine and sine coefficients $c_f$ and $s_f$. The amplitudes $c_f$ and $s_f$ correspond to the angular frequency $\omega_f$. In this study, the expansion term for the periodical term is chosen as 4. The periods of 161 days, 1, 2.5 and 3.7 years are taken into account. The 161-day period is included to consider effects from the insufficient ocean tide background model (Ray et al. 2003). Ray et al. (2003) showed that aliasing terms exist for the $S_2$, $K_2$, $K_1$ tide components that result in 161-day, 3.7-year and 7.4-year periods, respectively. Contribution of $K_1$ is not well retrievable due to its long aliasing period and the shorter time span of available GRACE monthly solutions. It is not considered further on. Schmidt et al. (2008) found a long-periodic wave of about 2.5 years on the global scale in GRACE data and the hydrological models for the GRACE period. They also showed that the 2.5-year period is also retrievable in hydrological models for a longer period of data. Seasonal and annual (and inter-annual) mass variations are usually related to the (sub-)surface water storage changes that can be considered as the largest contribution to the temporal gravity changes (Ray et al. 2003). Noise and other un-modeled terms are characterized by $\varepsilon$. Equation (1) can be solved by some standard least-squares adjustment. The quality of estimated secular trends and periodical terms strongly depend on the GRACE time span (interval) and the number of selected periodical terms. In other words, the modeled periodical terms should absorb all existing significant periodical signals in the (pseudo)-observations.

It should be mentioned that depending on ROI and SOI, appropriate post-processing techniques including filtering have to be used to separate the signals. From the several filter techniques published and used in the last years, the performance of the 1D isotropic Gaussian (Jekeli 1981), the 2D Fan (Zhang et al. 2009) and DDK filters (Kusche 2007) are tested for use in the permafrost region. In addition to these common filters, a de-striping filter has been applied to the spherical harmonics to minimize the effect of errors due to the north-south stripes in GRACE monthly solutions (Swenson and Wahr 2006).

## 3 GRACE Derived Mass Variations

After successful application of corrections due to the zonal coefficient, appropriate filtering and de-correlation, the EWT are computed at the nodes of an equiangular grid with $1° \times 1°$ taking into account the maximum degree of 90. Figure 2a–c show the performance of different filter techniques with different settings. Figure 2a shows the secular mass variations in the Siberian permafrost region by applying an isotropic 1D Gaussian filter with radius 350 km for GFZ monthly solutions. The double peak features in terms of minimum and maximum surface mass variation (i.e. EWT change) of 1.5 cm/a and 1.9 cm/a are visible in the basin of

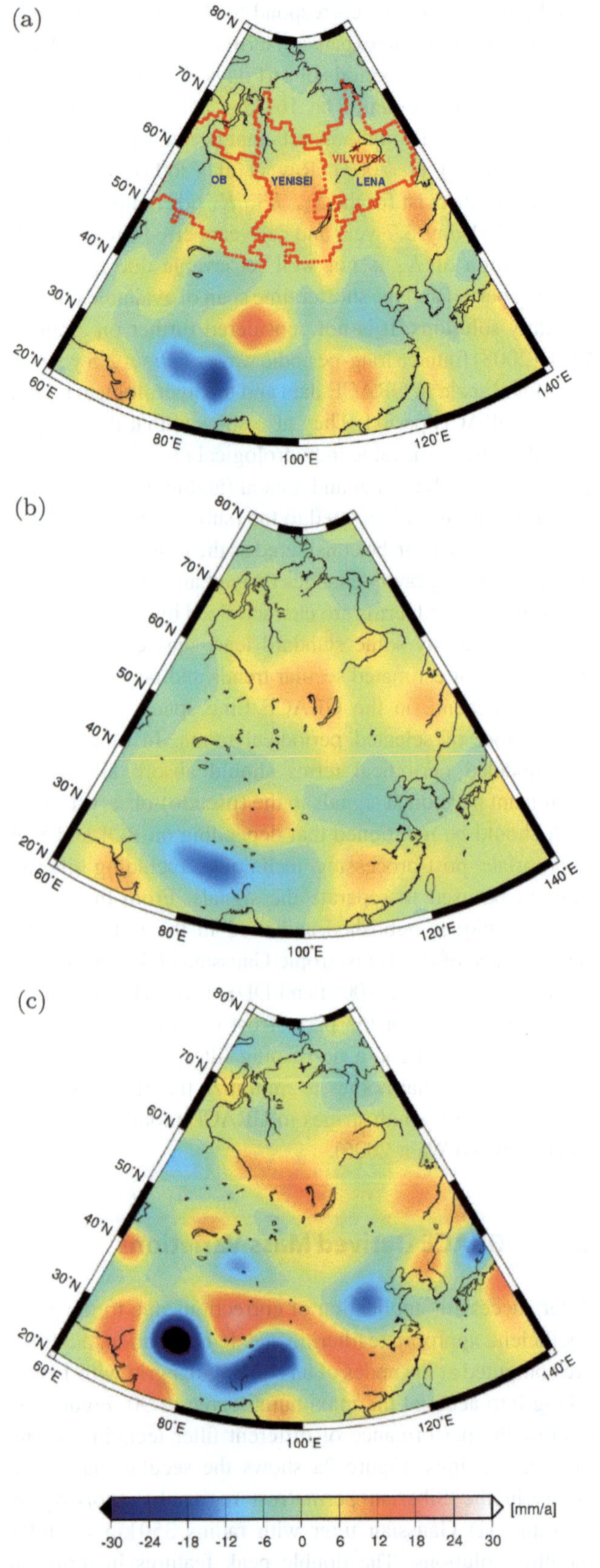


**Fig. 2** Secular trend estimation based on monthly gravity field solutions of GFZ-RL05a in the period of 2003–2013, zonal term $c_{2,0}$ replaced and various filters applied. (**a**): 1D Gaussian filter with radius 350 km (**b**): 2D Fan filter with radius 350 km (**c**): DDK3 filter

**Table 1** Statistical values of secular trend estimation for different filters using GFZ-RL05a data in Siberia covering the period of 2003–2013

| Case | Gaussian filter (350 km), de-striping + $c_{2,0}$ replaced | Fan filter (350 km), de-striping + $c_{2,0}$ replaced | DDK3 + $c_{2,0}$ replaced |
|---|---|---|---|
| Min. (cm/a) | −1.9 | −1.6 | −3.9 |
| Max. (cm/a) | 1.5 | 1.2 | 2.0 |
| RMS (cm/a) | 0.4 | 0.4 | 0.7 |
| Mean (cm/a) | 1.3 | 1.3 | 1.2 |

the Yenisei river. The minimum occurs around the Aral Sea and north-west of China. The basins of the Lena and Yenisei rivers (especially southern part) clearly show mass increase in the permafrost region. Figure 2b shows the secular mass variations for the same target region by applying the 2D-Fan filter with radius 350 km. The mass variation pattern is almost the same as for the Gaussian filter with radius 350 km. The double peaks (positive and negative secular trends) are almost the same as in the Gaussian case. But it seems that the signals, after applying the 2D-Fan filter, are smoothed more than by the Gaussian filter. Figure 2c shows the secular surface mass variations by applying the DDK3 filter (Kusche 2007) corresponding to a smoothing radius of approximately 350 km. Weaker smoothing (DDK3) obviously leaves an unrealistically strong variability in GRACE. The surface mass variations are amplified by a factor of 2 for the mass increasing rate and a factor 1.3 for the mass decreasing rate in this region. Table 1 shows the corresponding statistics for different filter settings. To study the estimated surface mass variations in the Siberian permafrost region in more detail, the station Vilyuysk located near the Vilyuy river on the left tributary of the Lena river is selected as an example. Velicogna et al. (2012) showed a mass increase for the whole Lena basin using the CSR solution in the period of 2002–2010. We also estimated a mass increase based on the GFZ-RL05a solution for the Vilyuysk region in the period of 2003–2008 (see Fig. 3a) and a slight mass decrease in the period of 2008–2013 (see Fig. 3b). Figure 4 shows a positive secular trend (mass increase) of 25 mm/a for the station Vilyuysk in the period of 2003–2008, and a negative secular trend (mass decrease) of −1 mm/a in the period of 2003–2013. To test the effect of selecting different time spans on estimated secular trends, we changed the periods, e.g., by taking the maximum peak in May 2008 for the first part trend estimation. The positive secular trend did not change significantly (from 2.5 to 2.2 cm/a). The secular trend of the second period changed from −0.1 to −0.4 cm/a. Such a shift would only slightly affect the major findings of the paper. If more data are available, we will test the negative secular trend estimation for the mass decrease in this region for the period beyond 2013.

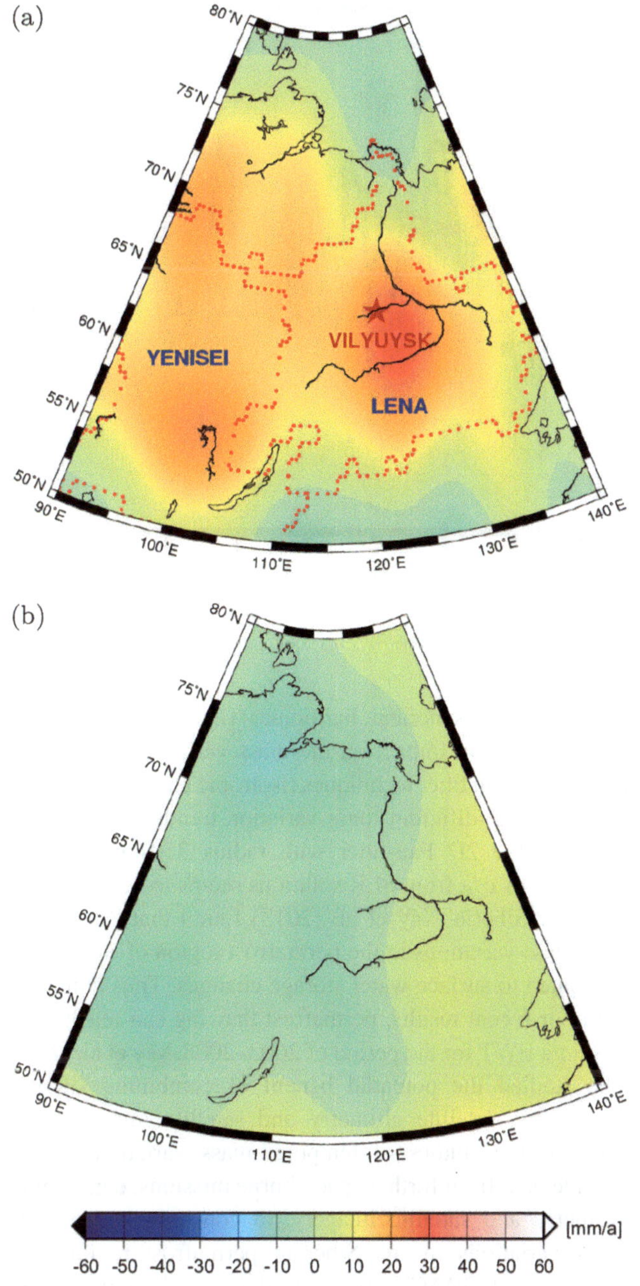


**Fig. 3** Secular trends of mass variations in Siberia based on GRACE for the period of (**a**): 2003–2008 and (**b**): 2008–2013 (Station Vilyuysk is marked with a *red* star)

## 4 Hydrological Mass Variations

Auxiliary hydrological models, e.g. GLDAS (Rodell et al. 2009) and precipitation data from GPCP centres (Adler et al. 2003) are used to compare those estimated GRACE secular trends. GLDAS provides hydrological parameters with a temporal resolution of 1 month and a spatial resolution of 1° × 1° for all land regions north of 60° South. GLDAS is generated by an optimal combination of all land (surface) data; e.g. soil-moisture in different depths or snow cover, and constrained by different satellite data. But permafrost contributions such as continuously frozen ground or thawing and freezing processes at the surface are not considered in GLDAS (Rodell et al. 2004). Hydro-climatic changes of permafrost layers in Siberia are complex and also include not stationary pattern changes (Milly et al. 2008). In this study, to be consistent with the procedure of GRACE data analysis, the same processing strategy and filter-techniques were applied to the GLDAS model. We then suppose that the remaining mass variation signals can be considered as permafrost layer changes.

Considering the hydrological model of GLDAS in the Siberian permafrost region for the period of 2003–2008, the maximum and minimum of soil moisture is found in the southern and western part of the Lena river. However, for the period of 2008–2013, soil moisture decreases in the same region which corresponds to the water mass decrease determined by GRACE. Therefore, the Lena basin may be affected by permafrost thawing. Figure 5a, b show several similarities to the GRACE results for the maximum and minimum in the Lena region, especially for the mass increase in the period of 2003–2008 and mass decrease in the period of 2008–2013. Figure 4 shows the time series of mass variations in terms of EWT at the selected station Vilyuysk based on GRACE solutions, the GLDAS model and differences between GRACE and GLDAS results. The times series as well as secular trend of GLDAS hydrological mass variations are similar to the GRACE results, but in some periods of time major differences are obvious. In addition, some studies show that the eastern part of Lena, especially the Kolyma river, experiences a strong mass increase during the period of 2003–2008 (Majhi and Yang 2008).

The differences between GRACE and GLDAS time series might be explained by permafrost thawing effects that were not modeled in GLDAS. The secular trends of 1.4 cm/a and 0.7 cm/a during the periods of 2003–2008 and 2008–2013 are estimated for the selected station. As GLDAS does not include permafrost contributions, these results may point to permafrost changes in the Lena basin. Therefore, the estimated positive secular trends for the GRACE minus GLDAS results during the period of 2003–2013 may show that permafrost thawing is proceeding in the target region. But permafrost thawing dynamics and geophysical processes behind it are very complex. By increasing the time span and accuracy of GRACE observations (e.g. with the launch of GRACE follow on in 2017) as well as the availability of more precise hydrological models, permafrost thawing processes in Siberia might be better determined and physically interpreted in near future.

There is some evidence that mass increase and decrease in the permafrost region, especially in the Lena basin, are

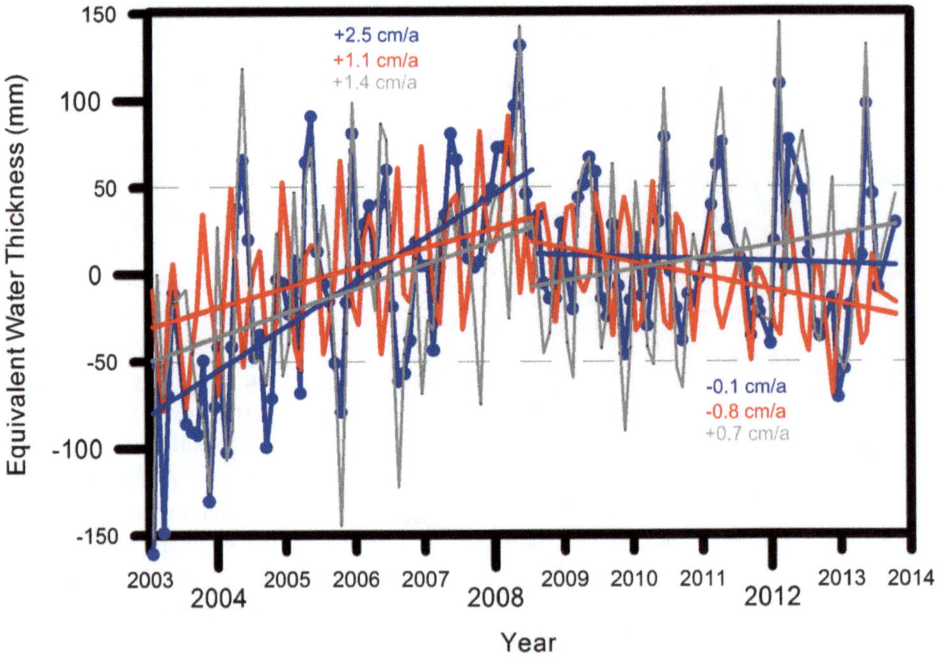


**Fig. 4** Time series of mass variations at the station Vilyuysk based on GFZ-RL05a solutions (*blue*), GLDAS model (*red*) and difference between GRACE and GLDAS results (*gray*) in the period of 2003–2013, and the corresponding estimated secular trends for the periods of 2003–2008 and 2008–2013

related to increase and decrease of precipitation. To test this assumption, we used GPCP data with a monthly temporal resolution for the period of 2003–2013. Figure 6a shows an increase of precipitation for the period of 2003–2008 that corresponds to soil moisture increase determined by GLDAS and surface mass increase estimated by GRACE. For the period of 2008–2013, a decrease of precipitation is visible in this region, that can be compared to the decrease of surface mass variations determined by GRACE. This means, in addition to permafrost thawing caused by climate warming, the precipitation rate has to be considered as an important second cause for mass variations in the Siberian permafrost region. Thus, with increasing permafrost thawing and consequently increasing permafrost active layers, which may absorb much water such as soil moisture, as well as an increasing precipitation rate (e.g. as can be seen in Fig. 5a), mass increase results in the period of 2003–2008. In the period of 2008–2013, based on the negative trend determined by GRACE and the negative precipitation trend, we expect that the permafrost thawing activities have slowed down in Siberia.

## 5 Conclusions and Outlook

Our investigations of mass variations in Siberia from GRACE and hydrological models mainly focused on the permafrost regime. It should be mentioned that the lower degree of the spherical harmonics (i.e. zonal term $c_{2,0}$) has a significant impact on the mass variation estimations. Using different filter techniques (isotropic and non-isotropic) gives slightly different mass variation trends in Siberia. In addition, the 2D Fan-filter with radius 350 km and after replacing of $c_{2,0}$ from SLR solutions seems to be an optimal filter for Siberia. Vey et al. (2013) found that 30–60% of total mass variations in the permafrost region of Siberia can be related to surface water storage changes. Thus combined with our recent results, permafrost thawing can reach up to 1.3 cm/a EWT for the period of 2003–2008. Vey et al. (2013) also studied the potential benefit of combining GRACE gravimetry, satellite altimetry and satellite imagery data. Thus, future studies of temporal mass variations should include data from further space-borne missions, e.g. satellite altimetry and satellite imagery, to constrain hydrological mass variations in the Siberian permafrost region. The separation of GRACE mass variations can be improved by assimilation of lake surface extent measurements and height variations from satellite altimetry (e.g. Jason-2) and hyper-spectral satellite (e.g. Landsat) data with different temporal-spatial resolutions. Based on these three techniques, the real (sub)-surface mass variation pattern in the Siberia might be better understood.

**Acknowledgements** We would like to thank the GeoForschungsZentrum (GFZ) in Potsdam and the German Space Operations Center (GSOC) of the German Aerospace Center (DLR) for providing continuously and nearly 100% of the raw telemetry and L2 data of the twin GRACE satellites.

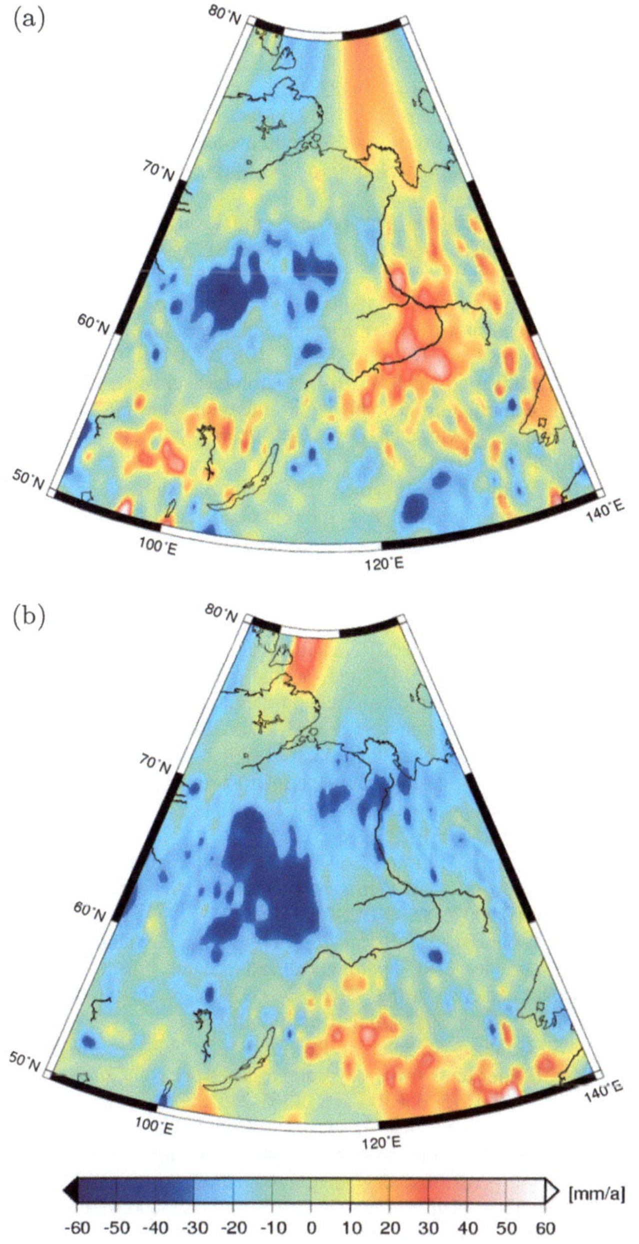


**Fig. 5** Secular trends of soil-moisture variations in Siberia from the GLDAS model for the periods of (**a**): 2003–2008 and (**b**): 2008–2013

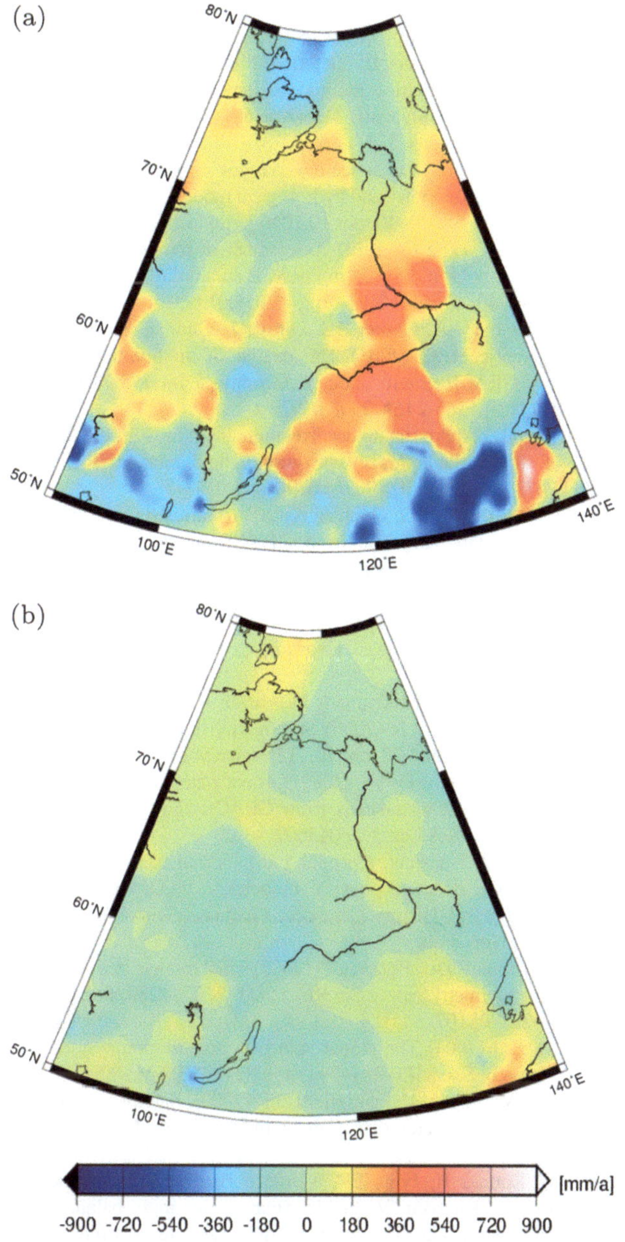


**Fig. 6** Precipitation trend estimation in Siberia based on the GPCP model for the period of (**a**): 2003–2008 and (**b**): 2008–2013

## References


Adler RF et al (2003) The version-2 global precipitation climatology project (GPCP) monthly precipitation analysis (1979–Present). J Hydrometeor 4(6):1147–1167

Chao B, Wu Y, Zhang Z, Ogawa R (2011) Gravity variation in Siberia: GRACE observation and possible causes. Terr Atmos Ocean Sci 22(2):149–155. doi:10.3319/TAO.2010.07.26.03(TibXS)

Cheng M, Tapley BD, Ries JC (2013) Deceleration in the Earth's oblateness. J Geophys Res Solid Earth 118(2):740–747. doi:10.1002/jgrb.50058

Davis JL, Tamisiea ME, Elsegui P, Mitrovica JX, Hill EM (2008) A statistical filtering approach for gravity recovery and climate experiment (GRACE) gravity data. J Geophys Res Solid Earth 113(B4). doi:10.1029/2007JB005043

Flechtner F, Dobslaw H (2014) AOD1B product description document for product release 05 (Rev. 4.2, May 20, 2014). Technical Note, GFZ German Research Centre for Geosciences Department 1: Geodesy and Remote Sensing

Frappart F, Ramillien G, Biancamaria S, Mognard NM, Cazenave A (2006) Evolution of high-latitude snow mass derived from the GRACE gravimetry mission (2002–2004). Geophys Res Lett 33(2). doi:10.1029/2005GL024778

Frappart F, Ramillien G, Famiglietti JS (2011) Water balance of the Arctic drainage system using grace gravimetry products. Int J Rem Sens 32(2):431–453. doi:10.1080/01431160903474954

Jekeli C (1981) Alternative methods to smooth the Earth's gravity field. Ohio State University, Department of Geodetic Science and Surveying, Report No. 327

Jin S, Feng G (2013) Large-scale variations of global groundwater from satellite gravimetry and hydrological models, 2002–2012. Glob Planet Change 106:20–30. doi:10.1016/j.gloplacha.2013.02.008

Jin S, Hassan A, Feng G (2012) Assessment of terrestrial water contributions to polar motion from GRACE and hydrological models. J Geodyn 62:40–48. doi:10.1016/j.jog.2012.01.009

Klees R, Revtova EA, Gunter BC, Ditmar P, Oudman E, Winsemius HC, Savenije HHG (2008) The design of an optimal filter for monthly GRACE gravity models. Geophys J Int 175(2):417–432. doi:10.1111/j.1365-246X.2008.03922.x

Kusche J (2007) Approximate decorrelation and non-isotropic smoothing of time-variable grace-type gravity field models. J Geodesy 81(11):733–749. doi:10.1007/s00190-007-0143-3

Landerer FW, Dickey JO, Güntner A (2010) Terrestrial water budget of the Eurasian pan-arctic from GRACE satellite measurements during 2003–2009. J Geophys Res Atmos 115(D23). doi:10.1029/2010JD014584

Majhi I, Yang D (2008) Streamflow characteristics and changes in Kolyma basin in Siberia. J Hydrometeorol 9(2):267–279. doi:10.1175/2007JHM845.1

Milly P, Betancourt J, Falkenmark M, Hirsch R, Kundzewicz Z, Lettenmaier D, Stouffer R (2008) Stationarity is dead: whither water management? Earth 4:20. doi:10.1126/science.1151915

Muskett RR, Romanovsky VE (2009) Groundwater storage changes in Arctic permafrost watersheds from GRACE and in–situ measurements. Environ Res Lett 4(4):045009

Ogawa R (2010) Transient, seasonal and inter-annual gravity changes from GRACE data: geophysical modelings. Ph.D. thesis, Department of Natural History Sciences, Graduate School of Science, Hokkaido University

Ray RD, Rowlands DD, Egbert GD (2003) Tidal models in a new era of satellite gravimetry. Space Sci Rev 108(1–2):271–282. doi:10.1023/A:1026223308107

Rodell M et al (2004) The global land data assimilation system. Bull Am Meteorol Soc 85(3):381–394. doi:10.1175/BAMS-85-3-381

Rodell M, Velicogna I, Famiglietti JS (2009) Satellite-based estimates of groundwater depletion in India. Nature 460(7258):999–1002

Sasgen I, Konrad H, Ivins ER, Van den Broeke MR, Bamber JL, Martinec Z, Klemann V (2013) Antarctic ice-mass balance 2003 to 2012: regional reanalysis of GRACE satellite gravimetry measurements with improved estimate of glacial-isostatic adjustment based on GPS uplift rates. Cryosphere 7(5):1499–1512. doi:10.5194/tc-7-1499-2013

Schmidt R, Petrovic S, Güntner A, Barthelmes F, Wünsch J, Kusche J (2008) Periodic components of water storage changes from GRACE and global hydrology models. J Geophys Res Solid Earth 113(B8). doi:10.1029/2007JB005363

Seoane L, Ramillien G, Frappart F, Leblanc M (2013) Regional GRACE-based estimates of water mass variations over Australia: validation and interpretation. Hydrol Earth Syst Sci 17(12):4925–4939. doi:10.5194/hess-17-4925-2013

Steffen H, Müller J, Peterseim N (2012) Mass variations in the Siberian permafrost region from GRACE. In: Kenyon S, Pacino MC, Marti U (eds) Geodesy for Planet Earth, vol 136. Springer, Berlin/Heidelberg, pp 597–603

Swenson S, Wahr J (2006) Post-processing removal of correlated errors in GRACE data. Geophys Res Lett 33(8):L08402. doi:10.1029/2005GL025285

Tapley BD, Bettadpur S, Watkins M, Reigber C (2004) The gravity recovery and climate experiment: mission overview and early results. Geophys Res Lett 31(9):L09607. doi:10.1029/2004GL019920

Tiwari V, Srinivas N, Singh B (2014) Hydrological changes and vertical crustal deformation in South India: inference from GRACE, GPS and absolute gravity data. Phys Earth Planetary In 231:74–80. doi:10.1016/j.pepi.2014.03.002

Treat CC, Wollheim WM, Varner RK, Grandy AS, Talbot J, Frolking S (2014) Temperature and peat type control co2 and ch4 production in Alaskan permafrost peats. Glob Chang Biol 20(8):2674–2686. doi:10.1111/gcb.12572

Velichko AA, Timireva SN, Kremenetski KV, MacDonald GM, Smith LC (2011) West Siberian plain as a late glacial desert. Quat Int 237(12):45–53. doi:10.1016/j.quaint.2011.01.013

Velicogna I, Wahr J (2013) Time-variable gravity observations of ice sheet mass balance: precision and limitations of the GRACE satellite data. Geophys Res Lett 40(12):3055–3063. doi:10.1002/grl.50527

Velicogna I, Tong J, Zhang T, Kimball JS (2012) Increasing subsurface water storage in discontinuous permafrost areas of the Lena river basin, Eurasia, detected from GRACE. Geophys Res Lett 39(9). doi:10.1029/2012GL051623

Vey S, Steffen H, Müller J, Boike J (2013) Inter-annual water mass variations from GRACE in central Siberia. J Geodesy 87(3):287–299. doi:10.1007/s00190-012-0597-9

Wahr J, Molenaar M, Bryan F (1998) Time variability of the Earth's gravity field: hydrological and oceanic effects and their possible detection using GRACE. J Geophys Res Solid Earth 103(B12):30205–30229. doi:10.1029/98JB02844

Werth S, Güntner A, Schmidt R, Kusche J (2009) Evaluation of GRACE filter tools from a hydrological perspective. Geophys J Int 179(3):1499–1515. doi:10.1111/j.1365-246X.2009.04355.x

Yang Z-p, Gao J-x, Zhao L, Xu X-l, Ouyang H (2013) Linking thaw depth with soil moisture and plant community composition: effects of permafrost degradation on alpine ecosystems on the Qinghai-Tibet plateau. Plant Soil 367(1–2):687–700. doi:10.1007/s11104-012-1511-1

Zhang Z-Z, Chao BF, Lu Y, Hsu H-T (2009) An effective filtering for grace time-variable gravity: fan filter. Geophys Res Lett 36(17). doi:10.1029/2009GL039459

Zhong M, Duan J, Xu H, Peng P, Yan H, Zhu Y (2009) Trend of China land water storage redistribution at medi- and large-spatial scales in recent five years by satellite gravity observations. Chin Sci Bull 54(5):816–821. doi:10.1007/s11434-008-0556-2

# Part V

# Solid Earth Investigations

# Comparative Study of the Uniform and Variable Moho Density Contrast in the Vening Meinesz-Moritz's Isostatic Scheme for the Gravimetric Moho Recovery

Robert Tenzer and Mohammad Bagherbandi

**Abstract**

In gravimetric methods for a determination of the Moho geometry, the constant value of the Moho density contract is often adopted. Results of gravimetric and seismic studies, however, showed that the Moho density contrast varies significantly. The assumption of a uniform density contrast thus might yield large errors in the estimated Moho depths. In this study we investigate these errors by comparing the Moho depths determined globally for the uniform and variable models of the Moho density contrast. These two gravimetric results are obtained based on solving the Vening Meinesz-Moritz's inverse problem of isostasy. The uniform model of the Moho density contrast is defined individually for the continental and oceanic lithosphere to better reproduce the reality. The global data of the lower crust and upper mantle retrieved from the CRUST1.0 seismic crustal model are used to define the variable Moho density contrast. This seismic model is also used to validate both gravimetric solutions. Results of our numerical experiment reveals that the consideration of the variable Moho density contrast improves the agreement between the gravimetric and seismic Moho models; the RMS of differences is 5.4 km (for the uniform density contrast) and 4.7 km (for the variable density contrast).

**Keywords**

Crust • Gravity • Isostasy • Mantle • Moho

## 1 Introduction

In methods for a gravimetric modeling of the crustal thickness a constant value of the Moho density contrast is often assumed (e.g., Moritz 1990; Čadek and Martinec 1991; Braitenberg and Zadro 1999; Braitenberg et al. 2006, 2010; Wienecke et al. 2007; Sjöberg 2009; Sampietro 2011; Tenzer and Bagherbandi 2012; Bagherbandi et al. 2013; Sampietro et al. 2013; Tenzer and Chen 2014). Different average values of the Moho density contrast can be found in literature. Heiskanen and Moritz (1967, p. 135) refer to the value of 600 kg m$^{-3}$. Dziewonski and Anderson (1981, Table 1) adopted the value of 480 kg m$^{-3}$ in the Preliminary Reference Earth Model (PREM) based on the analysis of global seismic data. Results of gravimetric studies also provided different estimates. Tenzer et al. (2009) reported the value of 520 kg m$^{-3}$ with more recent updates to 485 and 445 kg m$^{-3}$ (Tenzer et al. 2012a, 2014a). Sjöberg and Bagherbandi (2011) estimated the global average of the Moho density contrast based on combining gravimetric and seismic models. They reported the value of $448 \pm 187$ kg m$^{-3}$. They also provided the average values of $678 \pm 78$ and $334 \pm 108$ kg m$^{-3}$ for the continental and oceanic areas respectively and estimated that the Moho density contrast varies globally from 81.5 kg m$^{-3}$

R. Tenzer (✉)
The Key Laboratory of Geospace Environment and Geodesy, School of Geodesy and Geomatics, Wuhan University, Wuhan, China
e-mail: rtenzer@sgg.whu.edu.cn

M. Bagherbandi
Department of Industrial Development, IT and Land Management University of Gävle, Gävle, Sweden





S. Jin, R. Barzaghi (eds.), *IGFS 2014*, International Association of Geodesy Symposia 144,
DOI 10.1007/1345_2015_210

(in the Pacific region) up to 988 kg m$^{-3}$ (beneath the Tibetan Plateau). Tenzer et al. (2012b) reported a similar range of the Moho density contrast between 82 and 965 kg m$^{-3}$. It is important to note here that these Moho density contrast estimates (taken relative to the reference crustal density of 2,670 kg m$^{-3}$) are likely unrealistically large due to errors in used input models and partially also in applied numerical schemes. Large variations in the Moho density contrast were also confirmed from regional seismic studies. Niu and James (2002) and Jordi (2007), for instance, demonstrated that the density contrast regionally varies as much as from 160 kg m$^{3}$ (for the mafic lower crust) to 440 kg m$^{-3}$ (for the felsic lower crust), with an apparently typical value for the craton of about 440 kg m$^{-3}$.

The assumption of a constant compensation density in the Airy's (1855) local isostatic scheme is inconsistent with the findings of large density contrast variations at the Moho interface. Moreover, the regional/global compensation schemes according to Vening Meinesz (1931) and Moritz (1990), which adopted the Airy's compensation principle, have the same theoretical deficiency.

Reguzzoni et al. (2013) determined the gravimetric Moho model by assuming the lateral variations in the Moho density contrast, and Reguzzoni and Sampietro (2014) took into consideration also the radial density change within the Earth's crust. Tenzer et al. (2014b) investigated the effect of the upper mantle density structure on the Moho geometry. In this study we investigate this aspect in the context of the Vening Meinesz-Moritz (VMM) isostatic model (Sjöberg 2009, 2011; Sjöberg and Bagherbandi 2011). For this purpose we determine the VMM Moho depths for the uniform and variable models of the Moho density contrasts and validate these two gravimetric results using a global seismic model.

## 2 Method

Sjöberg and Bagherbandi (2011) formulated the generic expression for solving the VMM inverse problem of isostasy in the following form

$$-\mathrm{GR} \iint_{\sigma} \Delta\rho^{c/m}(\varphi', \lambda')\ K(\psi, s)\ d\sigma' = \Delta g^{i}(r, \varphi, \lambda), \tag{1}$$

where $\mathrm{G} = 6.674 \times 10^{-11}\mathrm{m^3kg^{-1}s^{-2}}$ is Newton's gravitational constant, $\mathrm{R} = 6361 \times 10^{3}\mathrm{m}$ is the Earth's mean radius, $K$ is the integral kernel function which relates the isostatic gravity anomalies $\Delta g^{i}$ with the Moho parameters (i.e., the Moho depth $D$ and the Moho density contrast $\Delta\rho^{c/m}$), $\sigma$ is the unit sphere, and $d\sigma' = \cos\varphi'\ d\varphi'\ d\lambda'$ is the surface integration element. The 3-D position is defined by the spherical coordinates with the radius $r$, latitude $\varphi$ and longitude $\lambda$. The integral kernel K in Eq. (1) reads (cf. Sjöberg 2009)

$$\mathrm{K}(\psi, s) = \sum_{n=0}^{\infty} \frac{n+1}{n+3}\left(1 - s^{n+3}\right)\ \mathrm{P_n}(\cos\psi), \tag{2}$$

where $\psi$ is the spherical distance $\psi$, $s = 1 - \tau = 1 - D/\mathrm{R}$, and $\mathrm{P_n}$ is the Legendre polynomial of degree $n$. The isostatic gravity anomaly $\Delta g^{i}$ is defined as follows (cf. Vening Meinesz 1931)

$$\Delta g^{i} = \Delta g^{B} + g^{c} = 0, \tag{3}$$

where $\Delta g^{B}$ is the Bouguer gravity anomaly, and $g^{c}$ is the isostatic compensation attraction (see also Moritz 1990). In global studies, the computation of the isostatic gravity anomaly $\Delta g^{i}$ is practically realized in a frequency domain according to the following expression

$$\Delta g^{i}_{n,m} = \frac{1}{4\pi}\begin{cases} 2\pi\ \mathrm{G}\ (\overline{\rho}^{c}\ \mathrm{H})_{0,0} - \tilde{g}^{c}_{0} & \text{if}\quad n = 0 \\ 2\pi\ \mathrm{G}\ (\overline{\rho}^{c}\ \mathrm{H})_{n,m} - \Delta g_{n,m} & \text{otherwise} \end{cases}. \tag{4}$$

As seen in Eq. (4), the zero-degree coefficient of the isostatic gravity anomaly $\Delta g^{i}_{0,0}$ is defined as the difference between the (zero-degree) spectral Bouguer gravity reduction term $2\pi\mathrm{G}(\overline{\rho}^{c}\ \mathrm{H})_{0,0}$ and the (nominal) compensation attraction $\tilde{g}^{c}_{0}$, which is computed approximately from (cf. Sjöberg 2009)

$$\tilde{g}^{c}_{0} \approx -4\pi\ \mathrm{G}\ \Delta\rho^{c/m}\ \mathrm{D_0}, \tag{5}$$

where $\mathrm{D_0}$ and $\Delta\rho^{c/m}$ are the nominal (mean) values of the Moho depth and density contrast respectively. The first- and higher-degree coefficients $\Delta g^{i}_{n,m}$ in Eq. (4) are obtained from $2\pi\mathrm{G}\ (\overline{\rho}^{c}\ \mathrm{H})_{n,m}$ after subtracting the coefficients of the gravity anomaly $\Delta g_{n,m}$. The spectral Bouguer gravity reduction term is expressed in terms of the coefficients of global topographic/bathymetric (density) spherical functions $(\overline{\rho}^{c}\ \mathrm{H})_{n,m}$. The density distribution function $\overline{\rho}^{c}$ is specified for the crustal density $\rho^{c}$ and the seawater density $\rho^{w}$ as follows

$$\overline{\rho}^{c}\ \mathrm{H} = \begin{cases} \rho^{c}\ H & H \geq 0 \\ (\rho^{w} - \rho^{c})\ H & H < 0 \end{cases}. \tag{6}$$

In order to formulate the linearized observation equation for the product $T\Delta\rho$, the integral term on the left-hand side of Eq. (1) is expanded into a Taylor series. The substitution of the first two terms of binomial series for $s^{n+3}$ from Eqs. (2) to (1) yields (cf. Sjöberg and Bagherbandi 2011)

$$\begin{aligned} &\mathrm{R}\iint_\sigma \Delta\rho(\varphi',\lambda')\,\mathrm{K}(\psi,s)\,d\sigma' \\ &= \mathrm{R}\sum_{n=0}^{\infty}(n+1)\iint_\sigma \Delta\rho^{c/m}(\varphi',\lambda')\left(\tau-\tau^2\frac{n+2}{2}\right)\mathrm{P_n}(\cos\psi)d\sigma' \\ &= 4\pi\sum_{n=0}^{\infty}\sum_{m=-n}^{n}\frac{n+1}{2n+1}\left[\left(\Delta\rho^{c/m}\mathrm{D}\right)_{n,m} - \frac{n+2}{2\mathrm{R}}\left(\Delta\rho^{c/m}\mathrm{D}^2\right)_{n,m}\right]\mathrm{Y}_{n,m}(\varphi,\lambda). \end{aligned} \tag{7}$$

and $\mathrm{Y}_{n,m}$ are the spherical harmonics. From Eq. (7), the linearized observation equation for the product $D\,\Delta\rho^{c/m}$ is found to be

$$D(\varphi,\lambda)\,\Delta\rho^{c/m}(\varphi,\lambda) = \sum_{n=0}^{\infty}\sum_{m=-n}^{n}\left[\frac{2n+1}{4\pi(n+1)}\Delta g^{i}_{n,m} - \frac{n+2}{2\mathrm{R}}\left(\Delta\rho^{c/m}\mathrm{D}^2\right)_{n,m}\right]\mathrm{Y}_{n,m}(\varphi,\lambda). \tag{8}$$

The solution to a system of observation equations in Eq. (8) was formulated for the (a priori known) uniform and variable models of the Moho density contrast. For the variable density contrast, the coefficients $(\Delta\rho^{c/m}\mathrm{D}^2)_{n,m}$ were calculated based on discretizing the following integral convolution

$$\left(\Delta\rho^{c/m}\mathrm{D}^2\right)_{n,m} = \frac{1}{4\pi}\sum_{m=-n}^{n}\iint_\sigma D^2(\varphi',\lambda')\,\Delta\rho^{c/m}(\varphi',\lambda')\,\mathrm{Y}_{n,m}(\varphi',\lambda')\,d\sigma'. \tag{9}$$

By analogy with Eq. (9), the coefficients $(\Delta\rho^{c/m}\mathrm{D}^2)_{n,m}$ for the uninform model were generated according to the following expression

$$\left(\Delta\rho^{c/m}\mathrm{D}^2\right)_{n,m} = \frac{1}{4\pi}\sum_{m=-n}^{n}\iint_\sigma D^2(\varphi',\lambda') \times \begin{Bmatrix}\Delta\rho_o^{c/m} & H<0 \\ \Delta\rho_c^{c/m} & H\ge 0\end{Bmatrix}\mathrm{Y}_{n,m}(\varphi',\lambda')\,d\sigma', \tag{10}$$

where the constant values $\Delta\rho_c^{c/m}$ and $\Delta\rho_o^{c/m}$ of the Moho density contrast were defined individually for the continents and oceans.

## 3 Results

We applied two schemes for a determination of the VMM Moho depths based on adopting the uniform and variable models of the Moho density contrast. The gravity field quantities were calculated globally on a $1\times1$ arc-deg surface grid. The Moho depths were determined on a $1\times1$ arc-deg global grid, which is identical to a position of gravity points. The gravity field quantities and the Moho geometry were determined with a spectral resolution complete to a spherical harmonic degree of 180 (which corresponds to a half-wavelength of 1 arc-deg, or about 100 km on equator).

### 3.1 Bouguer Gravity Anomalies

The gravity anomalies were generated from the GOCO03S coefficients (Mayer-Guerr et al. 2012). The spherical harmonic terms of the normal gravity field were computed according to the GRS-80 parameters (Moritz 2000). The Bouguer gravity reduction term (in Eq. 4) was calculated from the DTM2006.0 coefficients (Pavlis et al. 2007) of the Earth's solid topography (i.e., the topographic heights on land and the bathymetric depths offshore). The density distribution function in Eq. (6) was specified for the upper continental crust of 2,670 kg $m^{-3}$ (cf. Hinze 2003) and the seawater density of 1,030 kg $m^{-3}$. The global map of the Bouguer gravity anomalies is shown in Fig. 1. The values are between −483 and 771 mGal, with a mean of 329 mGal and a standard deviation of 202 mGal. The Bouguer gravity anomalies are typically positive over oceans while negative over land. The gravity maxima correspond with locations of the oceanic basins and trenches. The largest negative gravity values on land apply over orogens of the Tibetan Plateau and Himalaya.

The isostatic gravity anomalies (obtained from the Bouguer gravity anomalies after applying the isostatic compensation attraction) vary between −253 and 132 mGal, with a mean of −54 mGal and a standard deviation of 31 mGal (see Fig. 2).

### 3.2 VMM Moho Depths

The isostatic gravity anomalies (shown in Fig. 2) were used to determine globally the Moho depths based on solving the VMM inverse problem of isostasy. The gravimetric solutions

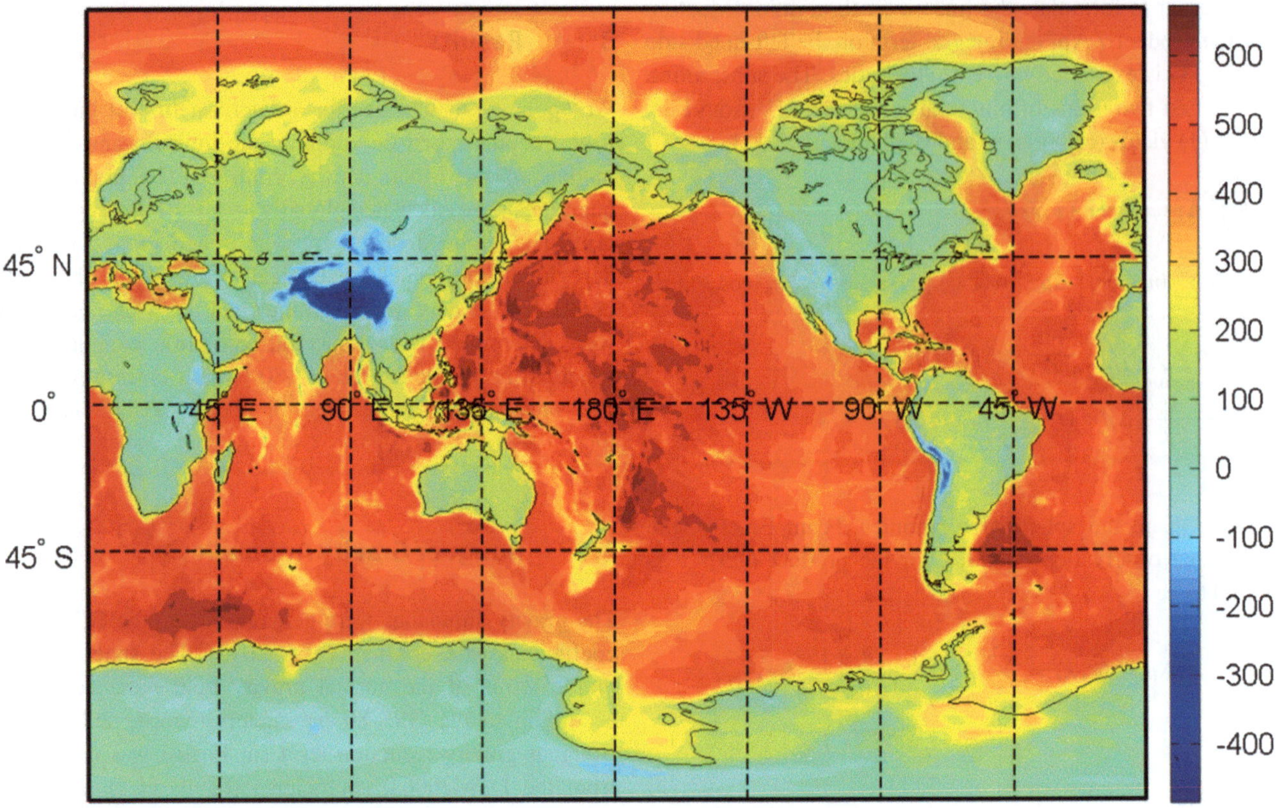


**Fig. 1** Bouguer gravity anomalies (in mGal)

were found for the uniform and variable models of the Moho density contrast. For the uniform model we adopted the constant values of the Moho density contrast of 678 kg m$^{-3}$ (for the continents) and 334 kg m$^{-3}$ (for the oceans) according to Sjöberg and Bagherbandi (2011). The variable model of the Moho density contrast was determined from the $1 \times 1$ arc-deg data of the CRUST1.0 (Laske et al. 2012) upper mantle and lower crust densities. The CRUST1.0 Moho density contrast varies at a relatively large interval between 10 and 610 kg m$^{-3}$. The density contrast minima are distributed along mid-oceanic ridges. The density contrast typically increases under oceanic basins. Under continental crustal structures, the density contrast minima are beneath continental margins, rift zones and sedimentary basins. The density contrast reaches maxima under cratonic and orogenic structures.

The gravimetrically determined Moho solutions are shown in Fig. 3. The VMM Moho depths for the uniform density contrast globally vary from 7.1 to 53.2 km, with a mean of 23.9 km and a standard deviation 9.0 km. The corresponding Moho depths for the variable density contrast vary from 5.0 to 61.1 km, with a mean of 23.7 km and a standard deviation of 11.3 km. The Moho depth differences between these two gravimetric models (shown in Fig. 4) are between −8.1 and 4.4 km, the mean of differences is 0.2 km and the RMS of differences is 2.1 km. The utilization of the variable model of the Moho density contrast thus changed only slightly the global mean of the Moho depths, while the range of Moho depths increased from 46.1 km (for a uniform model) to 56.1 km (for a variable model). As seen in the plot of the Moho depth differences between these two results (shown in Fig. 4), the largest absolute changes are detected under orogens of the Tibetan Plateau and Himalaya, where the maxima of the Moho depths increased locally even more than 5 km. Some large changes are also seen at locations of the Island hotspot and along continental margins. The application of the variable density contrast decreased the estimated values of the minimum Moho depths beneath some parts of the oceanic crust from 7.1 to only 5.0 km.

We note here that we did not apply an additional iterative computation of the variable Moho density contrast using the newly estimated gravimetric Moho depths, because of the presence of large uncertainties in the gravimetrically determined Moho depths as well as in the variable Moho density contrast values.

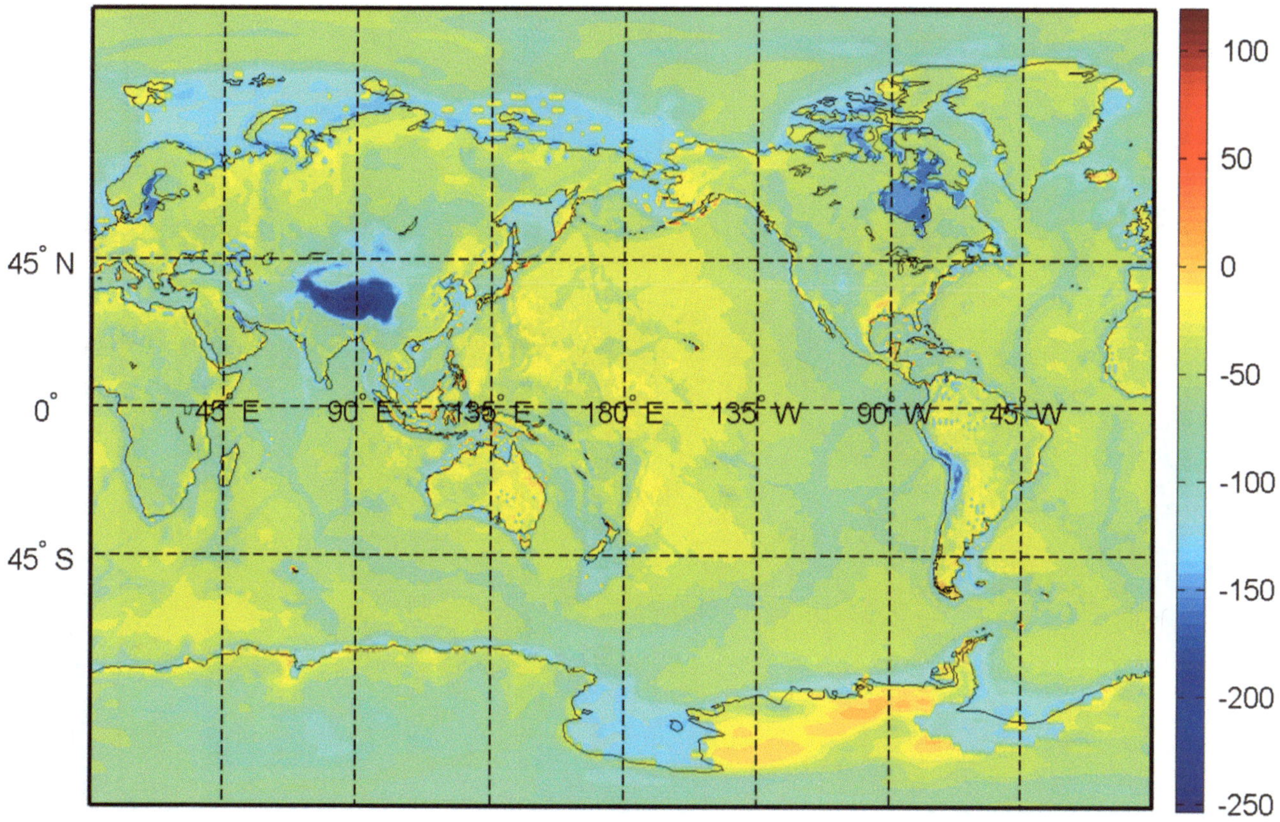


**Fig. 2** Isostatic gravity anomalies (in mGal)

### 3.3 Validation of Results

The gravimetric results (shown in Fig. 3) were compared with the CRUST1.0 seismic Moho model. The Moho depth differences between the gravimetric and seismic models are shown in Fig. 5. For the uniform Moho density contrast, these differences range between −21.6 and 29.2 km; the mean and RMS of differences are −1.0 and 5.4 km respectively. For the variable Moho density contrast, these differences are between −25.1 and 26.3 km; the mean and RMS of differences are −0.8 and 4.7 km respectively. As seen from this comparison, the gravimetric result for the variable density contrast better agrees with the CRUST1.0 Moho depths. Whereas, the systematic bias between the gravimetric and seismic models is very similar, the RMS fit improved from 5.4 km (for a uniform density contrast) to 4.7 km (for a variable density contrast). The significant improvement was found especially along the mid-oceanic ridges and under orogens of the Tibetan Plateau and Himalaya.

Both gravimetric solutions are slightly systematically biased from the CRUST1.0 seismic Moho model. The existence of this bias is explained by the long-wavelength gravitational contribution of the unmodelled mantle density heterogeneities. For more details we refer readers to studies, for instance, by Ishii and Tromp (2004), Chujkova and Maksimova (2010) and Chujkova et al. (2007, 2014).

## 4 Concluding Remarks

The facilitation of the variable Moho density contrast in solving the VMM inverse problem of isostasy substantially changed the Moho depth estimates. Large changes were detected, for instance, under orogenic formations of the Tibetan Plateau and Himalaya, where the application of the variable density contrast increased the estimated Moho depths 5 km or more compared to the uniform density contrast. A similar trend was found also under most of the

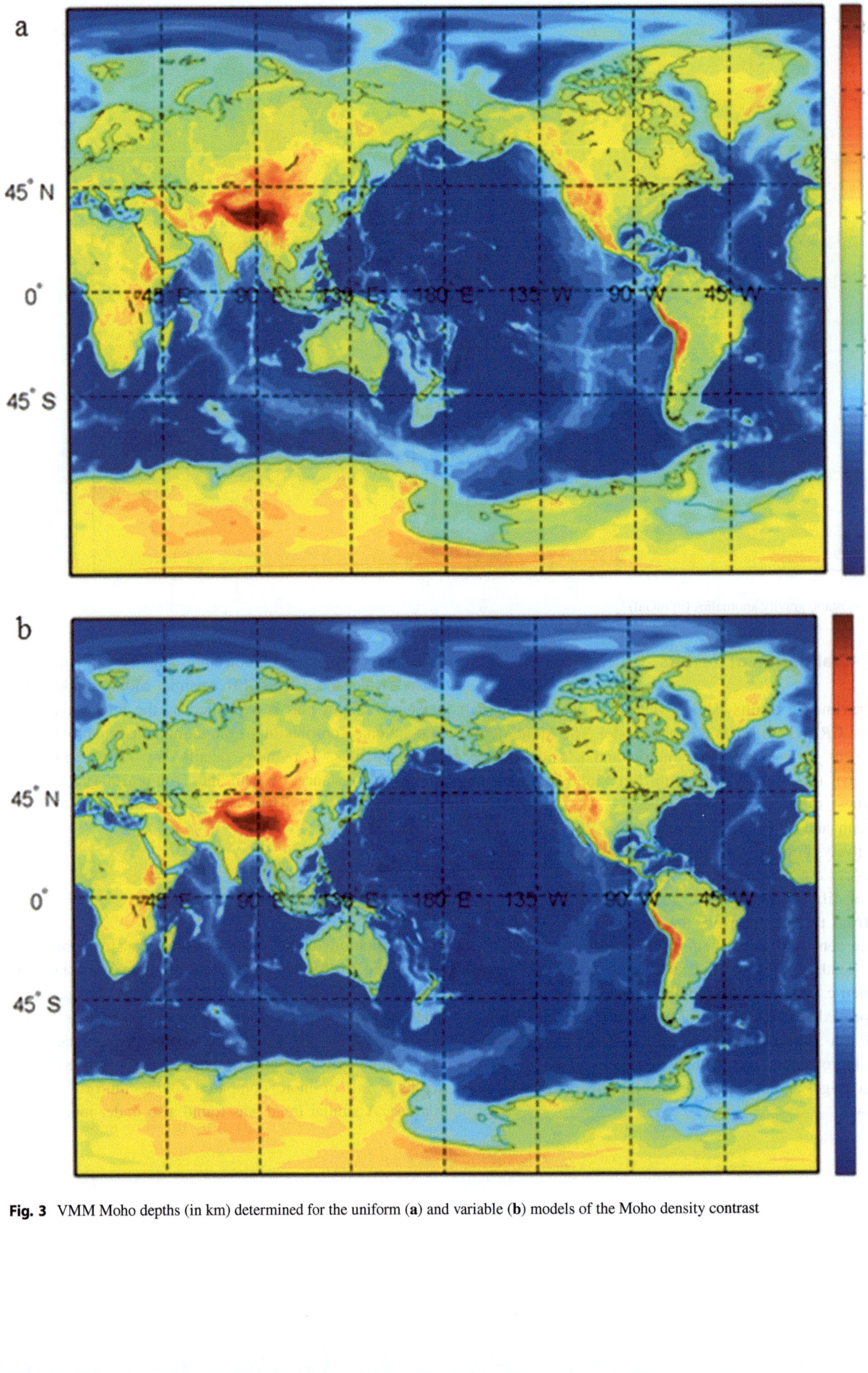


**Fig. 3** VMM Moho depths (in km) determined for the uniform (**a**) and variable (**b**) models of the Moho density contrast

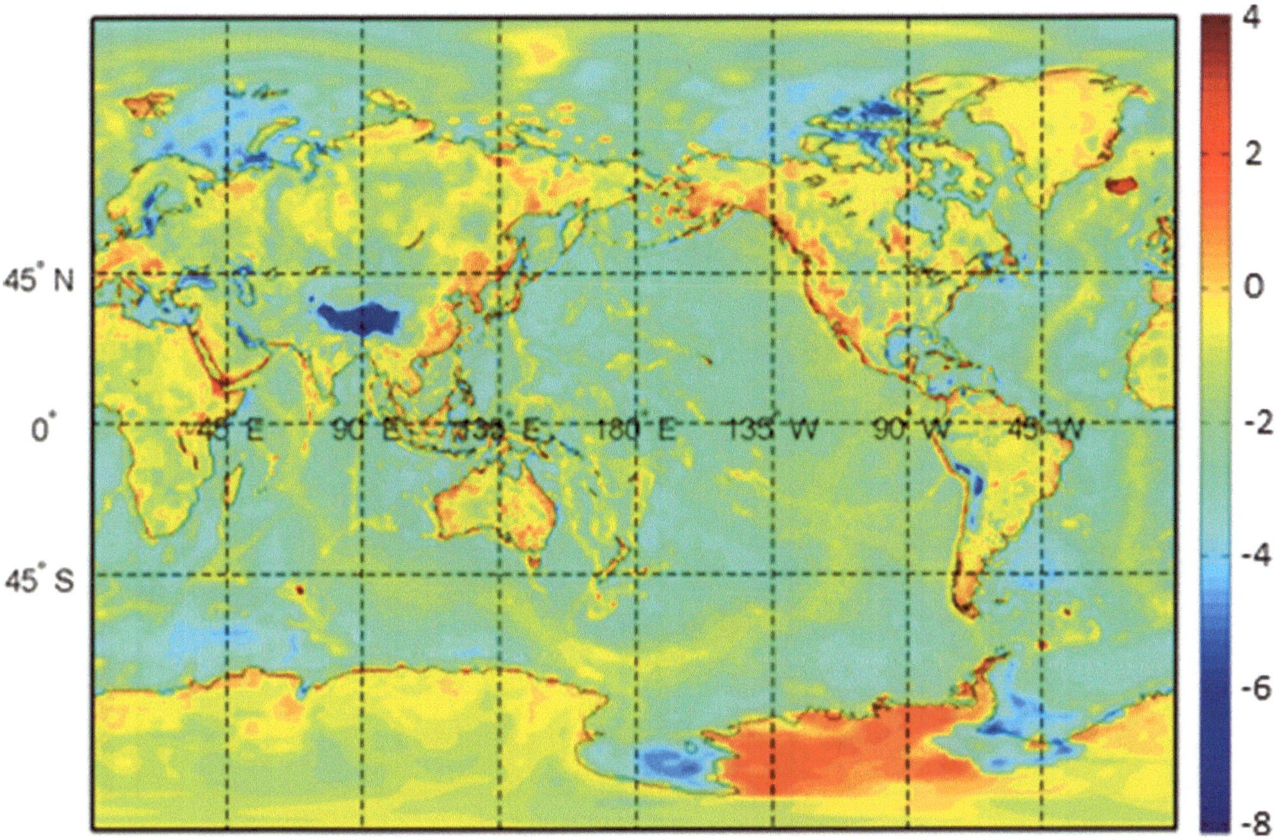


**Fig. 4** Moho depth differences (in km) between the gravimetric results obtained using the constant and variable models of the Moho density contrast

oceanic crustal structures. In contrast, continental crustal thickness increased under some parts of the continental margins. The largest increase (about 4 km) in estimated values of the Moho depths was detected at the Island hotspot in northern Atlantic, which is characterized by the largest thickness of the oceanic crust.

In overall, the application of the variable model of the Moho density contrast improved the agreement of the gravimetric result with the seismic model. The validation using the CRUST1.0 Moho depths have shown that the RMS fit improved from 5.4 km (for a uniform density contrast) to 4.7 km (for a variable density contrast).

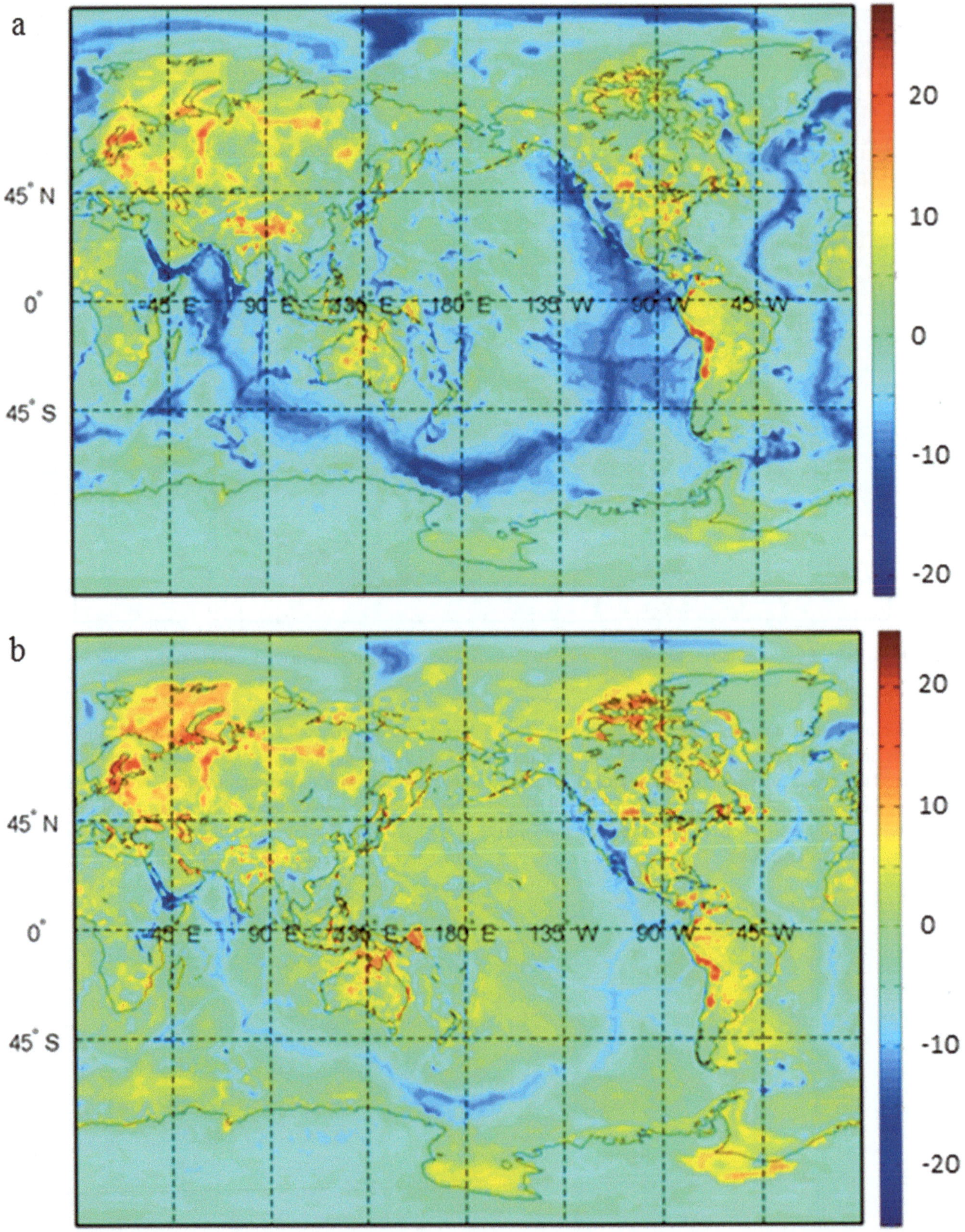


**Fig. 5** Differences (in km) between the VMM gravimetric results and the CRUST1.0 seismic Moho depths for the uniform (**a**) and variable (**b**) models of the Moho density contrast

## References


Airy GB (1855) On the computations of the effect of the attraction of the mountain masses as disturbing the apparent astronomical latitude of stations in geodetic surveys. Trans R Soc Lond B 145

Bagherbandi M, Tenzer R, Sjöberg LE, Novák P (2013) Improved global crustal thickness modeling based on the VMM isostatic model and non-isostatic gravity correction. J Geodyn 66:25–37

Braitenberg C, Zadro M (1999) Iterative 3D gravity inversion with integration of seismologic data. Boll Geofis Teor Appl 40(3/4):469–476

Braitenberg C, Wienecke S, Wang Y (2006) Basement structures from satellite-derived gravity field: South China Sea ridge. J Geophys Res 111:B05407

Braitenberg C, Mariani P, Reguzzoni M, Ussami N (2010) GOCE observations for detecting unknown tectonic features. In: Proceedings of the ESA living planet symposium, 28 June to 2, July 2010, Bergen, Norway, ESA SP-686

Čadek O, Martinec Z (1991) Spherical hermonic expansion of the earth's crustal thickness up to degree and order 30. Stud Geophys Geod 35:151–165

Chujkova NA, Maksimova TG (2010) Density inhomogeneities of the Earth's mantle end core. Mosc Univ Phys Bull 65:137–144

Chujkova NA, Nasonova LP, Maximova TG (2007) Gravity anomalies in the Earth's crust and upper mantle. Astron Astrophys Trans 26(4–5):391–399

Chujkova NA, Nasonova LP, Maximova TG (2014) Density, stress, and gravity anomalies in the interiors of the Earth and Mars and the probable geodynamical implications: comparative analysis. Izv Phys Solid Earth 50(3):427–443

Dziewonski AM, Anderson DL (1981) Preliminary reference earth model. Phys Earth Planet Inter 25:297–356

Heiskanen WA, Moritz H (1967) Physical geodesy. WH Freeman and Co, San Francisco

Hinze WJ (2003) Bouguer reduction density, why 2.67? Geophysics 68(5):1559–1560

Ishii M, Tromp J (2004) Constraining large-scale mantle heterogeneity using mantle and inner-core sensitive normal modes. Phys Earth Planet Inter 146:113–124

Jordi J (2007) Constraining velocity and density contrasts across the crust–mantle boundary with receiver function amplitudes. Geophys J Int 171:286–301

Laske G, Masters G, Ma Z, Pasyanos ME (2012) CRUST1.0: an updated global model of Earth's crust. In: Geophys Res Abs 14, EGU2012-3743-1, EGU General Assembly 2012

Mayer-Guerr T, Rieser D, Höck E, Brockmann JM, Schuh W-D, Krasbutter I, Kusche J, Maier A, Krauss S, Hausleitner W, Baur O, Jäggi A, Meyer U, Prange L, Pail R, Fecher T, Gruber T (2012) The new combined satellite only model GOCO03s. Abstract, GGHS2012, Venice

Moritz H (1990) The figure of the earth. Wichmann H, Karlsruhe

Moritz H (2000) Geodetic reference system 1980. J Geod 74:128–162

Niu F, James DE (2002) Fine structure of the lowermost crust beneath the Kaapvaal craton and its implications for crustal formation and evolution. Earth Planet Sci Lett 200:121–130

Pavlis NK, Factor JK, Holmes SA (2007) Terrain-related gravimetric quantities computed for the next EGM. In: Presented at the 1st international symposium of the International gravity service 2006, August 28 – September 1, Istanbul, Turkey

Reguzzoni M, Sampietro D (2014) GEMMA: an Earth crustal model based on GOCE satellite data. Int J Appl Earth Obs 35(A):31–43

Reguzzoni M, Sampietro D, Sansò F (2013) Global Moho from the combination of the CRUST2.0 model and GOCE data. Geophys J Int 195(1):222–237

Sampietro D (2011) GOCE exploitation for Moho modeling and applications. In: Proceedings of the 4th international GOCE user workshop, 31 March – 1 April 2011, Munich, Germany

Sampietro D, Reguzzoni M, Braitenberg C (2013) The GOCE estimated Moho beneath the Tibetan Plateau and Himalaya. In: International association of geodesy symposia. Earth on the edge: science for a sustainable planet, Proceedings of the IAG General Assembly, 28 June – 2 July 2011, Melbourne, Australia, vol 139. Springer, Berlin

Sjöberg LE (2009) Solving Vening Meinesz-Moritz inverse problem in isostasy. Geophys J Int 179:1527–1536

Sjöberg LE (2011) On the isostatic gravity anomaly and disturbance and their applications to Vening Meinesz-Moritz gravimetric inverse problem. Geophys J Int 193(3):1277–1282

Sjöberg LE, Bagherbandi M (2011) A method of estimating the Moho density contrast with a tentative application by EGM08 and CRUST2.0. Acta Geophys 59(3):502–525

Tenzer R, Bagherbandi M (2012) Reformulation of the Vening-Meinesz Moritz inverse problem of isostasy for isostatic gravity disturbances. Int J Geosci 3(5A):918–929

Tenzer R, Chen W (2014) Expressions for the global gravimetric Moho modeling in spectral domain. Pure Appl Geophys 171(8):1877–1896

Tenzer R, Hamayun K, Vajda P (2009) Global maps of the CRUST2.0 crustal components stripped gravity disturbances. J Geophys Res 114:B05408

Tenzer R, Hamayun K, Novák P, Gladkikh V, Vajda P (2012a) Global crust-mantle density contrast estimated from EGM2008, DTM2008, CRUST2.0, and ICE-5G. Pure Appl Geophys 169(9):1663–1678

Tenzer R, Bagherbandi M, Gladkikh V (2012b) Signature of the upper mantle density structure in the refined gravity data. Comput Geosci 16(4):975–986

Tenzer R, Chen W, Tsoulis D, Bagherbandi M, Sjöberg LE, Novák P, Jin S (2014a) Analysis of the refined CRUST1.0 crustal model and its gravity field. Surv Geophys. doi:10.1007/s10712-014-9299-6

Tenzer R, Chen W, Jin S (2014b) Effect of the upper mantle density structure on the Moho geometry. Pure Appl Geophys. doi:10.1007/s00024-014-0960-2

Vening Meinesz FA (1931) Une nouvelle methode pour la reduction isostatique regionale de l'intensite de la pesanteur. Bull Géod 29:33–51

Wienecke S, Braitenberg C, Götze H-J (2007) A new analytical solution estimating the flexural rigidity in the Central Andes. Geophys J Int 169(3):789–794

# The New Method to Find the Anomalous Internal Structure of Terrestrial Planets and Its Test on the Earth

N.A. Chujkova, L.P. Nasonova, and T.G. Maximova

**Abstract**

The original method for solving the inverse incorrect problem of gravimetry is developed. The method allows to find the anomalous density structure of terrestrial planets from space-based observations of the gravity field and topography. Previously we used this method to study the internal structure and geodynamic characteristics of the Earth and Mars for the purpose of comparison and interpretation. In this paper the distribution of the compensation's depths for the Earth's topographical harmonics and lateral distributions of anomalous masses in selected depths are found with the help of this method. This distribution of anomalous masses is compared with similar distribution obtained using of seismic data for Earth's Mohorovicic depths. The results obtained by these two methods give good coincidence. The maps of stresses' distribution in the Earth's lithosphere, obtained by our method, correspond to the map of earthquakes distribution. And the density distribution in the boundaries' layers of the Earth's core and mantle are in good agreement with the results of the analysis of free-oscillation data of the Earth. This gives reason to believe that our method provides a good model representation, which can be helpful as guidance in future studies on the surfaces of the terrestrial planets that are not available yet for seismic observations.

**Keywords**

Density anomalies • Depths of isostatic compensation • Earth • Inverse problem of gravimetry • Stresses

## 1 Introduction

Until now, an information about internal structure of the planets one can obtain only on the basis of space research of the gravity field and topography of the planets. However it is known that the inverse problem of gravimetry is ill-posed. Its solution is possible only when additional information about the internal structure is used, and based on some theoretical conclusions. These conclusions are based on cosmogonic scenarios of formation of terrestrial planets, on geophysical and geochemical information and high-energy physics data.

Extensive geophysical observations, available now only on the Earth, allow us to specify its internal structure (for the Moon there are only certain seismic and thermal monitoring).

The modern seismic and seismological studies show significant lateral velocity anomalies of seismic waves when passing through the transition layers upper-lower mantle (Gu et al. 2003; Lawrence and Shearer 2008), as well as on some depths of the upper and lower mantle (Antolik et al. 2003), associated with anomalies of temperature, density and elastic properties in these layers. In addition the significant velocity anomalies in the lowest layers of the mantle adjacent to the Earth's core are revealed (Lay and Garnero 2004), as well as the presence of anomalies in the boundary layers as in

N.A. Chujkova • L.P. Nasonova (✉) • T.G. Maximova
Sternberg Astronomical Institute, Moscow State University, Universitetskii pr. 13, Moscow 119991, Russia
e-mail: liudmila-nasonova@yandex.ru



S. Jin, R. Barzaghi (eds.), *IGFS 2014*, International Association of Geodesy Symposia 144,
DOI 10.1007/1345_2015_195

external (Soldati et al. 2003; Tanaka 2006), and in the inner Earth's core (Cormier 2007; Garcia et al. 2006; Bergman 2006; Krasnoshchekov et al. 2006; Zou et al. 2008). However the attempts to find contribution of laterally distributed density inhomogeneities based on the analysis of changes in seismic velocities cannot be considered successful, because for the deep Earth's interior even the sign of the relationship between the velocity change and density change is unknown. The relationship between the density anomalies and seismic velocities empirically investigated only for the surface layers of the Earth (the ratio of the Berch). It is not known whether this law is stored at high pressures and temperatures typical for the great depths. This is discussed in detail in the book (Brown and Musset 1993) and in the book ( Zharkov 1986), and this is known to all modern seismologists.

In recent years there are works on interpretation of observations of the free-oscillations of the Earth in order to determine the global lateral density variations in the different shells of the Earth (Ishii and Tromp 1999, 2004). However, similar studies of other authors (Kuo and Romanowicz 2002) have shown that the obtained density structures are unstable and a strong dependence on the assumed initial distribution model of seismic velocities and regularization schemes.

Unfortunately such data are not available to other planets now. Despite the absence of such observations for Mars, we made an attempt to determine its internal structure using the technique that we have developed earlier (Chuikova et al. 2011, 2012; Chujkova et al. 2014). The essence of this method is the determination of possible compensation's depths for different-order and different-degree harmonics for expansion of topographic heights relative to the equilibrium ellipsoid. Each topographic irregularity is characterized by a certain set of the harmonics. Therefore the maximal concentration of compensation of this set within a certain limited depth interval can testify the most probable depths of compensation of the considered topographic irregularity. Thus obtained compensation's depths allow one to solve the formulated problem of finding the lateral distribution of the compensating masses at selected depths. One can assume that if we apply the developed method to analyze the internal structure of the Earth, and the results are well consistent with the seismological and free-oscillation data of the Earth, it can be hoped that application of the same technique to analyze the internal structure of terrestrial planets will give reliable results. In order to ensure that the lack of seismic observations for the planets does not lead to significant errors in the interpretation of the internal structure, we performed a similar study with use of the most reliable data of the depth of the Moho surface (***M***).

## 2 Theory and Method of Calculation of Anomalous Structures of the Earth's Mantle and Core

Before calculating the abnormal structure of the Earth's mantle and core, it is necessary to subtract the contribution of topography or the total contribution of topography and density contrast on the boundary M from external gravity field harmonics. In order to determine the contribution of laterally distributed anomalous masses of relief and the density contrast on surface ***M*** to Stokes parameters of the gravity field, we represent these masses in the form of spherical expansion of the layers of constant density distributed along the height relative to the reference ellipsoid.

In the linear approximation, when the layer of anomalous masses is distributed over the spherical surface, there is a linear relationship between the coefficients $a_{nm}^{(s)}, b_{nm}^{(s)}$ of spherical expansion of the density of a simple layer $(s)$ and Stokes coefficients $\Delta C_{nm}^{(s)}, \Delta D_{nm}^{(s)}$, associated with the contribution of this layer at gravity field (Duboshin 1961):

$$\left\{ \begin{array}{c} \Delta C_{nm}^{(s)} \\ \Delta D_{nm}^{(s)} \end{array} \right\} = \frac{3}{2n+1} \left( \frac{R_s}{R_0} \right)^3 \frac{\Delta \sigma_s}{\overline{\sigma}} \left( \frac{R_s}{a} \right)^n \left\{ \begin{array}{c} a_{nm}^{(s)} \\ b_{nm}^{(s)} \end{array} \right\}, \quad (1)$$

here $R_s$, $\Delta\sigma_s$ – are the mean radius and mean density of the simple layer, and $R_0$, $\overline{\sigma}$ – are the mean radius and mean density of the entire planet, respectively, $a$ – is equatorial radius of the planetary ellipsoid. Here, the external gravitational potential of the Earth is represented as the following expansion in terms of spherical harmonics:

$$V(r, \varphi, \lambda) = \frac{fM_0}{r} \sum_{n=1}^{N} \left( \frac{a}{r} \right)^n \sum_{m=0}^{n} (C_{nm} \cos(m\lambda) + D_{nm} \sin(m\lambda)) P_{nm}(\sin\varphi),$$

where $M_0$ – is the mass of the Earth, $f$ is the gravitational constant, $r, \varphi, \lambda$ are the spherical coordinates of the observation point, $C_{nm}, D_{nm}$ are the Stokes constants, $P_{nm}(\sin\varphi)$ are the associated Legendre functions (i.e. spheroidal harmonics), $N = 18$.

The relative height $h_s(\varphi, \lambda)$ (relative to the average radius) for the layer number s is represented by the following expansion in spherical functions

$$h_s(\varphi, \lambda) = \sum_{n=1}^{N} \sum_{m=0}^{n} \left( a_{nm}^{(s)} \cos(m\lambda) + b_{nm}^{(s)} \sin(m\lambda) \right) P_{nm}(\sin\varphi)$$

In reality, the topographic masses do not form a simple spherical layer; instead, they are distributed along the height relative to the corresponding reference ellipsoid. In this case, taking into account the quadratic terms and ellipsoidal shape of the reference surface in the formula (1) for the coefficients $a_{nm}^{(s)}, b_{nm}^{(s)}$, one must use the following expression containing additional members (Chujkova et al. 2006; Chuikova et al. 2011), namely:

$$\begin{Bmatrix} a_{nm}^{(s)} \\ b_{nm}^{(s)} \end{Bmatrix} = \begin{Bmatrix} a_{nm} \\ b_{nm} \end{Bmatrix}_1 + \frac{n+2}{2}\begin{Bmatrix} a_{nm} \\ b_{nm} \end{Bmatrix}_2 + \alpha\,(n+2)\begin{Bmatrix} a_{nm} \\ b_{nm} \end{Bmatrix}_3. \tag{2}$$

Here the first term corresponds to the coefficients of expansion of relative (relative to the average radius of the layer $R_s$) heights for layer $h_s$, the term with index 2 corresponds to the coefficients of expansion of function $(h_s)^2$, and the term with index 3 corresponds to the coefficients of expansion of function $h_s \overline{P}_2(\sin\phi)$, $\alpha = 2/3e$, and $e$ – is oblateness of the reference ellipsoid. The formulas for expressing the coefficients $\{a_{nm}, b_{nm}\}_2$, $\{a_{nm}, b_{nm}\}_3$ in terms of $\{a_{nm}, b_{nm}\}_1$, were obtained in our previous work by mathematical modeling of the symbol calculations in computer algebra system (Nasonova and Chuikova 2007).

In order to determine the depth of compensation $d_{nm}$ for arbitrary harmonic of expansion for the anomalous masses of topography (or relief plus $M$), it is necessary to solve a system of two equations. The first equation represents the agreement between the contribution of topographic and compensating masses to the gravity and the observations. The second equation represents the fact that the pressures below the compensation depth are equal to the pressures of the equilibrium model. The solution of this system for the depth of compensation $d_{nm}$ for the arbitrary harmonic of the topography is defined by the relation:

$$d_{nm} = R_0 - R_M \left(a_{nm}^{M1}/a_{nm}^{M2}\right)^{1/n}. \tag{3}$$

Here $a_{nm}^{M1}, a_{nm}^{M2}$ – are coefficients of spherical expansion for the heights of the compensation surfaces ***M1*** (determined from the gravity field after the subtraction of the normal field and quadratic contribution of topographical masses) and ***M2*** (obtained from the hypothesis of the isostatic compensation of topographic masses relative to the corresponding equilibrium ellipsoid with a fixed radius of compensation $R_M$).

Therefore $a_{nm}^{M1}$ are determined from the first equation

$$C_{nm} - \Delta C_{nm}^{(s)} = \frac{3}{2n+1}\left(\frac{R_M}{R_0}\right)^3 \frac{\Delta\sigma_{com}}{\overline{\sigma}}\left(\frac{R_M}{a}\right)^n a_{nm}^{M1}, \tag{4}$$

were $\Delta\sigma_{com}$ is mean density contrast at $R_M$.

On the other hand, at a radius compensation $R_{com}$ the same condition should be performed with replacing $R_M$ on $R_{com} = R_0 - d_{nm}$, $a_{nm}^{M1}$ on $a_{nm}^{com}$.

$$C_{nm} - \Delta C_{nm}^{(s)} = \frac{3}{2n+1}\left(\frac{R_{com}}{R_0}\right)^3 \cdot \frac{\Delta\sigma_{com}}{\overline{\sigma}}\left(\frac{R_{com}}{a}\right)^n a_{nm}^{com}, \tag{5}$$

where

$$a_{nm}^{com} R_{com}^3 = a_{nm}^{M2} \cdot R_M^3. \tag{6}$$

$a_{nm}^{M2}$ is determined by the condition of the isostatic equilibrium on $R_M$ (i.e. from second equation): $a_{nm}^{M2} \cdot R_M^3 \cdot \Delta\sigma_{com} + a_{nm}^{(s)} \cdot R_s^3 \cdot \Delta\sigma_s = 0$

Substituting (6) into (5), and Eq. (5) into Eq. (4), we obtain $(R_{com})^n \cdot a_{nm}^{M2} = (R_M)^n \cdot a_{nm}^{M1}$.

From here the Eq. (3) follows. By replacing $a_{nm}$ by $b_{nm}$, $C_{nm}$ by $D_{nm}$ one obtain the similar formula for $b_{nm}$. As one can see, the solution is possible, that is $R_0 \geq d_{nm} \geq 0$ if $0 \leq a_{nm}^{M1}/a_{nm}^{M2} \leq (R_0/R_M)^n$.

The analysis of the solution both for the Earth and Mars (Chuikova et al. 2011, 2012; Chujkova et al. 2014) shows that compensation of the topographic masses on the single level is possible only for a certain set of the harmonics. For other harmonics we selected two variants of compensation, which lead to the smallest deviations from the equilibrium of the internal structure. In the first variant, compensation is carried out at two levels. The first of which is the Moho, likely depths of the second level are determined by analyzing the results for harmonics for which the solution (3) exists. The final choice is made by taking into account the weights, which are inversely proportional to the deviations from the equilibrium of the internal structure. In the second variant, the uncompensated topographical harmonics cause the stresses in the lithosphere such that the stresses does not exceed the strength's limit of the lithosphere.

After determining the depth of compensation, the main ranges of the depth of compensation were found. The maps of distribution of anomalous masses were constructed by recalculation of the density for a single layer to an average radius $\overline{R}_{com}$ for the selected range. Recalculation to the middle level was performed in such a way that for each harmonic the condition of equality of contribution to the external gravity field of all anomalous masses and pressure coincidence to the equilibrium model in the lower lying layers of the mantle must be satisfied. The recalculation was carried out according to the formula

$$\Delta m\left(\overline{R}, \varphi, \lambda\right) = \Delta\sigma_{com} \cdot \overline{R} \cdot \sum_{n=1}^{N}\sum_{m=0}^{n}\left(a_{nm}^{(s)}\alpha_{nm}\cos m\lambda + b_{nm}^{(s)}\beta_{nm}\sin m\lambda\right) \cdot P_{nm}(\cos\varphi),$$

where $\alpha_{nm}, \beta_{nm}$ are the recalculation's coefficients for anomalous mass at a depth $d_{nm}$ to the average radius $\overline{R}$ for each selected range. These coefficients must satisfy the given condition and were calculated separately for each selected range and for each depth $d_{nm}$.

The coefficients of spherical expansion for the stress as the vertical component of the tensor of the stress (in Pa) were calculated by the following formulas:

$$\left\{ \begin{matrix} a_{nm}^{st} \\ b_{nm}^{st} \end{matrix} \right\} = \left\{ \begin{matrix} a_{nm}^{M1} - a_{nm}^{M2} \\ b_{nm}^{M1} - b_{nm}^{M2} \end{matrix} \right\} \Delta\sigma_M . g_0 R_0, \qquad (7)$$

here $\Delta\sigma_M = \Delta\sigma_{com} = 0.31 \times 10^3$ kg/m$^3$ (mean density contrast on ***M***), $g_0 = 9.80$ m/s$^2$, $R_0 = 6{,}370 \times 10^3$ m. One can see from the formula (7) that the stresses are reduced to the surface of the planet of radius $R_M = R_0$. For calculation at the lower layers of radius $r$, formula (7) is changed,

$$\left\{ \begin{matrix} a_{nm}^{st} \\ b_{nm}^{st} \end{matrix} \right\}_r = \left\{ \begin{matrix} a_{nm}^{M1} \cdot \left(\frac{R_0}{r}\right)^n - a_{nm}^{M2} \\ b_{nm}^{M1} \cdot \left(\frac{R_0}{r}\right)^n - b_{nm}^{M2} \end{matrix} \right\} \cdot \Delta\sigma_M \cdot g_0 \cdot R_0 \cdot \left(\frac{R_0}{r}\right)^2 \qquad (8)$$

We see that coefficients $a_{nm}^{st}, b_{nm}^{st}$ are increased as $\left(\frac{R_0}{r}\right)^2$ and $a_{nm}^{M1}, b_{nm}^{M1}$ are increased as $\left(\frac{R_0}{r}\right)^n$.

## 3 Input Data

As the input data for the Earth's gravitational field we use the Stokes coefficients of degree $n \leq 18$, which are taken from the satellite-only model GOCO01S, based on data of the satellite gravity missions GOCE and GRACE. (Pail et al. 2010).

As the initial data for the relief of the Earth we use the spherical function expansion of the Earth's equivalent relief, which we obtained previously in (Chujkova et al. 1996). When calculating the heights of equivalent (i.e. reduced to the uniform density of 2.67 g/cm$^3$) relief of the Earth the following data were used: 1.03 g/cm$^3$ – is the average density of sea water; 0.91 g/cm$^3$ – is the average density of the ice sheet of Antarctica; 2.64 g/cm$^3$ – is the maximum density of the crust in the sedimentary basins of Eurasia. Detailed information about the used relief heights relative to the geoid heights, the depths of the seas and oceans, the thickness of sedimentary basins, as well as methods of reducing them to the heights of equivalent relief is given in (Chujkova et al. 1996).

For the depths of Moho the harmonic expansion of degree $n \leq 18$ is used. We obtain it from the analysis of seismic observations. The theory and methods of analysis of seismic data are givenin (Chujkova and Maximova 1996). The harmonic expansion for the topography and the Moho boundary are given on the website (http://lnfm1.sai.msu.ru/~chujkova).

In the study of anomalous internal structure of the Earth as an equilibrium model for the equilibrium ellipsoid for the viscoelastic Earth has been accepted (Shatina 2007). For this model the harmonic $C_{20}$ corresponds to measurements of the external gravity field.

## 4 Results and Main Conclusions

The Fig. 1a, b show the distribution histograms for the depths of compensation of the topographical harmonics of the Earth, for which compensation occurs at a single depth level.

It was found that only about 60% of the anomalous masses of crust are compensated at the single level. One can see that compensation almost entirely takes place in the crust and upper mantle (i.e. within the lithosphere). From the histograms can be seen that the compensation of various harmonics of the topography in the crust and mantle is carried out at various levels, indicating the stratification (layering) of crust and mantle. This result coincides with the results of the seismic studies also showing the stratification of the crust and upper mantle (Pavlenkova 1996). However, it can be seen from histogram that it is possible to allocate the basic compensation levels, for which can calculate the middle depth. The distribution of anomalous masses is calculated at this middle depth. The main compensation levels are clearly distinguished at the depths of 0–20 km, 20–55 km, 55–80 km, 90–165 km (Fig. 1a, c) and of 180–400 km, 600–900 km (Fig. 1b, d). This results consist well with the results of the seismic observations (Pavlenkova 1996; Rodkin 1993; Gu et al. 2003; Lawrence and Shearer 2008). The average values of depths of layers are calculated with weights corresponding to harmonic amplitudes. They are in good agreement with the depths of layers, obtained from the seismic data, at which the isostatic compensation of relief's masses may occur: $23 \pm 15$ km – the average depth of the Moho for the oceans and continents (Chujkova and Maximova 1996), in the range of 0–55 km; $92 \pm 27$ km – the depth of isostatic compensation, according to (Pavlenkova 1996; Thybo and Perchuć 1997) in the range of 55–165 km; $260 \pm 50$ km – the depth of the asthenosphere (Rodkin 1993; Gu et al. 2001) in the range of 180–400 km. In the lower mantle only some harmonics of low degree ($n = 2$–5) are compensated at depths of 600–900 km (average depth of $760 \pm 150$ km). For comparison, we performed a similar analysis taking into account the contribution to the gravity field of the density discontinuity on the boundary ***M***. The resulting histograms (Fig. 1c, d) well correspond to the results obtained without ***M*** (Fig. 1a, b). At lower mantle, only harmonics $a_{33}$ of relief and Moho are compensated at

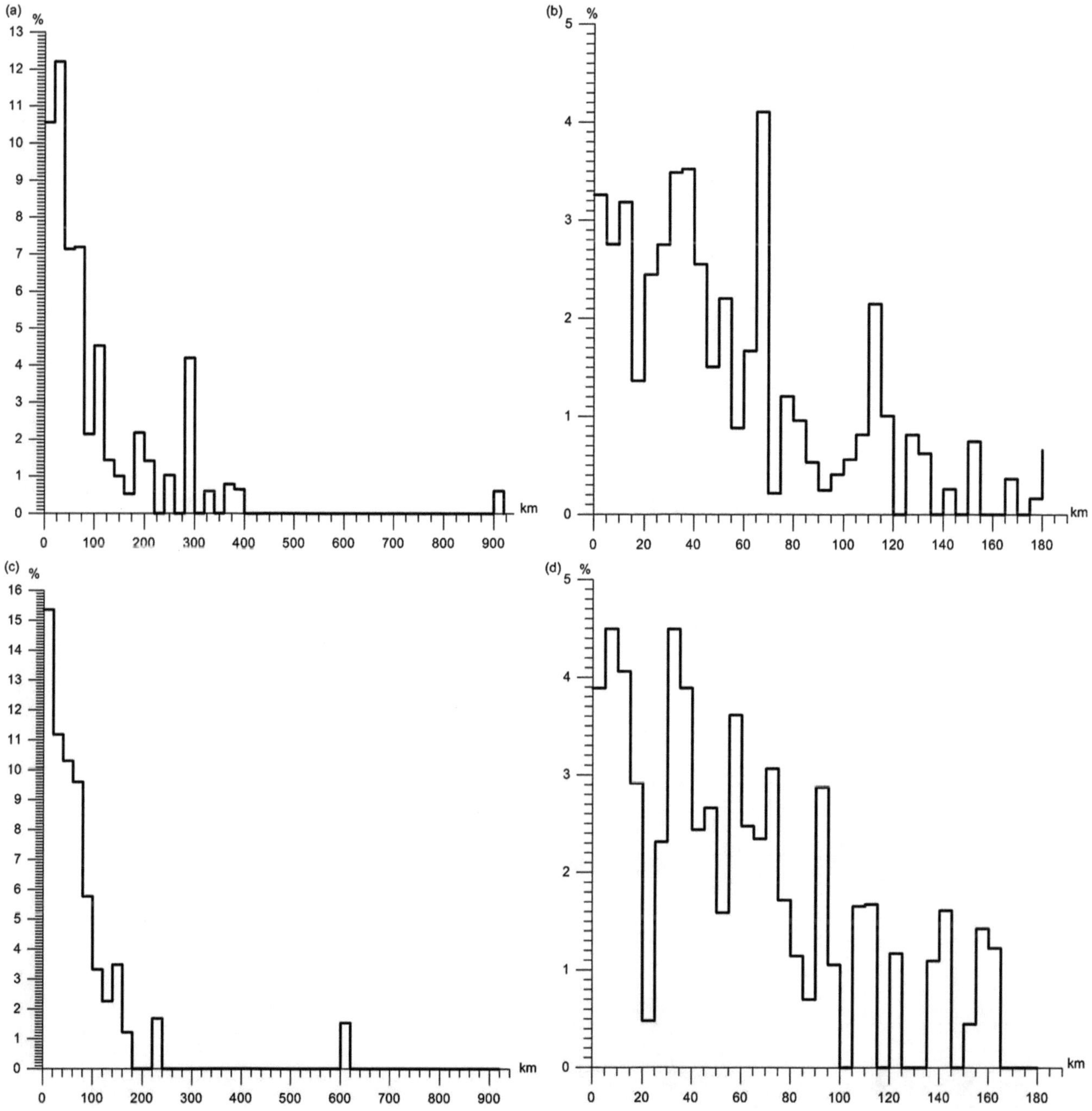


**Fig. 1** (**a–d**) The histograms for the depths of compensation of the topographical harmonics for the expansion of degree 18: with a distribution bin of (**a**, **c**) 20 km; (**b**, **d**) 5 km; (**a**, **b**) – taking into account only the relief, (**c**, **d**) – with an additional view of the density contrast on ***M***

one level (900 km for relief and 610 km at the joint account relief and Moho).

The Fig. 2a, b show the maps of the possible vertical stresses in the Earth, at depth 23 km, calculated as without taking into account the contribution of ***M*** in the gravitational field (Fig. 2a), and in view of it (Fig. 2b). At depth 92 km maximum values of stresses does not exceed 10 MPa. Distribution of the maximal stress gradients in the transition regions from the compressive stresses (positive values) to tensile stresses (negative values) is correlated with the distribution of the earthquakes (solarviews.com > cap/earthcyl2.htm). One can also notice the coincidence of areas of vertical tensile stresses with the regions of the present day uplifting of the Earth's surface (Fennoscandia, Canada, Antarctica).

The Figs. 3, 4, and 5 shows the maps of the lateral distribution of anomalous masses in lithosphere and upper mantle, an average depth in the selected range are obtained as by considering only the compensation relief's masses (Figs. 3a–5a) and taking into account the density contrast at the boundary

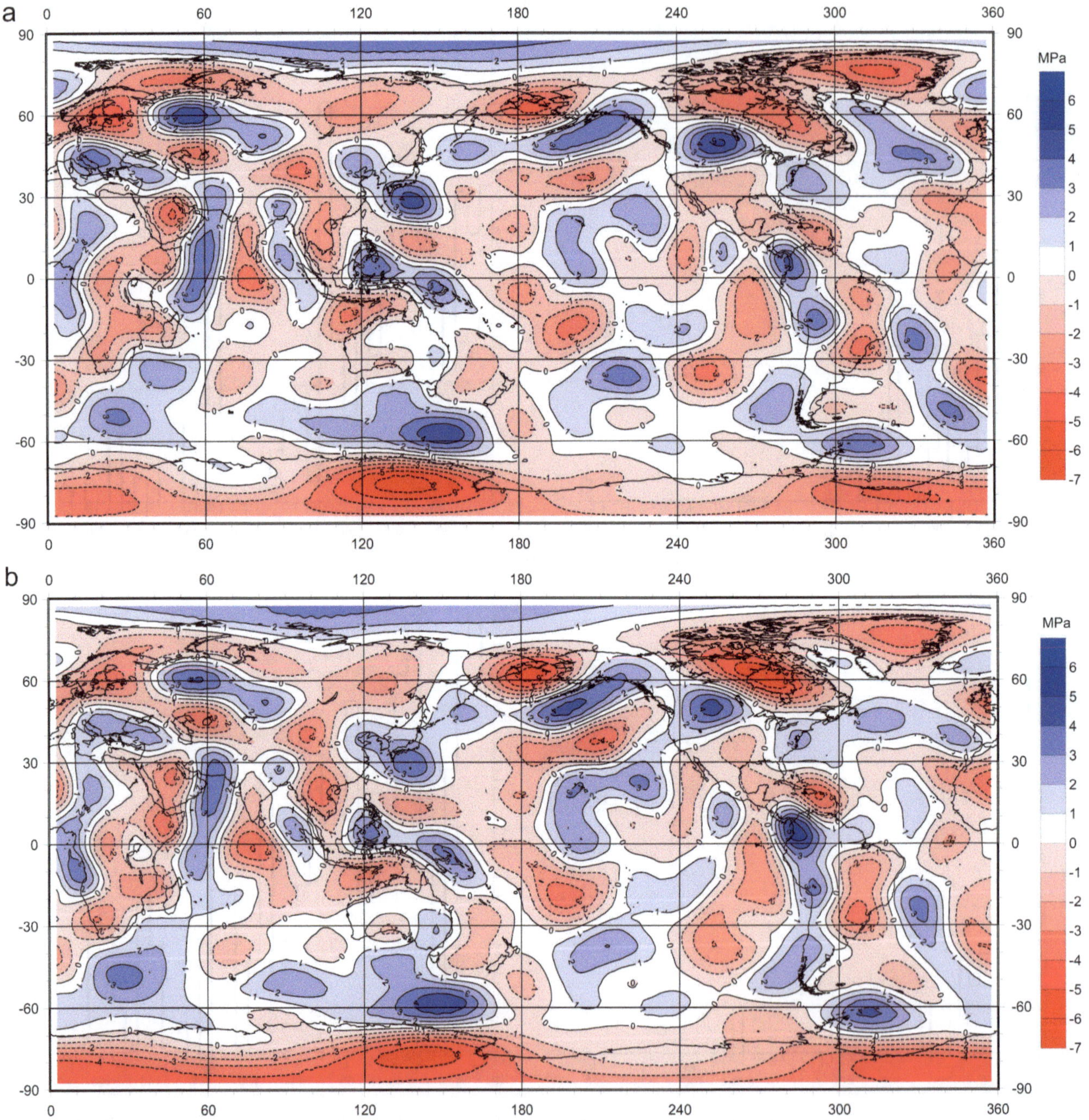


**Fig. 2** (**a**, **b**) The anomalies in the vertical nonisostatic stresses in the lithosphere of Earth for the expansion of degree 18: (**a**) with taking into account only compensation of topography masses, (**b**) with compensation of topography masses and density contrast on ***M***. Contour interval 1 MPa, range of variations is (**a**) (–6.55, 5.17) MPa, (**b**) (–6.01, 5.48) MPa

of ***M*** (0.31 g/cm$^3$). The difference of the density distributions in Fig. 3a, b corresponds to contribution of density anomalies caused by the density contrast at ***M*** (that has been taken into account in the data for Fig. 3b). It is obviously that Fig. 3b shows discrepancy of real value of the density contrast at the boundary ***M*** with the assumed constant value of 0.31 g/cm$^3$. On the Fig. 3a the particularly large deviations correspond to Indian and Australian geoid anomalies, as well as mountain areas in South America (Andes). All other maps (4–6) show a similar distribution pattern masses anomalous in upper and lower mantle as by taking into account, and without taking into account the density discontinuity on ***M***. It shows that neglect of density contrast at ***M*** does not introduce significant distortions.

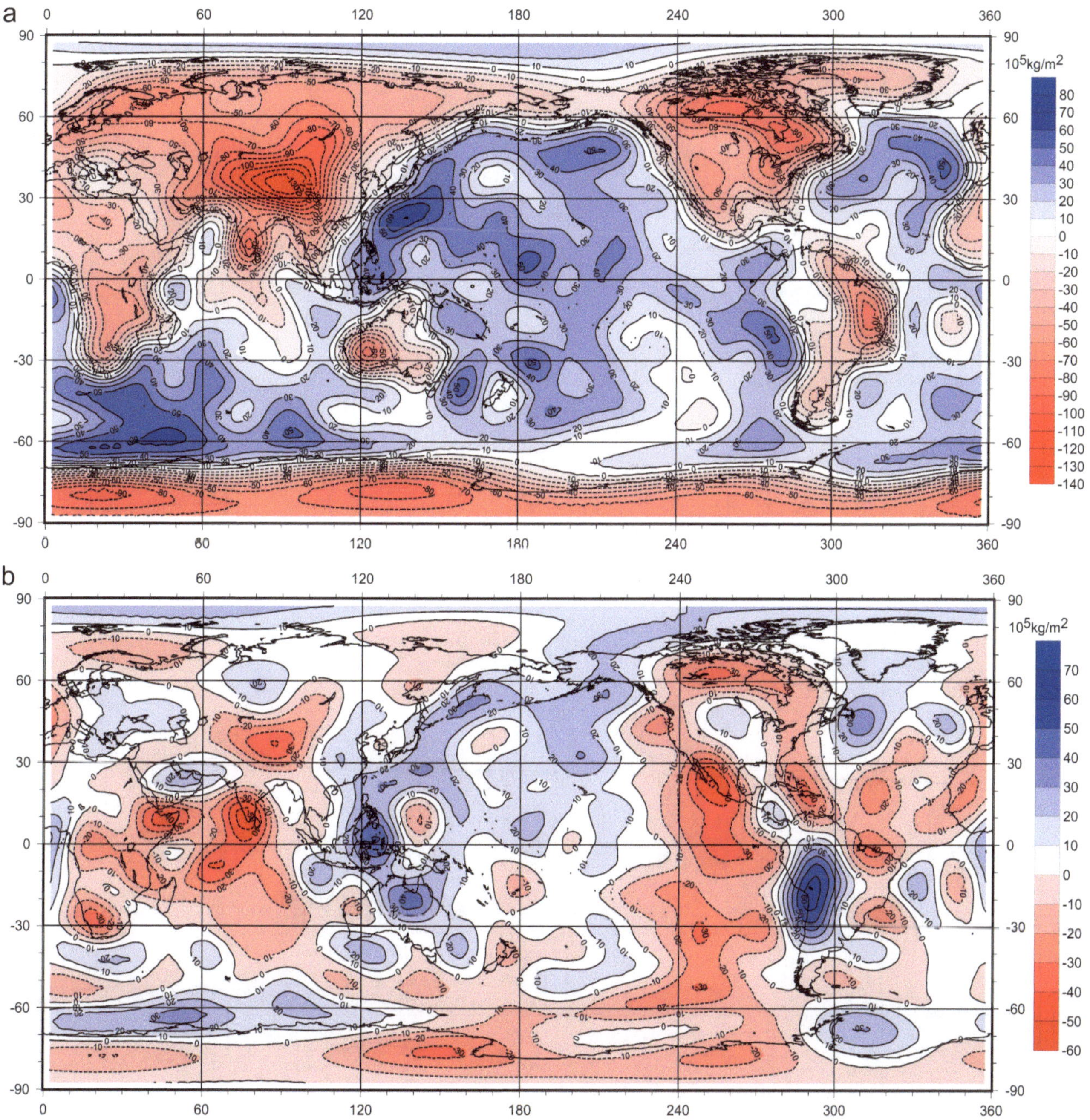


**Fig. 3** (**a**, **b**) Distribution of anomalous masses of the lithosphere at depths of 0–55 km, reduced to the simple-layer density at the average depth of 23 km: (**a**) include the contribution of mass relief, (**b**) with an additional view of the density contrast on ***M***. Contour interval $10 \times 10^5$ kg/m$^2$. The range of the variations is $(-139.68, 70.50) \times 10^5$ kg/m$^2$ (**a**); $(-56.12, 64.38) \times 10^5$ kg/m$^2$ (**b**)

The Fig. 6 shows the distribution of anomalous' masses at the top of lower mantle (Fig. 6a), at the upper boundary of the transition layer of the mantle-core (Fig. 6b), and in the upper layer of the outer core (Fig. 6c). They are obtained by taking into account the additional condition of hydrostatic equilibrium within the liquid core. These maps are substantially identical, both with ***M***, and without ***M***. These maps correspond well to the density distribution maps obtained from the analysis of the free-oscillations data of the Earth (Ishii and Tromp 2004, Fig. 8, depth at 600, 2,300, 2,850 km).

Although the obtained anomalies of density structures satisfy the pressure equilibrium condition below the depth of compensation, they, however, cause significant deviations of internal gravity from the gravity of the equilibrium planet in the mantle and inside the core. It can give rise to the

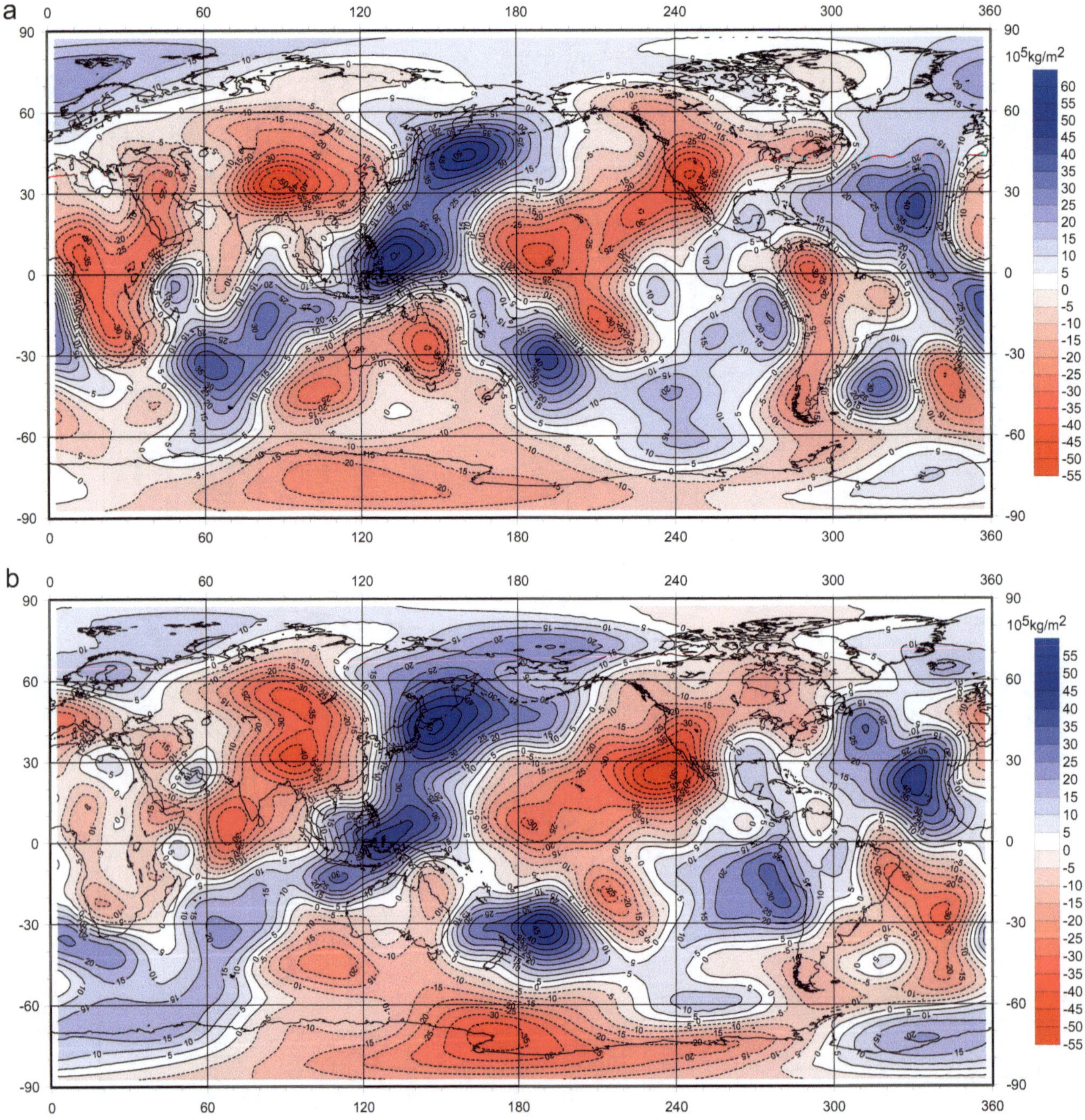


**Fig. 4** (**a**, **b**) Distribution of anomalous masses of the lithosphere at depths of 55–165 km, reduced to the density of the simple-layer at an average depth of 92 km: (**a**) with the contribution of mass relief, (**b**) an additional view of the density jump on ***M***. Contour interval $5 \times 10^5$ kg/m$^2$. The range of the variations is (–52.49, 55.50) $\times 10^5$ kg/m$^2$ (**a**); (–53.52, 51.87) $\times 10^5$ kg/m$^2$ (**b**)

convective motion in these regions. The absence of the own magnetic field at Mars can point to the equilibrium state of its core or to the possible presence of convective movements that are symmetrical about the polar axis. It, according to Cowling's theorem, cannot generate the global magnetic poloidal field. Testing of these conditions for Mars (Chuikova et al. 2012) shows that the anomalous gravity field in the Martian core corresponds to the second condition. For the Earth a structure of the anomalous field inside the core and in the lower mantle is more complex.

A detailed analysis of convective motions in the Earth's core and mantle, which may lead to existence and to variations of the magnetic field in the Earth, and its comparison with Mars, is given in (Chujkova et al. 2014).

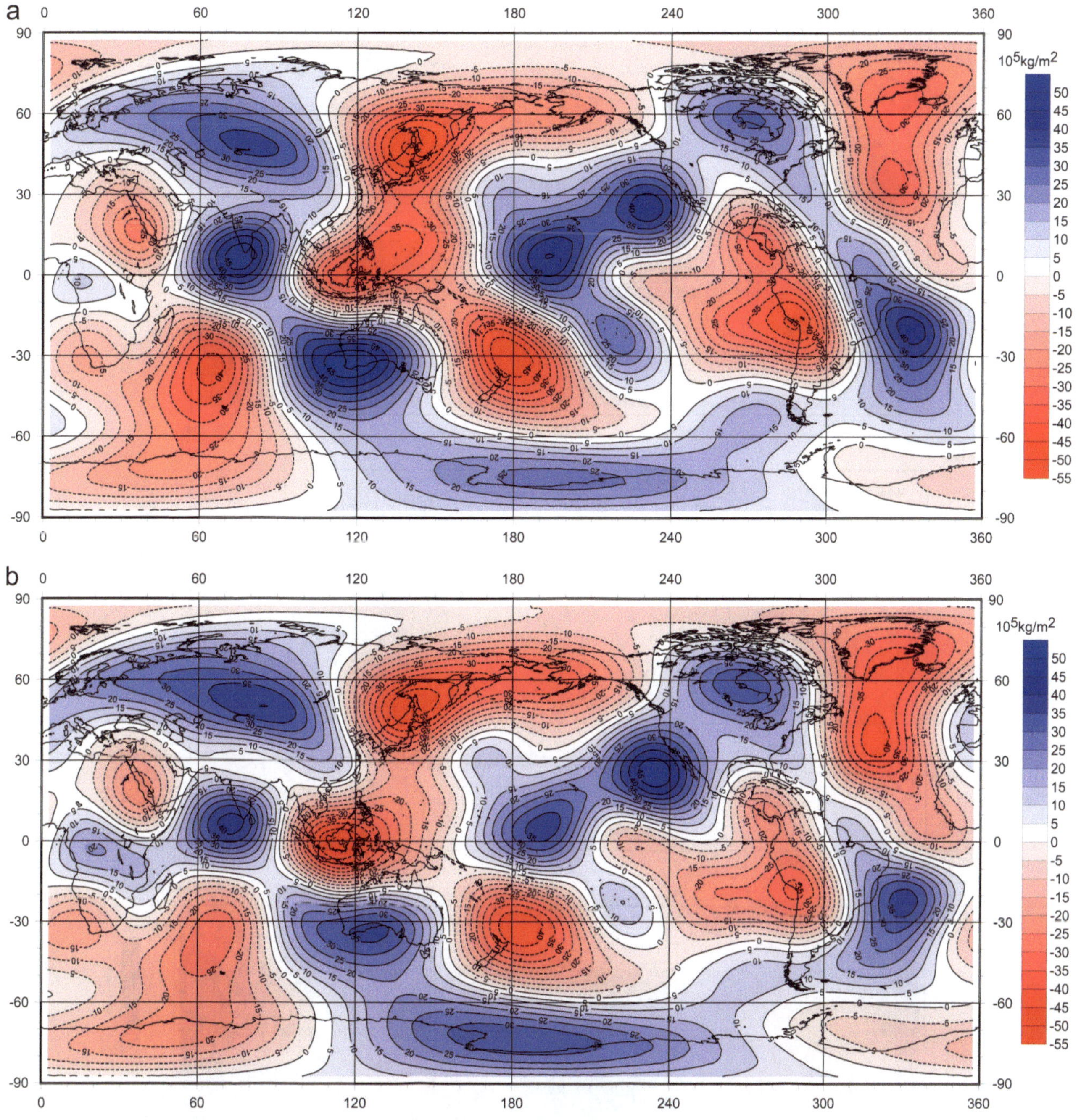


**Fig. 5** (**a**, **b**) Distribution of anomalous masses in the upper mantle at depths of 165–400 km, reduced to the density of the simple-layer at the average depth of 260 km: (**a**) with the contribution of the masses relief, (**b**) with an additional view of the density contrast on ***M***. Contour interval $5 \times 10^5$ kg/m$^2$. The range of the variations is (–51.07, 40.07) $\times 10^5$ kg/m$^2$ (**a**); (–51.62, 48.91) $\times 10^5$ kg/m$^2$ (**b**)

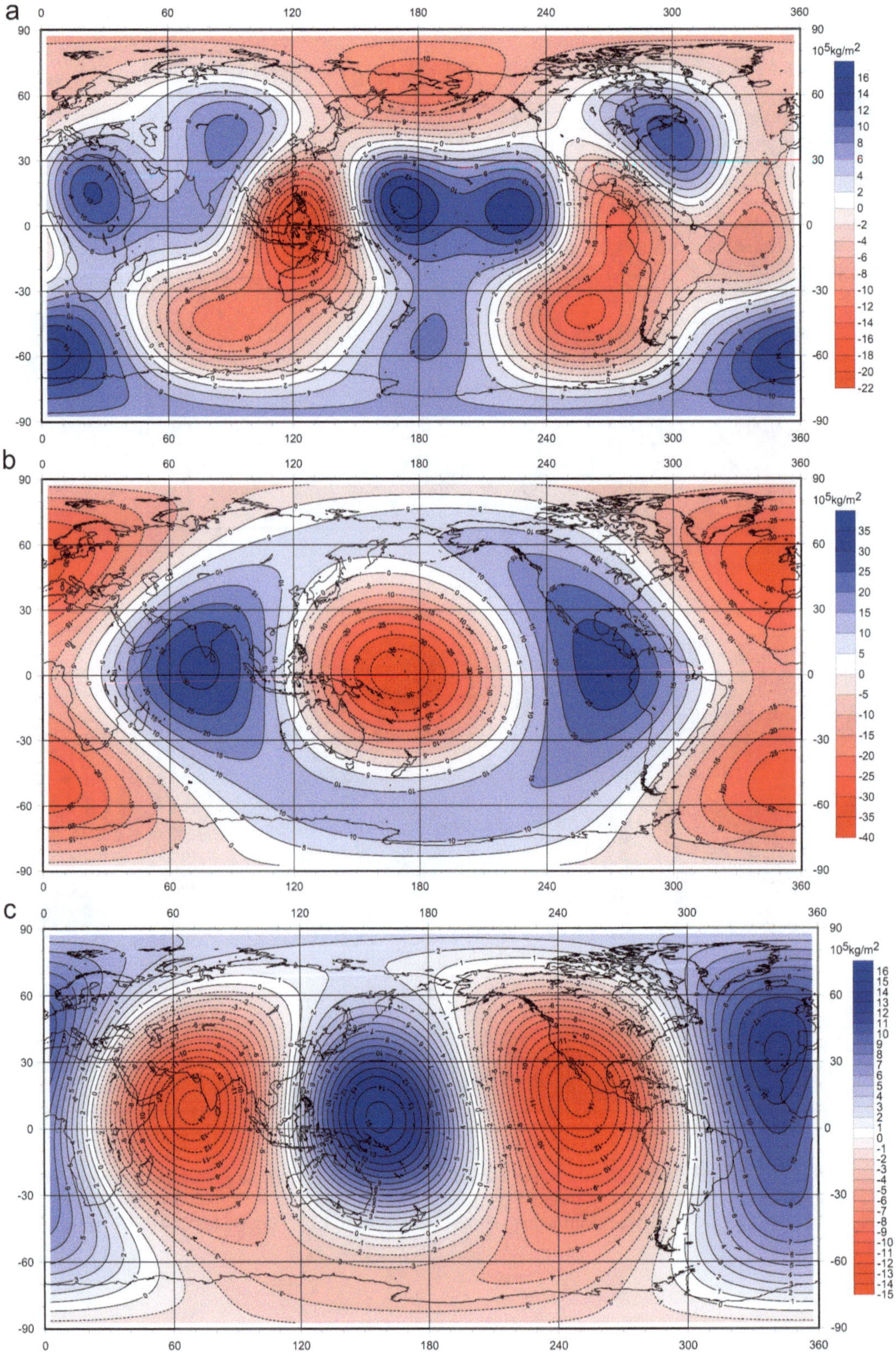


**Fig. 6** (**a–c**) Distribution of anomalous masses in the top layers of the lower mantle: (**a**) the upper boundary of the transition layer mantle – the core and the upper layer of the liquid core at depths of 600–900 km, reduced to the density of the simple-layer at the average depth of 750 km; (**b**) the upper boundary of the transition layer mantle – the core and the upper layer of the liquid core at an average depth of 2,370 km; (**c**) the upper boundary of the transition layer mantle – the core and the upper layer of the liquid core at an average depth of $2{,}970 \pm 80$ km. (**a**) Contour interval is $2 \times 10^5$ kg/m$^2$. The range of the variations is $(-21.23, 15.06) \times 10^5$ kg/m$^2$. (**b**) Contour interval is $5 \times 10^5$ kg/m$^2$. The range of the variations is $(-38.41, 31.50) \times 10^5$ kg/m$^2$. (**c**) Contour interval is $1 \times 10^5$ kg/m$^2$. The range of the variations is $(-14.37, 15.38) \times 10^5$ kg/m$^2$

## 5 Conclusion

It is shown that the developed method of determining the internal structure of density anomalies of the terrestrial planets from space-based observations of the gravitational field and topography, which was tested for the Earth, gives a good model representation. One can use it in future studies at the surfaces of planets. This problem was also considered in a recent papers (Tenzer et al. 2015; Jin and Zhang 2014; Jin et al. 2013).

## References

Antolik M, Gu YJ, Ekstrom G, Dziewonski AM (2003) J362D28: a new joint model of compressional and shear velocity in the Earth's mantle. Geophys J Int 153(2):443–466

Bergman MI (2006) Inner core dynamics. In: 10th Symposium on Study of the Earth's Deep Interior (SEDI), July 9–14, vol S6.2. Prague, p 56

Brown GC, Musset AE (1993) The inaccessible Earth. Chapman and Hall, 276 pp

Chujkova NA, Maximova TG (1996) A spherical harmonic and statistical analysis of the Moho surface (*M*) depths. Trudy GAISH LXV:33–50. http://lnfm1.sai.msu.ru/~chujkova/Trudi/chujkova.htm

Chujkova NA, Grushinsky AN, Maximova TG (1996) The spherical harmonic and statistical analysis of the earth equivalent rock and its isostatic compensation. Trudy GAISH, LXV:51–85.

Chujkova NA, Nasonova LP, Maximova TG (2006) Anomalies of the internal gravitaty field in the crust and upper mantle of the Earth. Vestnik Moscow Univ Phys Astron 4:48–56 (in Russian)

Chuikova NA, Nasonova LP, Maximova TG (2011) Anomalous structure of the crust and mantle of Mars. Moscow Univ Phys Bull 66(1):64–71

Chuikova NA, Nasonova LP, Maximova TG (2012) Anomalies of density, stresses, and the gravitational field in the interior of Mars. Moscow Univ Phys Bull 67(2):218–225

Chujkova NA, Nasonova LP, Maximova TG (2014) Density, stress, and gravity anomalies in the interiors of the Earth and Mars and the probable geodynamical implications: comparative analysis. Izvestiya Phys Solid Earth 50(3):427–443

Cormier VF (2007) Texture of the uppermost inner core from forward and back scattered seismic waves. Earth Planet Sci Lett 258:442–453

Duboshin GN (1961) The theory of attraction (Teoria pritiazgenia). GIF-ML, Moscow, pp 258–261 (in Russian)

Garcia R, Chevrot S, Tkalčić H (2006) New constraints on inner core structure from differential waveform analysis of PKP core phases at global scale. In: 10th Symposium on Study of the Earth's Deep Interior (SEDI), July 9–14, vol S2.10. Prague, p 25

Gu YJ, Dziewonski AM, Ekstrom G (2001) Preferential detection of the Lehmann discontinuity beneath continent. Geophys Res Lett 28:4655–4658

Gu YJ, Dziewonski AM, Ekstrom G (2003) Simultaneous inversion for mantle shear velocity and topography of transition zone discontinuities. Geophys J Int 154:559–583

Ishii M, Tromp J (1999) Normal-mode and free-air gravity constraints on lateral variations in velocity and density of the Earth's mantle. Science 285:1231–1236

Ishii M, Tromp J (2004) Constraining large-scale mantle heterogeneity using mantle and inner-core sensitive normal modes. Phys Earth Planet Inter 146:113–124

Jin SG, Zhang XG (2014) A Tikhonov regularization method to estimate Earth's oblateness variations from global GPS observations. J Geodyn 79:23–29. doi:10.1016/j.jog.2014.04.011

Jin SG, van Dam T, Wdowinski S (2013) Observing and understanding the Earth system variations from space geodesy. J Geodyn 72:1–10. doi:10.1016/j.jog.2013.08.001

Krasnoshchekov DN, Kaazik PV, Ovtchinnikov VM (2006) Heterogeneities of the Earth's solid core from PKiKP coda waves. In: 10th Symposium on Study of the Earth's Deep Interior (SEDI), July 9–14, vol S2.14. Prague, p 26

Kuo C, Romanowicz B (2002) On the resolution of density anomalies in the Earth's mantle using spectral fitting of normal-mode data. Geophys J Int 150:162–179

Lawrence JF, Shearer PM (2008) Imaging mantle transition zone thickness with SdS-SS finite-frequency sensitivity kernels. Geophys J Int 174:143–158

Lay T, Garnero EJ (2004) Core-mantle boundary structures and processec. In: Hawkesworth C, Sparks S (eds) State of the planet, vol 150. Geophysical monograph. American Geophysical Union, Washington, DC, pp 25–41

Nasonova LP, Chuikova NA (2007) The external and internal gravitational fields of the Earth's isostatically balanced crust in the quadratic approximation. Moscow Univ Phys Bull 62(6):397–405

Pail R, Goiginger H, Schuh WD, Höck E, Brockmann JM, Fecher T, Gruber T, Mayer-Gürr T, Kusche J, Jäggi A, Riese D (2010) Combined satellite gravity field model GOCO01S derived from GOCE and GRACE. Geophys Res Lett 37:L20314. doi:10.1029/2010GL044906

Pavlenkova NI (1996) General features of the uppermost mantle stratification from long-range seismic profiles. Tectonophysics 264(1996):261–278

Rodkin M (1993) Role of deep fluid regime in geodynamics and seismotectonics. Moscow 194 pp (in Russian)

Shatina AV (2007) Evolution of motion of mechanical systems with infinite number of the degrees of freedom. Extended abstract of Doct. Sci. (Phys. Math) Dissertation, Moscow State University, Moscow

Soldati G, Boschi L, Piersanti A (2003) Outer core density heterogeneity and the discrepancy between PKP and PcP travel time observations. Geophys Res Lett 30(4):39

Tanaka S (2006) A study on the seismic structure of the core-mantle boundary using global *SmKS* waveforms. In: 10th Symposium on Study of the Earth's Deep Interior, July 9–14, vol S1.17. Prague, pp 19–20

Tenzer R, Eshagh M, Jin SG (2015) Martian sub-crustal stress from gravity and topographic models. Earth Planet Sci Lett 425:84–92. doi:10.1016/j.epsl.2015.05.049

Thybo H, Perchuć E (1997) The seismic 8° discontinuity and partial melting in continental mantle. Science 275:1626–1629

Zharkov VN (1986) Interior structure of the Earth and planets. Char, Switzerland

Zou Z, Koper KD, Cormier VF (2008) The structure of the base of the outer core inferred from seismic waves diffracted around the inner core. J Geophys Res 113:B05314

# List of Reviewers

Hussein A. Abd-Elmotaal
Alberta Albertella
Matthew James Amos
Ole Baltazar Andersen
Jean-Pierre Barriot
Luisa M. C. Bastos
Oliver Baur
Battista Benciolini
Denizar Blitzkow
Johannes Bouman
Sean L. Bruinsma
Wei Chen
Vicki Childers
Nadezda Chujkova
Róbert Čunderlík
Silvio R.C. de Freitas
Maria Clara de Lacy
Olivier de Viron
Pavel Ditmar
Artu Ellmann
William Edward Featherstone
Vagner Goncalves Ferreira
Ehsan Forootan
Rene Forsberg
Oliver Francis
Frederic Frappart
Christian Gerlach
Antonio J. Gil
Richard Gross
Christian Hirt
Jianliang Huang
C. Hwang
Johannes Ihde
Shuanggen Jin
Ambrus Kenyeres
Norbert Kühtreiber
Marie-Francoise Lequentrec-Lalancette
Anno Löcher
Shaochong Luo
Jaakko Mäkinen
Urs Marti
Bruno Meurers
Paola Montone
Pavel Novak
Ove Christian D. Omang
Roland Pail
Vojtech Pálinkáš
Nikolaos Konstantinos Pavlis
Mirko Reguzzoni
Daniel R. Roman
Daniele Sampietro
Laura Sanchez
David Sandwell
Peter Schack
Wolf-Dieter Schuh
WenBin Shen
Christian Siemes
Dru Smith
Holger Steffen
Gabriel Strykowski
Robert Tenzer
Dimitrios Tsoulis
Ilias Tziavos
Michel Van Camp
Georgios S. Vergos
David Wiese
Herbert Wilmes
Jianguo Yan
Wei Zheng
M. Zhong
Leonid Zotov


S. Jin, R. Barzaghi (eds.), *IGFS 2014*, International Association of Geodesy Symposia 144,
DOI 10.1007/1345_2015

# Author Index




S. Jin, R. Barzaghi (eds.), *IGFS 2014*, International Association of Geodesy Symposia 144,
DOI 10.1007/1345_2015

MIX
Papier aus verantwortungsvollen Quellen
Paper from responsible sources
FSC® C105338

If you have any concerns about our products,
you can contact us on
**ProductSafety@springernature.com**

In case Publisher is established outside the EU,
the EU authorized representative is:
**Springer Nature Customer Service Center GmbH**
**Europaplatz 3, 69115 Heidelberg, Germany**

Printed by Libri Plureos GmbH
in Hamburg, Germany